# Multiple-conclusion Logic

# Multiple-conclusion Logic

**D. J. SHOESMITH**

*Director, Test Development and Research Unit, University of Cambridge*

*and*

**T. J. SMILEY**

*Lecturer in Philosophy, University of Cambridge, and Fellow of Clare College*

CAMBRIDGE UNIVERSITY PRESS

CAMBRIDGE

LONDON NEW YORK NEW ROCHELLE

MELBOURNE SYDNEY

CAMBRIDGE UNIVERSITY PRESS
Cambridge, New York, Melbourne, Madrid, Cape Town, Singapore, São Paulo, Delhi

Cambridge University Press
The Edinburgh Building, Cambridge CB2 8RU, UK

Published in the United States of America by Cambridge University Press, New York

www.cambridge.org
Information on this title: www.cambridge.org/9780521217651

© Cambridge University Press 1978

This publication is in copyright. Subject to statutory exception
and to the provisions of relevant collective licensing agreements,
no reproduction of any part may take place without the written
permission of Cambridge University Press.

First published 1978
Reprinted 1980
This digitally printed version 2008

*A catalogue record for this publication is available from the British Library*

*Library of Congress Cataloguing in Publication data.*

Shoesmith, D J
Multiple–conclusion logic
Bibliography:p.
Includes index
1. Many–valued logic.
I. Smiley, Timothy John, joint author. II Title.
BC126.S5  551′.3  79–300388

ISBN 978-0-521-21765-1 hardback
ISBN 978-0-521-09332-3 paperback

# Contents

PART II GRAPH PROOFS

PART III MANY-VALUED LOGIC

PART IV NATURAL DEDUCTION

# Preface

Logic is the science of argument, but ordinary arguments are lopsided: they can have any number of premisses but only one conclusion. In multiple-conclusion logic we allow any number of conclusions as well, regarding them (in Kneale's phrase) as setting out the field within which the truth must lie if the premisses are to be accepted. Thus we count a step in such an argument as valid if it is impossible for all its premisses to be true but all its conclusions false. Anything that can be said about premisses can now be said, mutatis mutandis, about conclusions. (For example, just as adding a member to a set of propositions makes more things follow from them, i.e. strengthens them as potential premisses, so it makes them follow from more things, i.e. weakens them as potential conclusions.) The subject owes much of its interest to the exploitation of this formal symmetry, while the contrasts between multiple- and single-conclusion calculi throw a fresh light on the conventional logic and its limitations.

Our subject is in its infancy. Its germ can be found in Gerhard Gentzen's celebrated *Untersuchungen über das logische Schliessen* (1934) if one is prepared to interpret his calculus of 'sequents' as a metatheory for a multiple-conclusion logic, but this is contrary to Gentzen's own interpretation, and it was Rudolf Carnap who first consciously broached the subject in his book *Formalization of logic* (1943). Carnap defined consequence and introduced rules of inference for multiple conclusions, but the first attempt to devise a proof technique to accommodate such rules was made by William Kneale in his paper *The province of logic* (1956).

Logicians in the subsequent decades appear to have ignored the lead of Carnap and Kneale. Doubtless this is because Kneale was mathematically an outsider and Carnap's jargon deterred even insiders; but doubtless too it reflects the prevalent conception of logic as the study of logical truth rather than logical consequence, and the property of theoremhood (deducibility from axioms) rather than the relation of deducibility in general. Such a climate is bound to be uncongenial to the development of multiple-conclusion logic, whose parity of treatment between premisses and conclusions calls for the more general approach from the outset. The reception given to Carnap's book by Church (1944) in his review is a notable case in point; but some recent Gentzen-inspired work on multiple and single conclusions by Dana Scott, which we cite at the appropriate places in the text, is one of a number of welcome signs that things are changing.

Our book is in four parts, of which the second and third are independent of one another. The aim of Part I is to redefine the fundamental logical ideas so as to take account of multiple conclusions. We begin by recalling four methods of defining consequence for single-conclusion calculi, and abstract from them to produce four different though ultimately equivalent criteria for a relation to be a possible consequence relation. We then devise the appropriate multiple-conclusion analogues, paying special attention to the definition of proof and the somewhat complex sense in which consequence with multiple conclusions is transitive. For both kinds of calculus we discuss the theory of axiomatisability, which surprisingly is less straightforward for single-conclusion calculi than it is either for theories (sets of theorems) or for multiple-conclusion calculi; the idea of consequence by rules of inference, where this is defined in advance of any ideas of proof and indeed is used as a criterion for their adequacy; and rules with infinitely many premisses or conclusions.

The other main theme of Part I is the connection between multiple- and single-conclusion logic. It turns out that each multiple-conclusion calculus has a unique single-conclusion part, but each single-conclusion calculus has a range of multiple-conclusion counterparts. We investigate the composition of these ranges, the extent to which the properties of calculi of one kind can be predicted from their counterpart or counterparts of the other kind, and the connection between multiple conclusions and disjunction.

Part II starts from the observation that for an argument to be valid it is not enough that each of its component steps is valid in isolation: they must also relate to one another properly. In order to discuss this generally overlooked ingredient of validity we need a way of formalising arguments that displays their steps explicitly and unambiguously, and for this purpose we introduce the idea of an argument as a graph of formulae. This representation makes it possible to define the form of an argument ('form' being construed not with reference to any particular vocabulary of logical constants but as something shared by arguments of altogether different vocabularies) in purely syntactic terms. On the other hand our criterion of validity is ultimately semantic, involving the idea of consequence by rules introduced in Part I. We therefore investigate the connection between form and validity, looking for syntactic conditions for validity, trying to determine the adequacy or otherwise of various syntactically defined classes of proof, and discussing such topics as conciseness and relevance. We do this first for multiple-conclusion arguments (showing incidentally that Kneale's definition of proof is inadequate), then for single-conclusion ones and finally for arguments by infinite rules.

Our treatment of the subject so far is less concerned with particular calculi than with features of consequence and proof common to them all, but the remainder of the book introduces two specific applications. In Part III we make a detailed comparison between the multiple- and single-conclusion treatment of a particular topic. We choose many-valued logic as our example because it is well known and accessible, but there is also a historical reason. For it was Carnap's discovery of 'non-normal' interpretations of the classical propositional calculus - interpretations which fit the calculus but not the normal truth-tables - that led him to advocate multiple conclusions as the only fully satisfactory means of capturing truth-functional logic. We therefore include two-valued calculi in our discussion of many-valued ones, and investigate the multiple-conclusion counterparts of the classical calculus along with those of many-valued calculi in general. We show that every finite-valued multiple-conclusion propositional calculus is finitely axiomatisable and categorical (though virtually no single-conclusion calculus is categorical); and we pose the many-valuedness problem - the problem of distinguishing many-valued calculi from the rest by some intrinsic feature of their consequence relations. In general it seems that distinctions which were blurred in the single-conclusion case become sharp in the multiple-conclusion one, and results which needed qualification become unconditional. On the other hand it appears that a single-conclusion calculus displays a stability when its vocabulary is enlarged which a multiple-conclusion one may not; and one result of this is that we have had to leave the many-valuedness problem open for multiple-conclusion calculi though it is solved for single-conclusion ones.

In Part IV we explore the possibility of replacing the indirect methods of ‘natural deduction’ by direct proofs using multiple-conclusion rules, and with it the possibility of obtaining within our theory such results as the subformula theorem. We illustrate these ideas for the classical predicate calculus in a purely rule-theoretic context, and in a proof-theoretic one (presupposing some of the ideas of Part II) for the intuitionist propositional calculus.

Our results are published here for the first time apart from a short abstract (1973), but we have worked together on and off over six years and Chapter 19 has its origin in Shoesmith's Ph.D. dissertation (1962). We are much indebted to the trustees of the Radcliffe Trust for giving one of us a welcome relief from the pressures of other work by their award of a Radcliffe Fellowship to Smiley for 1970 and 1971.

Cambridge, 1976

We have taken the opportunity provided by a reprinting to make some minor corrections, and to remedy our misuse of the term ‘recursive proof procedure’ in Chapter 4, replacing it by ‘recursive notion of proof’.

Cambridge, 1979

# Introduction

A multiple-conclusion proof can have a number of conclusions, say $B_1,\ldots,B_n$. It is not to be confused with a conventional proof whose conclusion is some one of the $B_j$, nor is it a bundle of conventional proofs having the various $B_j$ for their respective conclusions: none of the $B_j$ need be 'the' conclusion in the ordinary sense. This fact led Kneale to speak of the 'limits' of a 'development' of the premisses instead of the conclusions of a proof from them. We prefer to extend the sense of the existing terms, but hope to lessen one chance of misconstruction by speaking of a proof from $A_1,\ldots,A_m$ *to* $B_1,\ldots,B_n$ instead of a proof *of* $B_1,\ldots,B_n$ from $A_1,\ldots,A_m$.

The behaviour of multiple conclusions can best be understood by analogy with that of premisses. Premisses function collectively: a proof from $A_1,\ldots,A_m$ is quite different from a bundle of proofs, one from $A_1$, another from $A_2$ and so on. Moreover they function together in a conjunctive way: to say that B follows from $A_1,\ldots,A_m$ is to say that B must be true if $A_1$ and ... and $A_m$ are true. Multiple conclusions also function collectively, but they do so in a disjunctive way: to say that $B_1,\ldots,B_n$ follow from $A_1,\ldots,A_m$ is to say that $B_1$ or ... or $B_n$ must be true if all the $A_i$ are true.

It should not be inferred from this explanation that multiple conclusions can simply be equated with the components of a single disjunctive conclusion $B_1 \vee \ldots \vee B_n$. The objections to such a facile reduction of multiple- to single-conclusion logic are the same as the objections to reducing the conventional logic to a logic of single premisses. It is true that any finite set of premisses is equivalent to a single conjunctive one, $A_1, \ldots, A_m$ having the same joint force as $A_1 \& \ldots \& A_m$. But this equivalence is only established by appealing to the workings of the rule 'from A,B infer A&B', understood as involving two separate premisses (not one conjunctive one), and it would be circular to appeal to the equivalence to establish the dispensability of the rule. Infinite or empty sets of premisses could not in any case be treated in this way; nor do all calculi possess a conjunction. Moreover, the equivalence between a set of sentences and their conjunction is at best a partial one, for the conjunction, being a sentence itself, can be made a component of further sentences where a set cannot: contrast $\sim(A_1 \& A_2)$ and $\sim\{A_1, A_2\}$. Our remark is that considerations exactly analogous to these apply to multiple conclusions and disjunctions.

To see how multiple conclusions might invite the attention of the logician, imagine first a student assigned the modest task of devising axioms and rules for the propositional calculus. He sees that the truth-table for conjunction can be translated immediately into rules of inference, the stipulation that A&B is true when A and B are true producing the rule (1) 'from A,B infer A&B', and similarly for (2) 'from A&B infer A' and (3) 'from A&B infer B'. Not only are these rules justified by the truth-table, but they in turn dictate it: any interpretation of conjunction that fits the rules must fit the truth-table too, for by (1) A&B must be true if, and by (2) and (3) only if, A and B are both true. Encouraged by this start the student moves on to disjunction, where the three 'true' cells in the truth-table immediately produce the rules 'from A infer A∨B' and 'from B infer A∨B'. But when he comes to the remaining one - the entry 'false' when A and B are both false - the recipe fails. Moreover, even if he does find a complete set of rules, they cannot possibly dictate the intended interpretation of disjunction. For it is easy to show that all and only the tautologies and inferences of the propositional calculus are valid in the truth-tables below, where t stands for truth and $f_1$, $f_2$ and $f_3$ for subdivisions of falsity; yet A∨B can be true when A and B are both false. Our student has heard of the difficulties of excluding non-standard interpretations in the upper stories of mathematics; now he finds the same thing in the basement. He sees too that, if he could avail himself of it, the multiple-conclusion rule 'from A∨B infer A,B' would both translate the fourth cell of the original truth-table and serve to dictate the intended interpretation of disjunction in the same way as the rules for conjunction do.

| & | t | $f_1$ | $f_2$ | $f_3$ |
|---|---|---|---|---|
| t | t | $f_1$ | $f_2$ | $f_3$ |
| $f_1$ | $f_1$ | $f_1$ | $f_3$ | $f_3$ |
| $f_2$ | $f_2$ | $f_3$ | $f_2$ | $f_3$ |
| $f_3$ | $f_3$ | $f_3$ | $f_3$ | $f_3$ |

| ∨ | t | $f_1$ | $f_2$ | $f_3$ |
|---|---|---|---|---|
| t | t | t | t | t |
| $f_1$ | t | $f_1$ | t | $f_1$ |
| $f_2$ | t | t | $f_2$ | $f_2$ |
| $f_3$ | t | $f_1$ | $f_2$ | $f_3$ |

| ~ | |
|---|---|
| t | $f_3$ |
| $f_1$ | $f_2$ |
| $f_2$ | $f_1$ |
| $f_3$ | t |

Alternatively, consider the ambitious project of defining logic as advocated by Popper, Kneale and Hacking (for references see the historical note at the end of Section 2.1). It is proposed that a logical constant is one whose meaning can be explained by conventions governing its inferential behaviour. If the conventions are not merely to fix but to explain the meaning they must take the form of introduction rules, by which the behaviour of sentences containing the constant can be derived inductively from the behaviour of their constituents. (For example, using ⊢ to symbolise consequence and X for an arbitrary set of premisses, conjunction can be introduced by the rules 'if X ⊢ A and X ⊢ B then X ⊢ A&B' and 'if X,A ⊢ C or X,B ⊢ C then X,A&B ⊢ C'; but an elimination rule like 'if X ⊢ A&B then X ⊢ A' is ineligible, as is an appeal to the transitivity of ⊢.) Disjunction causes no difficulty this time, but material implication does. It turns out that the obvious introduction rules, 'if X,A ⊢ B then X ⊢ A⊃B' and 'if X ⊢ A and X,B ⊢ C then X,A⊃B ⊢ C', characterise intuitionist, not classical implication; and to introduce the latter it is necessary to have multiple conclusions. Indeed one would have to conclude that classical logicians, like so many Monsieur Jourdains, have been speaking multiple conclusions all their lives without knowing it.

No branch of mathematical logic relies exclusively on actual argumentative practice for its justification. We make use of an informal multiple-conclusion proof in Section 18.3, and note that the formalisation of a multiple-conclusion metacalculus provides a nice method of proving compactness (Theorem 13.1), but it can hardly be said that multiple-conclusion proofs form part of the everyday repertoire of mathematics. Perhaps the nearest one comes to them is in proof by cases, where one argues 'suppose $A_1$ ... then B, ... , suppose $A_m$ ... then B; but $A_1 \vee ... \vee A_m$, so B'. A diagrammatic representation of this argument exhibits the downwards branching which we shall see is typical of formalised multiple-conclusion proofs:

$$\frac{A_1 \vee A_2 \vee \ldots \vee A_m}{\begin{array}{cccc} A_1 & A_2 & \ldots & A_m \\ \vdots & \vdots & & \vdots \\ B & B & & B \end{array}}$$

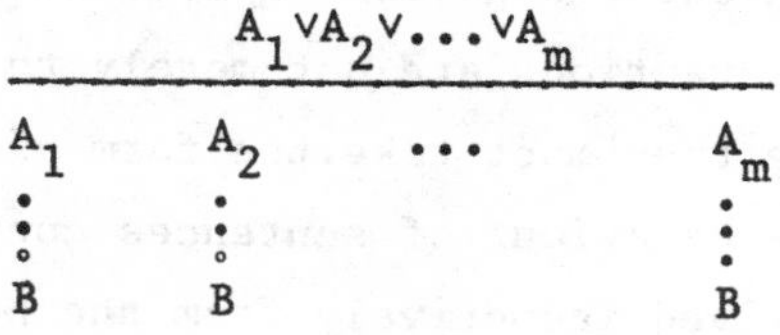

But the ordinary proof by cases is at best a degenerate form of multiple-conclusion argument, for the different conclusions are all the same (in our example they are all instances of the same formula B). Why are multiple-conclusion arguments not more prevalent?

One answer (suggested to us by Gareth Evans) concerns the difficulty of using multiple conclusions in an object-language as opposed to mentioning them in a metalanguage. We can assert a number of premisses as a series '$A_1$. $A_2$. ... $A_m$.' Each stage '$A_1$. ... $A_i$.' of this is complete in itself and independent of what may follow: the subsequent assertions merely add to the commitment represented by the previous ones. But if we tried to make a serial utterance '$B_1$. $B_2$. ... $B_n$.' in the way required for asserting multiple conclusions, as committing us to the truth of $B_1$ or of $B_2$ ... or of $B_n$, we should be withdrawing by the utterance of $B_2$ the unqualified commitment to $B_1$ into which we had apparently entered at the first stage, and so on. The utterance will therefore have to be accompanied by a warning (e.g. a prefatory 'Either') to suspend judgment until the whole series is finished, and we do not achieve a complete speech act until the utterance of the last $B_j$, duly marked as such. But this is as much as to admit that the various $B_j$ are functioning not as separate units of discourse but as components of a single disjunctive one.

To avoid Evans' point it is necessary to replace the unilinear flow of ordinary speech by discourse set in two or more dimensions, and all the notations we shall propose for multiple-conclusion proofs are of this kind. An example is our diagrammatic proof by cases above, where the convention would be that the inscription of sentences in different columns indicates a purely alternative commitment to their truth.

Another answer to the question (suggested to us by David Mayers) is that we usually focus on a desired conclusion - the *quod erat demonstrandum* - before devising an argument that leads to it, and if our interest were thus fixed on a (finite) set of multiple conclusions it would be a natural first step to replace them by a single disjunctive target. The situation is not symmetrical as between premisses and conclusions, since we do not usually know in advance what premisses we are going to need to prove a given conclusion. To redress the balance there would need to be more people engaged in arguing from given premisses in an attempt to refute them; i.e. more reductio ad absurdum proofs or more practising Popperians.

*Notation* We try to use the same terminology for single- and multiple-conclusion ideas, though since our principal concern is with the latter, terms like 'calculus' or 'rule' are to be understood in the multiple-conclusion sense unless explicitly qualified by the adjective 'single-conclusion' or unless the context is unequivocal, as in Chapter 1. Similarly we use a common notation whenever possible: for example $\vdash$ (used mainly for consequence relations) and R (for rules of inference) stand for binary relations between sets of premisses and individual conclusions in a single-conclusion context; between sets and sets in a multiple-conclusion one. Calculi of either kind are denoted by L, with or without a suffix or prime, and in this connection $\vdash$ stands for the consequence relation of L, $\vdash'$ that of L′ and so on. In addition the special suffix notations $\vdash_I$, $\vdash_R$ and $\vdash_M$ signify consequence relations determined respectively by a set I of interpretations, a set R of rules of inference and a matrix M: definitions are found in the text. We write $\nvdash$ for the complement of $\vdash$ so that $X \nvdash Y$ iff (if and only if) not $X \vdash Y$.

We use A, B, C, D to stand for formulae of a calculus and T, U, V, W, X, Y, Z for sets of them, V being reserved for the set of all formulae of the calculus and T and U for complementary sets (i.e. a pair of disjoint sets whose union is V). We adopt the lambda notation for functions and the braces notation for sets, by which $\lambda xy.x+y$ is the addition function, $\{A_1,\ldots,A_m\}$ the set whose members are $A_1,\ldots,A_m$ and $\{A: P(A)\}$ the set of formulae satisfying P. Other notation from set theory is $\langle x,y\rangle$ for an ordered pair, $\subset$ for set inclusion, X-Y for the set of formulae in X but not in Y, and $\Lambda$ for the empty set. (Some logicians use a symbol not unlike $\Lambda$ to stand for a particular 'absurd' formula. It will appear that there is some similarity between an absurd formula and an empty set qua conclusion, but none qua premiss.)

Wherever it is convenient to do so we adopt the usual elisions, suppressing braces and writing for example X-A for X-{A} and $A_1,\ldots,A_m$ for $\{A_1,\ldots,A_m\}$. When discussing multiple-conclusion calculi we therefore write $X \vdash B$ when we should strictly write $X \vdash \{B\}$, though we take care to retain the braces where (as in Chapter 5) the contrast between multiple- and single-conclusion calculi is the point at issue. We often use commas instead of the notation for set union, writing X,Y for $X \cup Y$; and we make regular use of Frobenius notation for sets of formulae, writing for example $X \supset B$ for $\{A \supset B: A \in X\}$ and, where s is a substitution, s(X) for $\{s(A): A \in X\}$. When the index set can be supplied from the context we write $\{X_i\}$ instead of $\{X_i\}_{i \in S}$ for a family indexed by S (if necessary adding the word 'family' to avoid confusion with the braces notation for a unit set or singleton), and similarly write $\bigcup X_i$ for the union of the various $X_i$. Greek letters stand for arrays - families of formulae or symbols - and in particular we use $\sigma$ for sequences, $\tau$ for trees and $\pi$ for graphs.

We call a member of a partially ordered set 'maximal' if no other member is greater than it, and 'maximum' if it is greater than every other member. 'Minimal' and 'minimum' are distinguished in the same way.

Results are numbered consecutively within each chapter, and in the course of proofs we cite them by number alone, writing 'by 2.12' instead of 'by Lemma 2.12' when appealing to the twelfth result of Chapter 2. For ease of reference, section and theorem numbers are also noted at the head of each pair of pages. Although we generally deal only with rules of inference with finite sets of premisses and conclusions, we discuss infinite rules and proofs in Chapters 6 and 12, and with this in view mark with a † those results in the previous chapters which fail to carry over to the infinite case. Definitions are not numbered but are easy to locate from the index, where the relevant page number appears first in the entry. References to the bibliography are by author and date, e.g. 'Carnap, 1943' or 'Carnap (1943)'.

# Part I · Multiple and single conclusions

# 1 · Single-conclusion calculi

## 1.1 *Consequence*

We take mathematical logic to be the mathematical study of consequence and proof by means of formalised languages or calculi. A *single-conclusion calculus* is constituted by a universe of formulae and a relation of consequence, construed as a binary relation between a set of premisses and a conclusion. We make no stipulations about the sort of entity that can be a formula or what sort of internal structure the formulae may possess. It is natural to assume that there is at least one formula, though no result of importance appears to depend on this. We make no other stipulation about the number of formulae, though since we use the axiom of choice we shall be assuming that they can be well-ordered. We write V for the relevant universe of formulae, and use T, U, W, X, Y, Z to stand for sets of formulae (i.e. subsets of V), and A, B, C, D for formulae. We write L for a calculus and ⊢ for its consequence relation, adding a prime or suffix when necessary, so that for example $\vdash_1$ stands for consequence in $L_1$. By a *single-conclusion relation* we mean any binary relation between sets of formulae on the one hand and individual formulae on the other. If R is a single-conclusion relation we write X R B to indicate that R holds between the set X and the formula B, and whenever X R B we say that the pair ⟨X,B⟩ is an *instance* of R.

Several ways of defining consequence for particular calculi are to be found in the literature. The 'semantic' method requires us to lay down the range of admissible interpretations of the formulae (typically by giving a fixed meaning to the so-called logical symbols and an arbitrarily variable one to the remainder), and to define X ⊢ B to mean that B is true on every interpretation on which all the members of X are true. At the opposite pole stands the definition of consequence as deducibility. Given certain formulae as axioms, and given rules by which a formula may be inferred from others, one first defines a proof (typically as a sequence in which the premisses are linked to the conclusion by steps conforming to the axioms and rules), and then defines X ⊢ B to mean that there exists a proof of B from X. Since the axioms and rules are usually specified in terms of the shapes, not the meanings, of the relevant formulae, this sort of definition is purely syntactic. Both methods are familiar to every student of the propositional calculus, the first being typified by the truth-table definition of tautological consequence and the second by any axiomatic presentation.

Two other methods are less well known. (For their origins see the note following this section.) The idea of 'consequence by rules', as we shall call it, mixes syntax with semantics by combining an appeal to axioms and rules with the idea of consequence as the transmission of truth. The meaning of even the logical symbols is taken to be arbitrarily variable, but attention is restricted to those interpretations that are sound in the sense that they make the axioms true and the rules truth-preserving, and X ⊢ B is defined to mean that B is true on every sound interpretation on which all the members of X are true. This is reminiscent of the well-known method of developing a mathematical theory in the abstract, whereby a sound interpretation is defined as one that makes the axioms true and the theorems are defined as the propositions that are true on every sound interpretation.

There is however one important difference. The development of an axiomatic theory presupposes some underlying logic, which limits the range of (sound plus unsound) interpretations. For example, in an abstract development of set theory the meaning of '$\in$' is free to vary but that of '=' or 'all' is circumscribed by the (formalised or informal) underlying logic. In the present case, by contrast, there is no need to invoke an underlying logic, and the range of interpretations from which the sound ones are selected is absolutely untrammelled. (If the reader insists on discerning an underlying logic even here, it can only be one that reflects the admission of every interpretation whatsoever, and since the wider the range of interpretations the fewer the formulae that follow from one another, it will be the weakest possible logic for the relevant universe. Such 'minimum' calculi are discussed further in Part III; and cf. Popper, 1947.)

Finally there is the 'implicit' method of defining consequence. The symbol $\vdash$ is introduced as a (metalinguistic) variable subject to constraints represented by postulates. If there is a unique (or perhaps a least) relation which satisfies these constraints it is said to be implicitly defined by the postulates. Typically, some of the postulates will express generic properties such as the transitivity of consequence, while the remainder will be special to the calculus in question (e.g. '$X \vdash A\&B$ iff $X \vdash A$ and $X \vdash B$' or '$X \vdash A \supset B$ iff $X,A \vdash B$').

Our own concern is less with the consequence relations of particular calculi than with what is common to them all, i.e. with the property of being a consequence relation. We therefore take each of the four methods and abstract from it so as to obtain a corresponding criterion for a single-conclusion relation to be a consequence relation. We shall show that these criteria are equivalent to each other provided rules of inference are allowed to have infinitely many premisses, but in the present chapter we confine the discussion to rules with a finite number of premisses.

Our treatment of the semantic method is prompted by the following line of argument.

(1) To say that a conclusion follows from a given set of premisses is to say that the conclusion must be true if the premisses are all true.

(2) The necessity with which conclusions follow is relative to the presuppositions of an argument, and different arguments may have different presuppositions. But whatever idea of necessity is involved there is a corresponding idea of possibility, and we can use this to reformulate (1) as

(3) To say that a conclusion follows from some premisses is to say that each possible state of affairs in which all the premisses are true is one in which the conclusion is true.

(4) Assuming the formulae of a calculus to be capable of truth and untruth, each relevant state of affairs may be represented by the set T of formulae which are true in it. Let I be the set of all such T, and let $X \vdash_I B$ mean that for all T in I, if $X \subset T$ then $B \in T$. Then by (3) to say that B follows from X is to say that $X \vdash_I B$.

(5) The only plausible way of generalising (4) to apply to an arbitrary universe of formulae is to be prepared to allow *any* set of sets of formulae to play the part of I in (4). This step is supported by considering the indefinite variety of argumentative presuppositions envisaged in (2).

(6) We therefore say that $\vdash$ is a consequence relation if it is of the form $\vdash_I$ for some set I of sets of formulae.

Without having any objection to (6) we prefer to reword it so as to bring out the analogy with our subsequent symmetrically worded definition of multiple-conclusion consequence. To do so we represent a state of affairs not simply by the set T of formulae which are true in it but, equivalently, by an ordered pair made up of the sets T and U of true and untrue formulae respectively. With this in mind we reserve the letters T and U to denote mutually complementary sets of formulae, and call any such pair $\langle T,U\rangle$ a *partition*. We also say that one set *overlaps* another if the two have a member in common. In these terms we can reformulate (4)-(6) as follows.

A partition $\langle T,U\rangle$ *satisfies* $\langle X,B\rangle$ if $B \in T$ or X overlaps U. Otherwise, i.e. if $X \subset T$ and $B \in U$, the partition *invalidates* $\langle X,B\rangle$. If $\langle X,B\rangle$ is satisfied by every partition in the set I we say it is *valid* in I and write $X \vdash_I B$. A relation $\vdash$ is a *consequence relation* if $\vdash\, = \,\vdash_I$ for some I, in which case we say too that $\vdash$ is *characterised* by I.

Turning to the implicit method, we formulate the following conditions on a relation $\vdash$.

| | |
|---|---|
| *Overlap* | If $B \in X$ then $X \vdash B$. |
| *Dilution* | If $X' \vdash B$ and $X' \subset X$ then $X \vdash B$. |
| *Cut for sets* | If $X,Z \vdash B$ and $X \vdash A$ for every A in Z, then $X \vdash B$. |

The title 'overlap' reflects our desire to use the same terminology in the single- as in the multiple-conclusion case, where it is clearly appropriate. 'Dilution' and 'cut' are translations of Gentzen's 'Verdünnung' and 'Schnitt', the former commonly but erroneously translated as 'thinning'.

Our first result will show that these three conditions provide an alternative criterion for a relation to be a consequence relation. In stating the result, and subsequently throughout the book, it is convenient to use the phrase '*is closed under* overlap' instead of 'satisfies the overlap condition', and similarly for dilution and cut.

*Theorem 1.1* A single-conclusion relation is a consequence relation iff it is closed under overlap, dilution and cut for sets.

*Proof* It is immediate from the definition that any consequence relation is closed under overlap and dilution. For cut, suppose that (i) $X \vdash A$ for each A in Z and (ii) $X,Z \vdash B$. Let $\langle T,U \rangle$ be any member of the set of partitions that characterises $\vdash$. If $X \subset T$ then $Z \subset T$ by (i) and so $B \in T$ by (ii); hence $X \vdash B$ as required. For the converse let $\vdash$ be closed under overlap, dilution and cut for sets, and let I be the set of all partitions of the form $\langle T(Y),U(Y) \rangle$, where $T(Y) = \{A: Y \vdash A\}$ and $U(Y) = \{A: Y \nvdash A\}$. If $X \nvdash B$ then $\langle T(X),U(X) \rangle$ invalidates $\langle X,B \rangle$, since by overlap $X \subset T(X)$. On the other hand if any partition $\langle T(Y),U(Y) \rangle$ invalidates $\langle X,B \rangle$, so that $Y \nvdash B$ although $Y \vdash A$ for every A in X, then by cut for sets it follows that $Y,X \nvdash B$ and so by dilution that $X \nvdash B$. Hence $\vdash$ is the consequence relation $\vdash_I$.

Our three conditions are suggested by rules devised by Hertz under the names 'triviality', 'immediate inference' and 'syllogism' (see note). He used the second to derive the first from a 'tautology' corresponding to $A \vdash A$, and similarly by using dilution $A \vdash A$ could replace overlap in our list. Hertz only envisaged finite sets of antecedents, and the straightforward generalisation of his syllogism rule to cover an arbitrary number of premisses would be 'if $X,Z \vdash B$ and $X_i \vdash A_i$ for each $A_i$ in Z, then $X, \bigcup X_i \vdash B$'. Formally speaking cut for sets is a special case of this, obtained by equating each $X_i$ with X, but the two are clearly equivalent given dilution. Yet another variant is 'if $Z \vdash B$ and $X \vdash A$ for each A in Z, then $X \vdash B$', and indeed this together with overlap alone could replace our original three conditions. Gentzen follows Hertz in most respects (see note), but his cut condition is the special case of Hertz's syllogism in which Z is a unit set. The two are equivalent in the finite case, but this ceases to be true in the infinite case. For Gentzen's version still allows only one formula at a time to be 'cut out', and even by iteration can deal with at most a finite number, whereas by taking Z to be infinite in Hertz's version infinitely many formulae can be cut out simultaneously. Theorem 1.3 will show that it is essential to adopt the latter rather than the former version if one is to obtain a condition strong enough to use in a discussion of consequence in general.

Given that arbitrary sets of premisses are permitted, a situation corresponding to the finite one treated by Hertz and Gentzen can be recreated by invoking the idea of compactness. A single-conclusion relation ⊢ is *compact* if whenever X ⊢ B there is a finite subset X′ of X such that X′ ⊢ B, and a *compact* calculus is one whose consequence relation is compact Consequence in a compact calculus is thus completely determined by consequence from finite sets of premisses, and Theorem 1.4 shows that in this case it makes no difference whether we formulate the cut condition so as to apply to formulae one at a time, or to a finite set, or to an infinite one.

In the theorems which summarise the preceding discussion *cut for finite sets* refers to the condition obtained from cut for sets by constraining Z (though not X) to range over finite sets only, and *cut for formulae* is the condition 'if X ⊢ A and X,A ⊢ B then X ⊢ B', obtained by constraining Z to range over unit sets {A}.

*Theorem 1.2* Cut for formulae is equivalent (granted dilution) to cut for finite sets.

*Theorem 1.3* Cut for finite sets is weaker (even granted overlap and dilution) than cut for sets.

*Theorem 1.4* For compact single-conclusion relations cut for formulae is equivalent (granted dilution) to cut for sets.

*Theorem 1.5* A compact single-conclusion relation is a consequence relation iff it is closed under overlap, dilution and cut for formulae.

*Proofs* For 1.2 we argue from cut for formulae to cut for finite Z by induction on the number of formulae in Z. The result is tautological when Z is empty. For the induction step it is sufficient to establish cut for $Z \cup \{C\}$ given cut for Z as follows: if $X,Z,C \vdash B$ and $X \vdash C$ and $X \vdash A$ for each A in Z then by dilution $X,Z \vdash C$, whence $X,Z \vdash B$ by cut for C and $X \vdash B$ by cut for Z, as required.

For 1.3 let $X \vdash B$ iff X is infinite or $B \in X$ or $B \neq C$, where C is some chosen formula. It is easy to verify that $\vdash$ is closed under cut for finite sets, but when V is infinite $\vdash$ is never closed under cut for sets, since $\Lambda \nvdash C$ although $V-C \vdash C$ and $\Lambda \vdash A$ for each A in V-C. Since $\vdash$ is closed under overlap and dilution this example shows that cut for finite sets cannot replace cut for sets in 1.1.

For 1.4 let $\vdash$ be compact and closed under cut for formulae. Suppose that $X,Z \vdash B$ and $X \vdash A$ for each A in Z. By compactness there exist finite sets X′ and Z′ such that $X' \subset X$ and $Z' \subset Z$ and $X',Z' \vdash B$. By dilution $X,Z' \vdash B$, and since by hypothesis $X \vdash A$ for each A in Z′ it follows by 1.2 and cut for Z′ that $X \vdash B$ as required. 1.5 is immediate from 1.4 and 1.1.

If $L_1$ and $L_2$ are single-conclusion calculi $L_1$ is said to be a *subcalculus* of $L_2$ if $\vdash_1 \subseteq \vdash_2$.

*Theorem 1.6* Every single-conclusion calculus has a maximum compact subcalculus.

*Lemma 1.7* Every single-conclusion relation $\vdash$ has a maximum compact subrelation $\vdash'$, and if $\vdash$ is closed under overlap, dilution and cut for formulae so is $\vdash'$.

*Proofs* For 1.7 let $\vdash'$ be such that $X \vdash' B$ iff $X \vdash B$ and $X' \vdash B$ for some finite subset $X'$ of $X$. It is then evident that $\vdash'$ is a compact subrelation of $\vdash$ and that any compact relation contained in $\vdash$ is contained in $\vdash'$, as required. If $\vdash$ is closed under dilution the definition of $\vdash'$ can be simplified to: $X \vdash' B$ iff $X' \vdash B$ for some finite subset $X'$ of $X$. Obviously $\vdash'$ too is closed under dilution; and it is closed under overlap if $\vdash$ is, for if $B \in X$ then $B \in X'$ for some finite subset $X'$ of $X$, whence $X' \vdash B$ and so $X \vdash' B$. If moreover $\vdash$ is closed under cut for formulae so is $\vdash'$, for if $X \vdash' A$ and $X,A \vdash' B$ then for some finite subsets $X_1$ and $X_2$ of $X$ we have that $X_1 \vdash A$ and (by dilution if necessary) $X_2,A \vdash B$, whence by hypothesis $X_1,X_2 \vdash B$ and so $X \vdash' B$ as required.

For 1.6 it is sufficient to observe that the maximum compact subrelation of any consequence relation is also a consequence relation by 1.7, 1.1 and 1.5.

*Historical note* In the second decade of the century C.I. Lewis proposed an axiomatic treatment of consequence in the guise of a binary connective $\dashv$, and when $A_1,\ldots,A_m \vdash B$ is equated with his $(A_1 \& \ldots \& A_m) \dashv B$ the finite cases of our three conditions can certainly be recovered from his system of strict implication (see e.g. Theorem 12.77 of Lewis and Langford, 1932). For an interesting reassessment of Lewis's project in the light of the present one see Scott, 1971 and 1974b.

Hertz (1923) developed a calculus with formulae of the form $A_1,\ldots,A_m \to B$, and when the arrow is read as $\vdash$ each of his rules becomes equivalent (with the qualifications mentioned in the text) to one of our conditions. For example, Hertz's assertion of the 'trivial' formula scheme $A_1,\ldots,A_m \to A_i$ becomes the assertion $A_1,\ldots,A_m \vdash A_i$ embodied in the overlap condition. This is not to say that Hertz himself interpreted his calculus in this way. On the contrary, his arrow is the Hilbert school's notation for a connective like $\supset$, and Hertz (1922) likens his own use of it to Russell's $\supset_x$, explaining that he restricts himself to formulae of this special sort in order to simplify his investigations into the independence of axiom systems.

Gentzen (1932) offers a variety of readings of Hertz's arrow as a relation symbol (e.g. taking the $A_i$ as standing for events and reading $\to$ as 'causes'), including taking the $A_i$ as standing for propositions and reading $\to$ as 'if the propositions ... are true then the proposition ... is true'. But in Gentzen, 1934, the arrow is read as a connective and $A_1,\ldots,A_m \to B$ equated with $(A_1\&\ldots\&A_m)\supset B$.

Tarski (1930a) is concerned with consequence in 'deductive disciplines', i.e. calculi characterised by rules of inference. He begins by defining the consequences of a set of formulae in any particular discipline to be the intersection of all the sets that include it and are closed under the relevant rules, where T is closed under R if $B \in T$ whenever X R B and $X \subset T$. This anticipates our definition of consequence by rules ($\vdash_R$) in the next section, since T is closed under R iff $\langle T,U\rangle$

satisfies R. Tarski rather abruptly abandons this approach with the remark that 'on account of the intended generality of these investigations we have no alternative but to regard the concept of consequence as primitive', but in fact such an alternative is readily available provided the mistaken first-level generalisations '$X \vdash B$ iff $(\exists R)(X \vdash_R B)$' or '$X \vdash B$ iff $(R)(X \vdash_R B)$' do not obscure the correct second-level one (cf. our Theorem 1.12): '$\vdash$ is a consequence relation in Tarski's sense iff $(\exists R)(\vdash = \vdash_R)$'. The implicit definition to which Tarski turns is couched in the language of functions - compare the function T( ) used in the proof of Theorem 1.1 above - but the relevant axioms correspond to overlap (Axiom 2), cut for sets (Axiom 3), and dilution and compactness (Axiom 4). To these Tarski (1930b) adds axioms relating consequence and the classical connectives. Tarski takes the justification for his Axiom 4, i.e. the compactness of consequence by rules, to be obvious from their finiteness; our Theorem 1.10 supplies a proof for those, like us, who do not find it obvious. Tarski's implicit definition is reconsidered from a multiple-conclusion standpoint by Scott (1974a), who covers some of the ground of our Sections 2.2 and 5.2 in the process.

## 1.2 *Rules of inference*

No apology is needed for considering non-compact calculi, if only because the classical logic of second- and higher-order quantification, which has as good a claim as any to represent the underlying logic of the working mathematician, is not compact. Nevertheless compact calculi have a special importance, reflected in the fact that only for a compact calculus is it possible to devise a finite proof technique. The conventional idea of a proof as a finite sequence of formulae governed by rules of inference presupposes that only a finite number of formulae can figure in a rule. We therefore consider single-conclusion *rules of inference* of the form 'from X infer B if X R B', where R is such that X R B can only hold if X is finite. In Chapters 6 and 12 we look at 'rules' with infinitely many premisses and 'proofs' designed to accommodate them, and with this in mind we mark our results with a † if they depend on the restriction to the finite case.

If X R B we call X the set of *premisses* and B the *conclusion* of the instance $\langle X,B \rangle$ of the rule; and we say that R defines the *scope* of the rule. It may or may not be possible to indicate the scope of a rule by means of formula schemes, as in the familiar formulation of modus ponens as 'from A,A⊃B infer B'. The essence of a rule evidently lies in the relation that determines its scope, and we therefore make free use of the terminology of relations - inclusion, intersection, closure, etc. - as applying to rules. Moreover a family of rules {from X infer B if X $R_i$ B} is always equivalent to a single rule 'from X infer B if X $\bigcup R_i$ B', and so we can talk indifferently of a rule, rules or a set of rules.

A partition $\langle T,U \rangle$ *satisfies* R if it satisfies every instance of R, and *invalidates* R if it invalidates some instance of it; and R is *valid* in a set I of partitions if it is satisfied by every member of I. We write $\vdash_R$ for the consequence relation characterised by the set of partitions that satisfy R, and say that B is a *consequence of* X *by* R if X $\vdash_R$ B, i.e. if every partition that satisfies R satisfies $\langle X,B \rangle$. As the proof of Theorem 1.9 shows, $\vdash_R$ is the least consequence relation containing R.

The *closure* of R under a number of conditions is defined as the intersection of all the relations that contain R and satisfy the conditions. Since if every one of a family of relations $\{\vdash_i\}$ is closed under overlap (or dilution or cut) then so is their intersection $\bigcap\vdash_i$, it follows that the closure of R under any combination of these conditions is the least relation that contains R and is closed under the conditions in question. In particular, R is closed under overlap (dilution, cut) iff it is identical with its closure, so that our use of the word 'closed' in this context is equivalent to the standard one.

*Theorem 1.8* $X \vdash_R B$ iff $\langle X,B\rangle$ is valid whenever R is valid.

*Theorem 1.9* $\vdash_R$ is the closure of R under overlap, dilution and cut for sets.

*Theorem 1.10*† $\vdash_R$ is compact.

*Theorem 1.11*† $\vdash_R$ is the closure of R under overlap, dilution and cut for formulae.

*Proofs* For 1.8, let $\langle X,B\rangle$ be valid in every set I in which R is valid; then if $\langle T,U\rangle$ is any partition that satisfies R we see by taking I to be the unit set of $\langle T,U\rangle$ that $\langle T,U\rangle$ satisfies $\langle X,B\rangle$. Hence $X \vdash_R B$, and the converse is immediate from the definitions.

For 1.9, since the various sets I in which R is valid characterise the various consequence relations $\vdash_I$ that contain R, it follows from 1.8 that $X \vdash_R B$ iff $X \vdash_I B$ for every such relation, whence the result by 1.1.

For 1.10 and 1.11, let $\vdash$ be the closure of R under overlap, dilution and cut for formulae, and let $\vdash'$ be the maximum compact subrelation of $\vdash$ (cf. 1.7). Since R itself is a compact subrelation of $\vdash$ we have $R \subset \vdash' \subset \vdash$; since $\vdash_R$ is closed under overlap, dilution and cut for formulae we have $\vdash \subset \vdash_R$; and since $\vdash'$ contains R and by

1.4 and 1.7 is closed under overlap, dilution and cut for sets we have by 1.9 that $\vdash_R \subset \vdash'$. Hence $\vdash' = \vdash = \vdash_R$ as required. An obvious corollary is that $\vdash_R$ is the intersection of the compact consequence relations that contain R. (Note that when 1.4 and 1.10 are presupposed, 1.9 follows immediately from 1.11 but not vice versa, since the results before 1.11 do not exclude the possibility of a non-compact relation lying between R and $\vdash_R$ and being closed under cut for formulae but not cut for sets.)

A set of rules R *characterises* a calculus if $\vdash_R$ is the consequence relation of the calculus. We show that the calculi which can be characterised by rules are precisely the compact ones.

*Theorem 1.12*† A single-conclusion calculus is characterised by rules of inference iff it is compact.

*Proof* A calculus characterised by rules is compact by 1.10. Conversely, if the consequence relation $\vdash$ of a calculus is compact, let R be the rule 'from X infer B if X is finite and $X \vdash B$'. If $X \vdash B$ then by compactness X′ R B for some finite subset X′ of X and so $X \vdash_R B$ by 1.9; but by 1.9 $\vdash_R$ is the least consequence relation containing R and so $\vdash = \vdash_R$.

We have not said anything so far about the validity of individual formulae, since we regard a calculus as a vehicle of inference, its governing idea being that of valid inference rather than valid formula. But the omission is easily rectified: a partition $\langle T,U \rangle$ *satisfies* B if it satisfies $\langle \Lambda,B \rangle$, i.e. if $B \in T$, and B is *valid* in a set I of partitions if it is satisfied by every member of I. Having said this, however, it is apparent that (as Carnap, 1937, observed) the validity of formulae is tantamount to a special case of the validity of inferences, for B is valid in I iff $\langle\Lambda,B\rangle$ is, i.e. iff $\Lambda \vdash_I B$. There is thus no need to continue with a separate account of the validity of formulae, and we can use the familiar notation $\vdash B$ as an abbreviation for $\Lambda \vdash B$. Similarly we have not yet treated the use of formulae as *axioms* in conjunction with rules of inference. The definition of $\vdash_R$ can be extended in an obvious way to allow for the presence of axioms as well as rules, but when this is done it becomes apparent that postulating an axiom B is tantamount to postulating a rule of inference with zero premisses, viz. 'from Λ infer B'. There is thus no need to provide separately a systematic treatment of axioms.

## 1.3 *Sequence proofs*

A (*sequence*) *proof of* B *from* X *by* R is a finite sequence of formulae $C_1,\ldots,C_n$ such that $C_n = B$ and each $C_i$ is either a member of X or the conclusion of an instance of R all of whose premisses occur in $C_1,\ldots,C_{i-1}$. We say that B is *deducible from* X *by* R if there exists a proof of B from X by R.

This definition assumes that axioms have been assimilated to rules in the way described above, and so does not contain the usual clause allowing $C_i$ to be an axiom. We do not define 'proof of B' separately, since it can be subsumed under the present definition as a proof of B from Λ. Similarly theoremhood can be equated with deducibility from Λ and will not be given a separate treatment.

*Theorem 1.13* Deducibility and consequence by the same single-conclusion rules are the same relation.

*Proof* Let $C_1,\ldots,C_n$ be a proof $\sigma$ of B from X by R, and let $\langle T,U\rangle$ be any partition which invalidates $\langle X,B\rangle$. Then $C_n \in U$, so let $C_i$ be the first member of $\sigma$ to belong to U. Since X and U are disjoint $C_i \notin X$, so $C_i$ must be the conclusion of an instance of R whose premisses precede it in $\sigma$ and so are all in T. Hence $\langle T,U\rangle$ invalidates R. This shows that deducibility by R is contained in $\vdash_R$. For the converse it is sufficient by 1.11 to show that deducibility by R contains R and is closed under overlap, dilution and cut for formulae. For cut, if there are proofs $\sigma$ of A from X and $\sigma'$ of B from X,A, then the sequence $\sigma,\sigma'$ obtained by juxtaposing $\sigma$ and $\sigma'$ is a proof of B from X; for any occurrence of A in $\sigma'$ may be justified in $\sigma,\sigma'$ in the same way as the final occurrence of A in $\sigma$. The other properties are immediate.

The definition of a proof as a sequence of formulae is not the only possible one, and we discuss a number of alternatives, including the Hilbert school's idea of a proof as a tree, in Chapter 11. Each idea of proof carries with it a corresponding idea of deducibility, and we shall say that a proposed definition of proof is *adequate* if, for all R, deducibility by R coincides with $\vdash_R$. Since deducibility by finite proofs is necessarily a compact relation (e.g. any sequence proof from X is a proof from those members of X that actually occur in it), Theorem 1.10 becomes significant as a precondition of the adequacy of any sort of finite proof technique, while Theorem 1.13 shows that in particular

*Theorem 1.14* Sequence proofs are adequate for single-conclusion rules.

'Adequacy' is sometimes encountered in the literature as a name for a property of rules of inference. We may say that a set of rules is adequate for a calculus if it is both sound and complete, where R is sound with respect to L if B is deducible from X by R only when X ⊢ B and complete if, conversely, B is deducible from X by R whenever X ⊢ B. A typical adequacy theorem for rules thus resembles an adequacy theorem for proofs in that each equates a syntactic idea with a (wholly or partly) semantic one, but where the first theorem concerns a particular set of rules the second is entirely general with regard to them. And although the adequacy of rules is invariably discussed with reference to sequence proofs, which as we have seen are adequate, it is possible for rules to be adequate for a calculus when - perhaps only when - they are coupled with an inadequate proof technique. (Theorem 20.7 can perhaps be seen as a case in point.)

Theorem 1.13 can also be regarded in a purely abstract light as the analogue of a historically significant result about the ancestral. The straightforward way of defining the ancestral *R of a binary relation R is to say that x *R y if there exists a finite sequence $z_1, \ldots, z_n$ such that $z_1$ is x and $z_n$ is y and $z_i$ R $z_{i+1}$ for each i. Frege's way is to say that x *R y if y belongs to every set which contains x and is closed under R; and there is an obvious parallel between the first of these definitions and that of deducibility, and between the second and the idea of consequence by rules.

# 2 · Multiple-conclusion calculi

## 2.1 *Consequence*

A *multiple-conclusion calculus* is constituted by a universe V of formulae and a relation ⊢ of multiple-conclusion consequence. As in Section 1.1 we assume that V is not empty, though curiously enough with an empty universe of formulae there would still be two distinct multiple-conclusion calculi, in one of which $\Lambda \vdash \Lambda$ and in the other $\Lambda \nvdash \Lambda$. By a *multiple-conclusion relation* we mean a binary relation between sets of formulae, and when such a relation R holds between sets X and Y we write X R Y and call the pair ⟨X,Y⟩ an *instance* of R. We shall generally take the adjective 'multiple-conclusion' for granted and write simply 'calculus' to contrast with 'single-conclusion calculus', and similarly with 'relation' and other terms.

Our task in this and the succeeding chapter is to develop the appropriate multiple-conclusion analogues of the various definitions of consequence that we have produced for single conclusions. First we say that a partition ⟨T,U⟩ *satisfies* ⟨X,Y⟩ if X overlaps U or Y overlaps T. Otherwise, i.e. if $X \subset T$ and $Y \subset U$, we say that ⟨T,U⟩ *invalidates* ⟨X,Y⟩. If ⟨X,Y⟩ is satisfied by every partition in the set I we say it is *valid* in I and write $X \vdash_I Y$. We say that ⊢ is a *consequence relation* if $\vdash = \vdash_I$ for some I, in which case we say too that ⊢ is *characterised* by I. If T and U consist of the true and untrue formulae on some interpretation, ⟨T,U⟩ satisfies ⟨X,Y⟩ iff at least one member of X is untrue or at least one member of Y is true, this being the relation of 'involution' defined by Carnap (see the note at the end of this section). An important difference between multiple- and single-conclusion satisfaction has to do with the possibility of amalgamating a number of satisfying partitions: if for each i $\langle T_i, U_i \rangle$ satisfies $\langle X, B \rangle$ then so does $\langle \bigcap T_i, \bigcup U_i \rangle$, but there is no comparable result for the satisfaction of a pair $\langle X, Y \rangle$. The difference shows itself in connection with argumentative structure in Chapter 11 and categoricity in Chapter 17.

In following out the second approach the only point of difficulty is the formulation of the cut condition. Had we construed multiple conclusions conjunctively we would have postulated transitivity in the literal sense – 'if $X \vdash Y$ and $Y \vdash Z$ then $X \vdash Z$'; but this is clearly inappropriate if the members of Y are to function conjunctively in $Y \vdash Z$ but disjunctively in $X \vdash Y$, and it turns out that we need a version that takes account of all the different ways of partitioning Y into a pair of subsets. Consider, therefore, the following conditions, where a *partition of* Z is any pair $\langle Z_1, Z_2 \rangle$ such that $Z_1 \cup Z_2 = Z$ and $Z_1 \cap Z_2 = \Lambda$. (The word 'partition' on its own, as opposed to 'partition of Z', will be reserved for partitions of V, as in Section 1.1.)

*Overlap* If X overlaps Y then $X \vdash Y$.

*Dilution* If $X' \vdash Y'$, where $X' \subset X$ and $Y' \subset Y$, then $X \vdash Y$.

*Cut for sets* If $X, Z_1 \vdash Z_2, Y$ for each partition $\langle Z_1, Z_2 \rangle$ of Z, then $X \vdash Y$.

As before, a relation is *closed under* these conditions if it satisfies them, and its *closure* under them is the intersection of all the relations containing it and satisfying them. It is again evident that a relation is closed iff it is identical with its closure.

*Theorem 2.1* A relation is a consequence relation iff it is closed under overlap, dilution and cut for sets.

*Proof* Closure under overlap and dilution is immediate from the definition of a consequence relation. For cut, let $\vdash_I$ be any consequence relation and suppose that $X \nvdash_I Y$, so that some partition $\langle T,U \rangle$ in I invalidates $\langle X,Y \rangle$. For any Z let $Z_1 = Z \cap T$ and $Z_2 = Z \cap U$. Then $\langle Z_1,Z_2 \rangle$ is a partition of Z and $\langle T,U \rangle$ invalidates $\langle X \cup Z_1, Z_2 \cup Y \rangle$, i.e. $X,Z_1 \nvdash_I Z_2,Y$ as required.

For the converse let $\vdash$ be closed under overlap, dilution and cut for sets: we show that $\vdash = \vdash_I$, where I is the set of partitions $\langle T,U \rangle$ such that $T \nvdash U$. If $X \nvdash Y$ then by cut for V there exists a partition $\langle T,U \rangle$ such that $X,T \nvdash U,Y$. Appealing to the overlap condition it follows that $X \subset T$ and $Y \subset U$, and hence $\langle T,U \rangle$ both belongs to I and invalidates $\langle X,Y \rangle$. On the other hand if $X \vdash Y$ then by dilution $X,T \vdash U,Y$ for every partition $\langle T,U \rangle$ in I, and since $T \nvdash U$ this means that either X overlaps U or Y overlaps T; i.e. $\langle X,Y \rangle$ is satisfied by every member of I.

In formulating cut for sets a condition is asserted for all X, Y, Z. By confining it to a particular Z we obtain *cut for* Z; by confining it to finite Z we obtain *cut for finite sets*; and by confining it to singletons Z = {A} we obtain *cut for formulae*, viz. if $X,A \vdash Y$ and $X \vdash A,Y$ then $X \vdash Y$. Cut for formulae stands to the cut rule of Gentzen, 1934, in the same way as the corresponding single-conclusion condition does to that of Gentzen, 1932 (cf. Section 1.1 and the note at the end of this section). The following results show how the present cut conditions are related. Theorems 2.3 and 2.4 have single-conclusion analogues (Theorems 1.2 and 1.3), but Theorem 2.2 does not, since every single-conclusion relation is trivially closed under single-conclusion cut for V.

*Theorem 2.2* Cut for sets is equivalent (granted dilution) to cut for V.

*Theorem 2.3* Cut for formulae is equivalent (granted dilution) to cut for finite sets.

*Theorem 2.4* Cut for finite sets is weaker (even granted overlap and dilution) than cut for sets.

*Proofs* For 2.2, if $\vdash$ is closed under cut for V then whenever $X \nvdash Y$ there exists a partition $\langle T,U\rangle$ such that $X,T \nvdash U,Y$. For any Z let $Z_1 = Z \cap T$ and $Z_2 = Z \cap U$; then $\langle Z_1,Z_2\rangle$ is a partition of Z such that $Z_1 \subset T$ and $Z_2 \subset U$, whence by dilution $X,Z_1 \nvdash Z_2,Y$ as required.

For 2.3 suppose that $\vdash$ is closed under cut for formulae and that $X \nvdash Y$. Let S be the set of pairs $\langle W_1,W_2\rangle$ of subsets of a given finite set Z such that $X,W_1 \nvdash W_2,Y$. Then S is nonempty (since by hypothesis $\langle \Lambda,\Lambda\rangle \in S$), finite and partially ordered by the subpair relation $\subset$ such that $\langle W_1,W_2\rangle \subset \langle W_1',W_2'\rangle$ iff $W_1 \subset W_1'$ and $W_2 \subset W_2'$. Hence S has a maximal member $\langle Z_1,Z_2\rangle$, and $Z_1 \cup Z_2 = Z$ since if $A \in Z-(Z_1 \cup Z_2)$ then by cut for A either $X,Z_1,A \nvdash Z_2,Y$ or $X,Z_1 \nvdash A,Z_2,Y$, contradicting the maximality of $\langle Z_1,Z_2\rangle$. By dilution, $X,Z_1-Z_2 \nvdash Z_2,Y$, and as $\langle Z_1-Z_2, Z_2\rangle$ is a partition of Z it follows that $\vdash$ is closed under cut for Z, as required. (Obviously any other division of the overlap of $Z_1$ and $Z_2$ would serve the same purpose.)

For 2.4, let V be infinite and let $X \vdash Y$ iff X or Y is infinite or X overlaps Y. It is easily verified that $\vdash$ is closed under overlap, dilution and cut for finite sets; but it is not closed under cut for sets, for $\Lambda \nvdash \Lambda$ although $T \vdash U$ for every partition $\langle T,U\rangle$.

Overlap and dilution have no plausible rivals as analogues

of the corresponding single-conclusion conditions. This is not the case with cut, as the following series shows:

*Cut$_1$* If X ⊢ A,Y for each A in W and X,W ⊢ Y then X ⊢ Y.

*Cut$_2$* If X,B ⊢ Y for each B in W and X ⊢ W,Y then X ⊢ Y.

*Cut$_3$* If X ⊢ A,Y for each A in $W_1$ and X,B ⊢ Y for each B in $W_2$ and X,$W_1$ ⊢ $W_2$,Y, then X ⊢ Y.

*Theorem 2.5* Cut$_1$ and cut$_2$ are independent, but each is stronger than cut for formulae.

*Theorem 2.6* Cut$_3$ is equivalent (granted dilution) to cut$_1$ and cut$_2$ together, but stronger than either separately.

*Theorem 2.7* Cut for sets is stronger (granted dilution) than cut$_3$.

*Proofs* For 2.5 we note that cut for formulae is the special case of cut$_1$ or cut$_2$ in which W is taken to be a unit set. To show that there are no other implications between the three conditions, let V be infinite and let X ⊢ Y iff X is infinite or Y nonempty. It is easily verified that ⊢ is closed under overlap, dilution, cut$_2$ and cut for formulae, but that cut$_1$ fails when X is finite, Y empty and W infinite. Similarly the relation ⊢ such that X ⊢ Y iff Y is infinite or X nonempty is closed under overlap, dilution, cut$_1$ and cut for formulae, but not under cut$_2$.

For 2.6 we obtain cut$_3$ from cut$_1$, cut$_2$ and dilution thus: if X ⊢ A,Y for each A in $W_1$ then by dilution X ⊢ A,$W_2$,Y for each A in $W_1$. If too X,$W_1$ ⊢ $W_2$,Y then X ⊢ $W_2$,Y by cut$_1$; and if also X,B ⊢ Y for each B in $W_2$ then X ⊢ Y by cut$_2$. Conversely cut$_1$ is the special case of cut$_3$ in which $W_2$ is empty, and cut$_2$ is the case in which $W_1$ is empty. Since by 2.5 neither cut$_1$ nor cut$_2$ implies the other, it follows that neither implies cut$_3$.

For 2.7 let $Z = W_1 \cup W_2$ and note that every partition $\langle Z_1, Z_2 \rangle$ of Z is such that either $Z_1$ contains a member B of $W_2$ or $Z_2$ contains a member A of $W_1$ or (only if $W_1$ and $W_2$ are disjoint) $Z_1 = W_1$ and $Z_2 = W_2$. Using dilution, $\mathrm{cut}_3$ is obtained as a corollary of cut for this Z. On the other hand the counterexample of 2.4 shows that $\mathrm{cut}_3$ does not imply cut for sets.

Since all the counterexamples used in proving Theorems 2.4-7 are of relations closed under overlap and dilution, they establish the crucial difference between cut for sets and its rivals, namely that only it is strong enough to sustain Theorem 2.1.

*Historical note* Gentzen (1934) makes Hertz's notation symmetrical by introducing 'sequents' $A_1, \ldots, A_m \rightarrow B_1, \ldots, B_n$. On what may be called the material interpretation of sequents the arrow is a connective (material implication) belonging to the same object-language as the $A_i$ and $B_j$, the sequent is a notation for a common or garden formula $(A_1 \& \ldots \& A_m) \supset (B_1 \vee \ldots \vee B_n)$, and the calculus of sequents LK is a variant of the conventional predicate calculus. On the 'metalinguistic' interpretation the arrow is a relation symbol, the sequent is the statement $A_1, \ldots, A_m \vdash B_1, \ldots, B_n$, and the calculus of sequents is a (single-conclusion) metacalculus for a multiple-conclusion object-calculus. On either interpretation Gentzen's rules are single-conclusion rules: it is their content, not their form, that may involve multiple conclusions. The choice between the two interpretations has been a matter not so much of dispute as tacit disagreement. Church (1956) speaks for one side in saying categorically that 'Gentzen's arrow is not comparable to our syntactical notation $\vdash$ ', while Scott, 1971, (if one discounts his attempt to assimilate the metalinguistic and the Fregean use of ⊢) speaks for the other side by taking the metalinguistic interpretation for granted. It seems clear enough that Gentzen himself interpreted his calculus in the material way. He says as much (e.g. 1934, I.2.4), but the decisive evidence is his treatment of the equivalence between the sequent calculus and the conventional one (1934, V.1 and 6). (Passages pointing the other way are 1936, II.5.1 - though the effect has been enhanced in translation - and 1938, 1.2 'Explanatory remarks' on sequents without succedents; but in each case the material interpretation is reaffirmed on the same page of the original.) If, however, one considers the question as a logical rather than a historical one, the answer seems to be that although ⊢ and ⊃ differ both in grammar and strength, neither difference can be resolved inside the calculus of sequents. For example, a disjunction of sequents like '$A \rightarrow B$ or $A \rightarrow \sim B$', which could settle the matter of strength, is simply inexpressible. Similarly the arrow is both non-iterable, like ⊢, and only capable of holding between finitely many formulae, like ⊃. If one set out deliberately to encapsulate just the principles that are common to ⊃ and ⊢ (not to mention Lewis's ⥽ ), one could hardly do it better than the sequent calculus. But, granted this, the metalinguistic interpretation is surely the more fruitful one. Our only cavil (apart from the artificiality of the restriction to a finite number of premisses and conclusions) is that studying multiple conclusions exclusively through a metacalculus of sequents is rather like waiting for Godot. The time is profitably spent (in establishing metametatheorems such as Gentzen's Hauptsatz), but the actual multiple-conclusion calculus remains offstage; and one could be forgiven for exploring, as Carnap and Kneale do, more mundane but less allusive ways of tackling the subject.

Carnap (1943) introduces multiple conclusions by taking as primitive the idea of a 'junctive', where a pair of junctives goes with each set of formulae; a 'conjunctive' being, roughly speaking, a set construed conjunctively and a 'disjunctive' one construed disjunctively. But we regard this as a false start, and in the final section of the book Carnap himself briefly reworks his ideas in terms of plain sets. Here, after defining involution in the way mentioned in the text, he defines 'L-involution' as the relation which holds between X and Y iff X involves Y in every 'L-state' or possible state of affairs, thereby anticipating our $\vdash_I$. He then introduces rules of inference (including infinite ones) and, in the absence of any conception of deducibility by multiple-conclusion proofs, defines 'C-involution' as the relation which holds between X and Y iff Y overlaps every T such that (a) $X \subset T$ and (b) for all W and Z, if $W \subset T$ and W R Z then Z overlaps T. C-involution is thus equivalent to consequence by R as defined in Section 2.3 below.

To the best of our knowledge the first person to put forward the inferential definability of the logical constants as the basis for a demarcation of logic was Popper (1947ff.), in a series of ambitious but confused papers: 'these bad and ill-fated papers', as he was to call them. (See Popper, 1974, and reviews indexed in the Journal of Symbolic Logic, vol.14, 1949). The project was revived by Kneale, 1956, though he, like Popper, assigns no special significance to introduction rules. The origin of this is Gentzen's remark (1934, II.5.13) that the introduction rules of his natural-deduction system NJ 'represent, as it were, the definitions of the symbols concerned, and the eliminations are no more, in the final analysis, than the consequences of these definitions'. It is made part of the demarcation of logic by Hacking (1979), who makes the whole project his own by pursuing it at a new level of sophistication and precision.

## 2.2 *Compactness*

A relation ⊢ is *compact* if whenever X ⊢ Y there exist finite subsets X′ and Y′ of X and Y such that X′ ⊢ Y′; and a calculus is *compact* if its consequence relation is compact. The term entered the literature of single-conclusion logic from that of topology, and in passing it is worth seeing what analogy there is between compactness in multiple-conclusion logic and in topology. Let $\vdash_I$, then, be any consequence relation, let I(X,Y) be the set of those members of I that satisfy ⟨X,Y⟩, and let [I] be the set of all such sets. The *topology generated by* [I] is the topology on I whose open sets are the unions of finite intersections of members of [I], including I as I(V,V) and Λ as I(Λ,Λ). A set of subsets of I *covers* I if the union of its members is I, and a topology on I is *compact* if every set of open sets that covers I has a finite subset that also covers I.

*Theorem 2.8* $\vdash_I$ is compact iff the topology generated by [I] is compact.

*Proof* By a classical result (see e.g. Theorem 5.6 of Kelley, 1955) the topology generated by [I] is compact iff every subset $\{I(X_i,Y_i): i \in S\}$ of [I] that covers I has a finite subset that also covers I. This is equivalent to the condition that if $\bigcup X_i \vdash_I \bigcup Y_i$ when the union is taken over S then $\bigcup X_i \vdash_I \bigcup Y_i$ when the union is taken over some finite subset of S. Evidently this condition holds if $\vdash_I$ is compact, and conversely we can demonstrate the compactness of $\vdash_I$ by applying the condition in the case where $X_i \cup Y_i$ is a singleton for each i in S.

For Theorem 2.8 to be more than a topological gloss the topology generated by [I] ought to have some independent logical significance. This is most plausible in the case of calculi incorporating the multiple-conclusion version PC of the two-valued propositional calculus (cf. Chapter 18), for which the topology generated by [I] reduces to [I]. For in this case the union of any members of [I] is always in [I] since $\bigcup I(X_i,Y_i) = I(\bigcup X_i, \bigcup Y_i)$, and the intersection of any two members is also in [I] since $I(X_1,Y_1) \cap I(X_2,Y_2) = I(\Lambda, (\sim X_1 \vee Y_1) \& (\sim X_2 \vee Y_2))$. The open sets of the topology are therefore the members of [I].

All this relates only to multiple-conclusion logic. A single-conclusion calculus is never characterised by a unique set of partitions, as Theorem 5.2 will show, and even in the case of the classical calculus some characteristic sets of partitions give rise to compact topologies (cf. Lyndon, 1964) and others to non-compact ones (cf. Section 18.3). The most we have in general is, by Theorems 2.8, 5.8 and 5.12, that a single-conclusion consequence relation ⊢ is compact iff the topology generated by [I] is compact for at least one I that characterises ⊢.

*Theorem 2.9* For compact relations cut for formulae is equivalent (granted dilution) to cut for sets.

*Theorem 2.10* A compact relation is a consequence relation iff it is closed under overlap, dilution and cut for formulae.

*Proofs* 2.10 is immediate from 2.9 and 2.1. The proof of 2.9 follows that of 2.3 except that we need to show that S has a maximal member even for infinite Z. For this we note that a pair belongs to S iff all its finite subpairs do so, since by compactness and dilution $X,W_1 \vdash W_2,Y$ iff $X,W_1' \vdash W_2',Y$ for some finite subsets $W_1'$ of $W_1$ and $W_2'$ of $W_2$. The result then follows by applying Tukey's lemma, which states that if a nonempty set S of sets is of 'finite character', i.e. if a set belongs to S iff all its finite subsets do so, then S has a maximal member (cf. Kelley, 1955). The application is achieved by considering a domain of objects '⟨A' and 'A⟩' for each A in V, and observing that by our general notational conventions $\langle W_1, W_2 \rangle$ may be read as short for $\{\langle A: A \in W_1\} \cup \{A\rangle: A \in W_2\}$, whereupon the subpair relation becomes straightforward set-inclusion. Alternatively the result can be proved by applying Zorn's lemma; thus Scott (1974a), making the further assumption of closure under overlap, proves what is in effect our 2.10. The result can also be proved directly, by an inductive argument whose strategy is well known from its use by Lindenbaum (see Theorem 56 of Tarski, 1930a) and by Henkin (1949). All the versions rely on the axiom of choice unless V is already well-ordered.

Leaving aside these variations, the use made of the argument as a whole is interestingly different in the multiple- and single-conclusion cases. It is not needed for the single-conclusion analogue of the present result (1.4), and therefore is not needed in order to show the adequacy of sequence proofs for single-conclusion rules; but it is of course needed for any Henkin-style proof of the completeness of a particular set of rules. Contrariwise, the argument appears to be essential in the multiple-conclusion case if we are to show the compactness of $\vdash_R$ and hence the adequacy of any finite technique of proof, but may not otherwise be needed when proving the completeness of particular rules (as for example in 18.1).

A calculus $L_1$ is a *subcalculus* of $L_2$ if $\vdash_1 \subset \vdash_2$.

*Theorem 2.11* Every calculus has a maximum compact subcalculus.

*Lemma 2.12* Every relation $\vdash$ has a maximum compact subrelation $\vdash'$, and if $\vdash$ is closed under overlap, dilution and cut for formulae so is $\vdash'$.

*Proofs* 2.11 is a corollary of 2.12, using 2.1 and 2.10. The proof of 2.12 follows that of 1.7, taking $X \vdash' Y$ iff ($X \vdash Y$ and) $X' \vdash Y'$ for some finite subsets $X'$ and $Y'$ of X and Y. In particular, to show that $\vdash'$ is closed under cut for formulae, suppose that $X,A \vdash' Y$ and $X \vdash' A,Y$. Then for some finite subsets $X_1$ and $X_2$ of X and some finite subsets $Y_1$ and $Y_2$ of Y we have (by dilution if necessary) that $X_1,A \vdash Y_1$ and $X_2 \vdash A,Y_2$, whence by cut $X_1,X_2 \vdash Y_1,Y_2$ and so $X \vdash' Y$ as required.

## 2.3 *Rules of inference*

We consider rules of the form 'from X infer Y if X R Y', where R is a relation such that if X R Y then X and Y are finite. If X R Y we call the members of X and Y the *premisses* and *conclusions* of the instance ⟨X,Y⟩ of the rule, and in practice we identify the rule with the relation R that determines its scope. We consider rules with infinitely many premisses and conclusions in Chapters 6 and 12, and meanwhile continue to mark theorems with a † if they hold only in the finite case. To illustrate the idea of a rule, including the use of formula schemes to convey its scope, we give a set of rules for the two-valued propositional calculus PC (cf. Chapter 18).

| | | |
|---|---|---|
| From A,A⊃B infer B | From B infer A⊃B | From Λ infer A,A⊃B |
| From A&B infer A | From A&B infer B | From A,B infer A&B |
| From A infer A∨B | From B infer A∨B | From A∨B infer A,B |
| From A,~A infer Λ | From Λ infer A,~A | |

These *rules for PC* are taken from Kneale, 1956, except that where we have rules with an empty set of premisses or conclusions he has rules permitting inference from or to any formula. Thus his rules for negation are 'from A,~A infer B' and 'from B infer A,~A', and his third rule for material implication is 'from C infer A,A⊃B'. It is clear from Kneale, 1962, that he does this not because of some settled policy of excluding the empty case, but because the distinction is not important to him. For some of its implications, however, see Section 3.1 and the discussion of the sign of calculi in Sections 5.4 and 16.1.

A partition ⟨T,U⟩ *satisfies* a relation R if it satisfies every instance of R and *invalidates* R if it invalidates some instance of it, and R is *valid* in a set I of partitions if it is satisfied by every member of I. We write $\vdash_R$ for the consequence relation characterised by the set of partitions which satisfy R, and say that Y *is a consequence of* X *by* R if X $\vdash_R$ Y, i.e. if every partition that satisfies R satisfies ⟨X,Y⟩.

*Theorem 2.13* X $\vdash_R$ Y iff ⟨X,Y⟩ is valid whenever R is valid.

*Theorem 2.14* $\vdash_R$ is the closure of R under overlap, dilution and cut for sets.

*Theorem 2.15*† $\vdash_R$ is compact.

*Theorem 2.16*† $\vdash_R$ is the closure of R under overlap, dilution and cut for formulae.

*Proofs* follow those of 1.8-11, using the corresponding multiple-conclusion results.

*Immediate consequence* by R is defined as the closure of R

under overlap and dilution, i.e. Y is an immediate consequence of X iff they overlap or there exist subsets X′ of X and Y′ of Y such that X′ R Y′. Immediate consequence is generally a much weaker relation than consequence by the same rules, but the two are equivalent in the special case where the premisses and conclusions form a partition of V:

*Lemma 2.17* $T \vdash_R U$ iff U is an immediate consequence of T by R.

*Proof* If $T \vdash_R U$ every partition that invalidates ⟨T,U⟩ invalidates R. Since every partition invalidates itself, it follows that ⟨T,U⟩ invalidates R and from this that U is an immediate consequence of T. The converse is obvious by 2.14.

A set of rules R *characterises* a calculus if the consequence relation of the calculus is $\vdash_R$. As in the single-conclusion case the calculi that can be characterised by rules of inference are precisely those that are compact.

*Theorem 2.18*† A calculus is characterised by rules of inference iff it is compact.

*Proof* follows that of 1.12, using the rule 'from X infer Y if X and Y are finite and $X \vdash Y$'.

# 3 · Tree proofs

## 3.1 *Definition*

Our first definition of proof for multiple-conclusion logic generalises the idea of a sequence proof in a manner suggested by the informal technique of proof by cases. A proof will be cast in the form of a tree of formulae: it will begin like a sequence but each time the application of a rule of inference introduces a number of conclusions the tree will split into a number of fresh branches. The conclusions of the rule are treated as if they represented alternative possibilities or cases, and the proof is therefore continued along each of the resulting branches independently. In writing out such a proof it is easiest to make it run down the page, adding horizontal lines where necessary to clarify the branching structure. In this way, for example, Figure 3.1 shows how (A⊃B)⊃B can be proved from A∨B by the rules for PC set out in Section 2.3; and Figure 3.2 shows how ~A,B can be proved from A⊃B by the rules for PC.

```
        ______________________
        A⊃B       (A⊃B)⊃B
        A∨B
      ____________
      A          B
      B       (A⊃B)⊃B
   (A⊃B)⊃B
```

Figure 3.1

```
      A⊃B
    _______
    A    ~A
    B
```

Figure 3.2

A tree of formulae, like a sequence, is a special case of what may be called an array. An array is not simply a set of formulae: once we recognise that the same formula can occur at different places in (say) a sequence, we must distinguish its occurrences from each other and so from the formula itself. This is not to be confused with the 'type-token' distinction between a symbol considered as an abstract entity or type and its occurrences (physical particulars or tokens). In a sequence consisting exclusively of symbol types the type-token distinction would be irrelevant but we should still need to draw the present one. As usual there is more than one way of treating the matter systematically: occurrence can be taken as an independent idea or it can be defined in other set-theoretic terms, and an array can be construed as something whose occurrences consist of occurrences of formulae or as something which itself consists of occurrences of formulae. We adopt the second alternative in each case, defining an *occurrence* of a formula A to be an ordered pair $\langle i,A \rangle$ for arbitrary index i, and an *array* to be a set of occurrences. If the use of the same index with different formulae is forbidden an array becomes identical with a family of formulae, given the customary treatment of a family as a function from a set of indices. A structured (e.g. well-ordered) array of formulae is a *replica* of another if there is an isomorphism between them in which corresponding elements are occurrences of the same formula. We use Greek letters for arrays, but for the moment use A, B, C to stand alike for formulae and their occurrences. We include the word 'occurrence' whenever it is needed to remove an ambiguity, but in general we avail ourselves of the customary elliptical language; e.g. in the following definition the clause '$\phi$ is a branch ending in a formula in Y' is short for '$\phi$ is a branch ending in an occurrence of a formula in Y'.

A *sequence* of formulae is a well-ordered array. A *tree* of formulae is a partially ordered array such that for each occurrence A the set $\hat{A}$ of occurrences that precede A is a sequence, and a *branch* is a maximal well-ordered subset of a tree (cf. e.g. Jech, 1971). We define a *file* to be an initial segment of a branch, and the *rank* $\rho(\phi)$ succeeding a file $\phi$ is the set of occurrences A such that $\hat{A} = \phi$. A (*tree*) *proof from* X *to* Y *by* R is a finite tree of formulae such that for each file $\phi$ either (i) $\rho(\phi)$ consists of an occurrence of a formula in X or (ii) $\phi$ is a branch ending in a formula in Y or (iii) $\rho(\phi)$ consists of the conclusions of an instance of R whose premisses occur in $\phi$. We say 'a proof from X to Y' in preference to 'a proof of Y from X' partly to avoid the suggestion that the members of Y are being proved individually and partly to provide a phrase in which premisses and conclusions are mentioned in that order, as they are in 'X ⊢ Y'.

It follows immediately from the definition that any replica of a proof from X to Y by R is also a proof from X to Y by R, and we illustrate the other ideas involved by verifying the claim that Figure 3.2 represents a proof from A⊃B to ~A,B by the rules for PC. The tree in question has two branches, namely $\phi_1$ = {A⊃B,A,B} and $\phi_2$ = {A⊃B,~A}. Besides itself $\phi_1$ contains the three files $\phi_3 = \Lambda$, $\phi_4$ = {A⊃B}, and $\phi_5$ = {A⊃B,A}. The files in $\phi_2$ are itself, $\phi_3$ and $\phi_4$. Like all branches, $\phi_1$ and $\phi_2$ are succeeded by the empty rank, while the ranks succeeding the other files are $\rho(\phi_3)$ = {A⊃B}, $\rho(\phi_4)$ = {A,~A}, and $\rho(\phi_5)$ = {B}. It is now easy to see that each file satisfies one of the conditions in the definition of proof: $\phi_1$ and $\phi_2$ satisfy (ii) since they end in one of the conclusions ~A and B; $\phi_3$ satisfies (i) since $\rho(\phi_3)$ is the premiss A⊃B; and $\phi_4$ and $\phi_5$ satisfy (iii) since $\rho(\phi_4)$ consists of the conclusions of the rule 'from Λ infer A,~A' and $\rho(\phi_5)$ is the conclusion of the rule 'from A,A⊃B infer B'.

The difference in style between the definitions of sequence proof and tree proof calls for an explanation. Certainly it is easy to bring the former into line with the latter, for any sequence is a one-branch tree and so we have: a finite sequence of formulae is a proof of B from X by (single-conclusion) R if for each file $\phi$ either (i) $\rho(\phi)$ consists of a formula in X or (ii) $\phi$ is the entire sequence and ends in B, or (iii) $\rho(\phi)$ consists of the conclusion of an instance of R whose premisses are in $\phi$. Moreover the entire theory of sequence proofs can be reproduced in a multiple-conclusion context. For to every single-conclusion rule R there corresponds a *singleton-conclusion* rule $R_1$ such that X $R_1$ {B} iff X R B; proofs by $R_1$ may branch (as on p.4 when $A_1 = \ldots = A_m$), but we have

*Theorem 3.1* A sequence proof of B from X by single-conclusion R is a tree proof from X to {B} by the corresponding singleton-conclusion rules $R_1$, and every branch of a tree proof from X to Y by $R_1$ is a sequence proof of B from X by R for some B in Y.

> *Proof* Since there are no instances of $R_1$ with an empty set of conclusions, every branch in a proof from X to Y by $R_1$ must end in a member of Y, and the result follows immediately from the definition of tree proof and the new definition of sequence proof.

If we had excluded rules with zero conclusions, as Kneale does, we could alternatively have brought the definition of tree proof into line with that of sequence proof. For in this case conditions (i) and (iii) cannot apply to a branch anyway, so nothing is lost by ignoring the empty rank associated with it. The remaining ranks are simply the equivalence classes determined by the relation 'has the same predecessors as', and a proof from X to Y by R is a finite tree of formulae such that every branch ends in a member of Y and each new-style 'rank' consists either of an occurrence of a member of X or of the conclusions of an instance of R whose premisses are among the preceding formulae. The original definition of sequence proof corresponds to the special case of this where there is only one branch and so each 'rank' consists of exactly one occurrence. When rules with zero conclusions are allowed, however, the empty rank following a branch of a tree proof may be justified by appeal to a zero-conclusion rule, so that the branch need not end in a conclusion: indeed it may even end in a premiss, as Figure 3.4 shows. In such a case the device of treating branches separately from proper files is no longer any help in harmonising the styles of the two definitions.

In writing out proofs that use zero-conclusion rules it is helpful to add a horizontal line below branches which terminate by virtue of such a rule. In this way Figure 3.3 represents a proof from ~(A&B) to ~Av~B by the rules for PC, and Figure 3.4 represents a proof from ~~B to B.

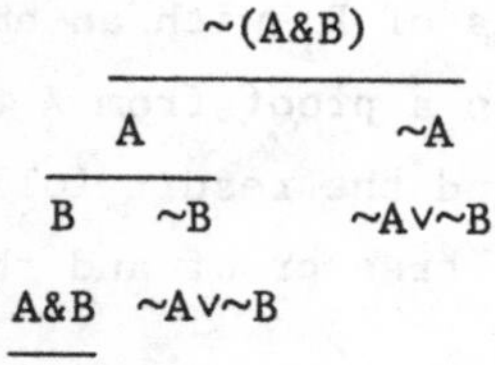

~(A&B)<br>
A    ~A<br>
B  ~B    ~Av~B<br>
A&B  ~Av~B

Figure 3.3

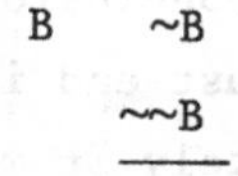

B  ~B<br>
~~B

Figure 3.4

Zero-conclusion rules draw some support from the informal practice of proof by cases. In proving B from ~~B by a consideration of the alternative cases (i) B and (ii) ~B, the argument in case (ii) would, if we used Kneale's rule 'from A,~A infer C', have to run '~B, but ~~B, so B'. In practice mathematicians argue more like this: '~B, but ~~B; this is a contradiction and so we need not pursue this case.' It is just this style of argument which is formalised by means of a zero-conclusion rule, as in Figure 3.4.

## 3.2 *Extended proofs*

There are several ideas of multiple-conclusion proof that have a claim to provide the natural analogue of the sequence proof, and we have chosen perhaps the most restrictive of them. It can be relaxed in any or all of three respects.

In the first place there is no provision in our existing definition for the presence of a rank $\rho(\phi)$ to be justified on the ground that it repeats one of the formulae in $\phi$. Admittedly no similar provision is explicit in the definition of sequence proof, but there it would be otiose: in whatever way the first occurrence of a formula in a sequence proof is justified (i.e. whether as a premiss or an axiom or as being inferred by a rule from one or more preceding formulae), every subsequent occurrence of it can be justified in exactly the same way. Our demonstration that single-conclusion deducibility is closed under cut (Theorem 1.13) relied on this implicit provision for repetition to justify the claim that the juxtaposition of a proof of A from X with one of B from X,A produces a proof of B from X. For tree proofs, on the other hand, a provision for repetition is not otiose, and the juxtaposition of proofs does not necessarily produce a proof without it. For example, taking X to be {A⊃B} and Y to be {~A,B}, $\tau_1$ of Figure 3.5 is a proof from X to A,Y and $\tau_2$ a proof from X,A to Y by the rules for PC. The tree $\tau$ is obtained by juxtaposing $\tau_1$ and $\tau_2$ in the obvious way, but $\tau$ is not the expected proof from X to Y since there is no way of justifying the rank consisting of the second occurrence of A.

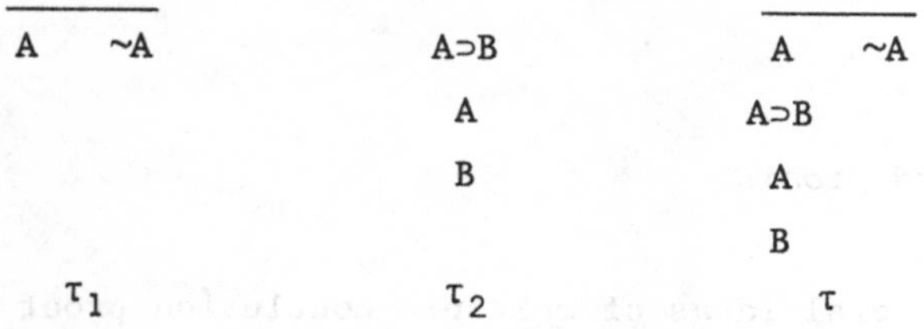

$\tau_1$: $\overline{A \quad {\sim}A}$

$\tau_2$: $A \supset B$, $A$, $B$

$\tau$: $\overline{A \quad {\sim}A}$, $A \supset B$, $A$, $B$

Figure 3.5

Secondly we might allow additional formulae to occur in a rank alongside a premiss of the proof, or alongside the conclusion of a rule, by replacing 'consists of' by 'contains' in the relevant clauses (i) and (iii) of the definition. The motive for this proposal is the fact that a sequence proof can contain formulae which have nothing to do with the deduction of its conclusion, provided only that they follow from the premisses in the prescribed way. For example, the sequence A&B,A,B is a perfectly acceptable sequence proof of B from A&B by the rules 'from A&B infer A' and 'from A&B infer B', although its middle formula represents a dead end or digression as far as the deduction of B is concerned. Since we already allow this sort of digression in the files of a tree proof, why not also allow its ranks to contain formulae which are not deducible from the premisses, provided they lead on to the conclusions in the prescribed way? Since this 'prescribed way' has the feature just mentioned, that not all the formulae in a file need be used in the subsequent deduction, the proposal has the somewhat curious consequence that a formula can occur in a proof without playing any part in it, whether as a premiss or conclusion of the proof as a whole or of any inferential step in it. For example, the tree shown in Figure 3.6 would on the present proposal be a proof from A&B

to B by the rules for PC, in which the formula C occurs in this wholly extraneous way. The effect on proofs using zero-conclusion rules is especially striking; e.g. the tree shown in Figure 3.4 would continue to be a proof from ~~B to B even when an arbitrary tree of formulae was grafted onto the tip of its right-hand branch.

$$\frac{A\&B}{B \qquad C}$$
$$\qquad\quad B$$

Figure 3.6

Thirdly, in the absence of zero-conclusion rules, our existing definition does not allow a file to be continued past a conclusion unless en route to a subsequent one. In this it resembles the definition of sequence proof, which does not allow A&B,B,A to count as a proof of B from A&B even when the final formula duly follows from the premiss. Where, however, a file contains the premisses of a zero-conclusion rule, the existing definition does allow the file to be continued past them, provided only that the subsequent material follows from the premisses (i.e. conforms to clause (i) or (iii) of the definition). This being so, it would seem reasonable to allow similarly for the continuation of files justified by rules with non-zero conclusions. We might therefore relax the definition by changing clause (ii) from '$\phi$ is a branch ending in a formula in Y' to '$\phi$ is a branch overlapping Y' - or even to '$\phi$ overlaps Y', which would permit an arbitrary continuation in the spirit of the previous variation.

These variations can be adopted or rejected independently. The connection with sequence proofs expressed by Theorem 3.1 is unaffected by the first of the suggested changes; to match the second we should replace 'every' by 'some' in the statement of the theorem, and to match either form of the third we should replace 'branch' by 'branch ... contains a file which'. Each notion of proof carries with it a corresponding relation of deducibility, such that Y is *deducible* from X by R iff there is a proof of the relevant kind from X to Y by R. We show that all the variant definitions of tree proof lead to the same deducibility relation. We do this by considering the broadest of them, which incorporates all three variations and exhibits an attractive verbal symmetry between rank and file. Namely we say that a finite tree of formulae is an *extended* (*tree*) *proof* from X to Y by R if for each file $\phi$ either (i) $\rho(\phi)$ overlaps X or (ii) $\phi$ overlaps Y or (iii) $\rho(\phi)$ is an immediate consequence of $\phi$ by R.

*Theorem 3.2* A proof from X to Y by R is an extended proof from X to Y by R.

*Theorem 3.3* An extended proof from X to Y by R contains as a subtree a proof from X to Y by R.

*Theorem 3.4* Tree proofs and extended proofs (and hence any intermediate variety) determine the same relation of deducibility.

*Proofs* 3.2 is immediate from the definitions, and 3.4 from 3.2 and 3.3. For 3.3, let $\tau$ be an extended proof from X to Y by R, and let $\tau_1$ be obtained by omitting everything subsequent to a formula in Y. Then $\tau_1$ is an extended proof from X to Y by R in which no formula in Y occurs in a file other than finally. We may therefore designate in each nonempty rank $\rho(\phi)$ of $\tau_1$ one occurrence of a formula in X or of a formula occurring in $\phi$, or failing that one occurrence of each conclusion of an instance of R whose premisses all occur in $\phi$. Let $\tau_2$ be obtained from $\tau_1$ by pruning out all undesignated occurrences and anything subsequent to them. Then $\tau_2$ is a proof from X to Y by the rule R′ obtained by adding 'from A infer A' to R, and it has the property that no formula occurs more than once in each rank, that no formula in Y occurs more than once in each file, and that any occurrence preceded by another occurrence of the same formula is the sole member of a rank. Let $\tau_3$ be obtained by omitting all repetitions of each formula subsequent to its first occurrence in each branch. If $\phi$ is a branch of $\tau_3$ it is contained in a branch $\phi'$ of $\tau_2$. But $\phi'$ either ends in an occurrence of a member of Y (which must be in $\phi$ also), or contains the premisses of an instance of R′ with zero conclusions. Since the same formulae occur in $\phi$ and $\phi'$, the justification of $\phi'$ in $\tau_2$ carries over in either case to justify $\phi$ in $\tau_3$. If on the other hand $\phi$ is a file of $\tau_3$ succeeded by a nonempty rank $\rho(\phi)$, let $\phi'$ be the subtree of $\tau_2$ which is the union of the predecessors of members of $\rho(\phi)$. Then $\phi'$ is a file, since $\phi = \phi' \cap \tau_3$ and no member of any multi-membered rank of $\phi'$ is omitted in constructing $\tau_3$. So $\rho(\phi)$ is the rank succeeding $\phi'$ in $\tau_2$, and its justification in $\tau_2$ can be carried over to $\tau_3$ since the same formulae occur in $\phi$ and $\phi'$. Hence $\tau_3$ is a proof from X to Y by R′ in which no formula occurs more than once in any file, and is therefore a proof by R as required.

## 3.3 *Adequacy*

By virtue of Theorem 3.4 we can use 'deducibility' ambiguously with reference to any of the varieties of proof we have considered, and in establishing the following results about deducibility we can appeal to whichever variety is most convenient for the purpose at hand.

*Theorem 3.5* Deducibility and consequence by the same rules are the same relation.

*Proof* If there is an extended proof $\tau$ from X to Y by R, let $\langle T,U\rangle$ be any partition that satisfies R. There exist files $\phi$ (e.g. the empty file) such that $\phi \subset T$, so let $\phi$ be a maximal file with this property. Then $\rho(\phi) \subset U$. Since $\langle T,U\rangle$ satisfies R, the rank $\rho(\phi)$ is not an immediate consequence of $\phi$, so either $\rho(\phi)$ overlaps X or $\phi$ overlaps Y. Hence either U overlaps X or T overlaps Y, so that $\langle T,U\rangle$ satisfies $\langle X,Y\rangle$. Accordingly $X \vdash_R Y$. For the converse it is sufficient by 2.16 to show that deducibility by R (a) contains R and is closed under (b) overlap, (c) dilution and (d) cut for formulae. For (a) it is easily verified that Figure 3.7 represents a proof from $\{A_1,\ldots,A_m\}$ to $\{B_1,\ldots,B_n\}$ by R if $\{A_1,\ldots,A_m\}$ R $\{B_1,\ldots,B_n\}$. For (b), if A is common to X and Y, the array consisting of a single occurrence of A is a proof from X to Y by any R; and for (c) any proof from X′ to Y′ by R is a fortiori a proof from X to Y by R whenever $X' \subset X$ and $Y' \subset Y$. For (d), suppose that there is a proof $\tau_o$ from X to A,Y by R, and let $\phi_1,\ldots,\phi_n$ be the branches of $\tau_o$ that end in A ($n \geq 0$). Suppose also that there is a proof from X,A to Y by R, and let $\tau_1,\ldots,\tau_n$ be n replicas of it, disjoint from each other and from $\tau_o$. Let $\tau$ be obtained by grafting $\tau_i$ onto $\phi_i$ for each i; i.e. let $\tau$ be the union of $\tau_o,\ldots,\tau_n$, with an occurrence B preceding B′ in $\tau$ iff B precedes B′ in one of $\tau_o,\ldots,\tau_n$ or, for some i, $B \in \phi_i$ and $B' \in \tau_i$. We show that $\tau$ is an extended proof from X to Y by R, distinguishing two categories of file. The first category comprises the files of $\tau_o$ other than $\phi_1,\ldots,\phi_n$; these are justified in $\tau$ as they are in $\tau_o$. The second consists of the files of the form $\phi_i,\phi$, where $\phi$ is a file of $\tau_i$. Here the justification of $\phi$ in $\tau_i$ carries over to justify the new file in $\tau$, except where $\phi$ is succeeded in $\tau_i$ by an occurrence of the premiss A of $\tau_i$. In such a case, however, A occurs both in the file $\phi_i,\phi$ and in the rank succeeding this in $\tau$, and the necessary justification is supplied by clause (iii) of the definition of extended proof.

We have already noted (Figure 3.5) that $\tau$ is not in general a proof from X to Y by R as defined in Section 3.1. Similarly the juxtaposition of extended proofs does not in general produce the required extended proof. For example, by the rules for PC the sequence A&A,A,B is an extended proof from A&A to A, and A,A∨A is an extended proof from A to A∨A, but A&A,A,B,A,A∨A is not an extended proof from A&A to A∨A. To ensure that the juxtaposition of two proofs does produce another of the same sort it is necessary to adopt an intermediate definition by incorporating the first variation discussed in Section 3.2 but not the third.

$$\frac{\begin{array}{c} A_1 \\ A_2 \\ \cdot \\ \cdot \\ \cdot \\ A_m \end{array}}{B_1 \ \ldots \ B_n}$$

Figure 3.7

As in the single-conclusion case, we say that a notion of proof is *adequate* if the corresponding relation of deducibility by R coincides, for each R, with $\vdash_R$. So Theorems 3.4 and 3.5 show that all the varieties of tree proof are adequate; in particular

*Theorem 3.6* Tree proofs are adequate.

*Theorem 3.7* Extended proofs are adequate.

We conclude by defining a special subclass of 'universal' trees, one such tree $\tau_Z$ being associated with each finite set Z. We show that the universal trees form an adequate class of all-purpose proofs in the sense that $\tau_Z$ is an extended proof from X to Y whenever Y is deducible from X using only members of Z, where 'deducible' refers to any kind of proof considered in this book; i.e. if $\pi$ is a tree proof or graph proof or whatever, and Z is the set of formulae appearing in $\pi$, then $\tau_Z$ can prove whatever $\pi$ proves.

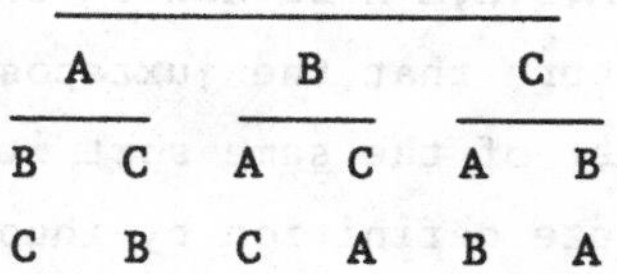

| A | | B | | C | |
|---|---|---|---|---|---|
| B | C | A | C | A | B |
| C | B | C | A | B | A |

Figure 3.8

The idea behind the construction of $\tau_Z$ is that each rank $\rho(\phi)$ should consist of one occurrence of every formula in $Z-\phi$, as illustrated by Figure 3.8 for the case where $Z = \{A,B,C\}$. It is easy to show by induction that any two trees with this property are replicas of each other. For any finite Z, then, a *universal tree* $\tau_Z$ is to consist of an occurrence $A_\sigma$ of each A in Z for each sequence $\sigma$ of distinct members of Z-A, these occurrences being partially ordered by the stipulation that $A_\sigma$ is to precede $A'_{\sigma'}$ iff the sequence formed by $\sigma$ followed by A is an initial segment of $\sigma'$. It is easy to see that the predecessors of each $A_\sigma$ form a sequence, in fact a replica of $\sigma$; thus $\tau_Z$ is a finite tree of formulae. Moreover the various files of $\tau_Z$ are replicas of the various sequences $\sigma$ of distinct formulae in Z, the rank following such a file consisting of an occurrence $A_\sigma$ of each formula A in $Z-\sigma$. In other words each rank $\rho(\phi)$ is $Z-\phi$, and

*Lemma 3.8* The set of pairs $\langle\phi,\rho(\phi)\rangle$ of files and ranks of a universal tree $\tau_Z$ is the set of all partitions of Z.

A *universal proof* from X to Y by R is a universal tree which is an extended proof from X to Y by R. The result about deducibility using only members of Z is best derived from one couched in terms of consequence. Let R′ comprise just those instances of R whose premisses and conclusions are all in Z, let L′ be the calculus with universe Z and characterised by R′ and let $X \vdash_R^Z Y$ iff there exist subsets X′ and Y′ of X and Y such that $X' \vdash' Y'$. Then for any adequate concept of proof $\vdash_R^Z$ coincides with deducibility by R using only members of Z. Note that it is closed under dilution but not overlap (and is thus not a consequence relation): Y may be deducible from X using only members of Z even if X or Y contains a formula A not in Z, provided A does not figure in the proof; but it is not in general possible to deduce A from A without invoking A.

*Theorem 3.9* $\tau_Z$ is a universal proof from X to Y by R iff $X \vdash_R^Z Y$.

*Theorem 3.10* Universal proofs are adequate.

*Proofs* For 3.9, suppose that $X \vdash_R^Z Y$ and let R′ and L′ be as described in the definition of $\vdash_R^Z$, whence $X' \vdash' Y'$ for some subsets X′ and Y′ of X and Y. For each partition $\langle Z_1,Z_2\rangle$ of Z either (a) X′ overlaps $Z_2$ or Y′ overlaps $Z_1$ or (b) $X' \subset Z_1$ and $Y' \subset Z_2$. In the latter case $Z_1 \vdash' Z_2$ by dilution, and applying 2.17 to L′ it follows that $Z_2$ is an immediate consequence of $Z_1$ by R′. Hence by 3.8 $\tau_Z$ is an extended proof from X′ to Y′ by R′ and so from X to Y by R. Conversely, if $\tau_Z$ is an extended proof from X to Y by R it is one from X′ to Y′ by R′, where $X' = X \cap Z$ and $Y' = Y \cap Z$. By 3.7 $X' \vdash' Y'$ and hence $X \vdash_R^Z Y$.

For 3.10, the adequacy of universal proofs follows from that of extended proofs in general (3.7), since by 3.7 and 3.9 there is a universal proof from X to Y by R iff there is an extended proof from X to Y by R.

# 4 · Axiomatisability

## 4.1 *Multiple-conclusion calculi*

The informal axiomatic method tries to relegate everything controversial in a theory to its axioms, and so requires that deduction from them shall be uncontroversial: given a putative proof of something from the axioms it must be an effective matter to decide whether it really is one or not. The analogous condition for a calculus governed by rules of inference is that there should be an effective procedure for deciding, for any X and Y, whether a putative proof from X to Y by the rules really is one or not. Naturally this decision procedure presupposes that information about X and Y is forthcoming. Thus we envisage a set of instructions which, when presented with a finite tree of formulae, determines a computation that proceeds algorithmically except that from time to time questions of the form ‘is A in X?’ or ‘is A in Y?’ may be asked, whereupon the correct answer is assumed to be supplied by some external agency or ‘oracle’ (cf. Rogers, 1967, Sections 9.2 and 15.1). The computation is to end with a verdict as to whether or not the given tree is a proof from X to Y.

To make this condition precise in terms of the classical theory of recursiveness we need to assume a countable (i.e. finite or countably infinite) universe of formulae, and an effective numbering - a Gödel numbering - of them. We write $A_n$ for the formula with Gödel number n. Gödel numbers can then be assigned to finite sets of formulae in the standard way, and we write

$X_n$ or $Y_n$ for the finite set with Gödel number n. Finite trees and sequences can similarly be numbered so that two arrays have the same number iff they are replicas, and we write $\tau_n$ and $\sigma_n$ for a finite tree and finite sequence with Gödel number n.

A condition on formulae is then said to be *recursive* if the corresponding numerical condition is recursive. A condition on sets of formulae may similarly be called *recursive* if the corresponding condition on sets of numbers is recursive. Where, however, a condition P applies exclusively to finite sets, there is a weaker sense in which it may plausibly be called *recursive*, namely as meaning that $\{n: P(X_n)\}$ is recursive. We shall distinguish these two senses notationally by speaking of P(X) being recursive in the one case and $P(X_n)$ being recursive in the other. Thus where the recursiveness of P(X) implies the existence of a procedure for deciding whether an arbitrary set satisfies P, the recursiveness of $P(X_n)$ only implies a procedure for deciding whether an arbitrary finite set satisfies P. Moreover where the first procedure necessarily relies on an oracle for information about the set being tested, the second takes for granted a complete enumeration of its members (its Gödel number merely being a cipher for this). It could happen that in the course of applying a decision procedure of the first kind to a particular finite set we might unwittingly enumerate all its members but, assuming an infinite universe of formulae, no finite number of questions of the form ‘is A in X?’ could ever assure us that we had done so. As a result P(X) is never recursive (except in the trivial cases of a finite universe or an unsatisfiable P) when P is true of finite sets only. For suppose that P(X) is recursive, where P is a satisfiable condition true of finite sets only. Then P(Y) for some Y, and we let Z consist of all formulae save those for which the question ‘is A in Y?’ is put to the oracle and answered negatively in the process of deciding that

P(Y). Then exactly the same process of questioning and computation decides that P(Z), so that Z is finite and hence the universe too must be finite.

Since by definition rules of inference have finite sets of premisses and conclusions, it follows that there are no non-trivial rules which are recursive in the stronger sense, and only the weaker sense is appropriate. We therefore say that a *rule* R *is recursive* if $X_m$ R $Y_n$ is recursive. Similar considerations apply to tree proofs, which are by definition finite, though not to their sets of premisses and conclusions, which need not be finite. We therefore say that *tree proof by* R *is recursive* if the relation '$\tau_j$ is a tree proof from X to Y by R', holding between j, X and Y, is recursive; and we say that *tree proof is recursive* if the relation '$\tau_j$ is a tree proof from X to Y by R', holding between j, X, Y and R, is recursive.

To say generally that there is a recursive notion of proof for a calculus L is to say that there is a notion of proof such that (i) the relation 'j is the Gödel number of a proof from X to Y' is recursive, and (ii) X ⊢ Y iff there exists a proof from X to Y. More particularly, therefore, we say that there is a *recursive notion of tree proof* for a calculus if it is characterised by rules R such that tree proof by R is recursive. We shall see that this condition is equivalent to its being characterised by recursive rules, but it should be observed that there is no direct equivalence between the recursiveness of any given rules R and the recursiveness of tree proof by the same R. For example, the rule 'from $X_m$ infer $Y_n$ if $P(X_m)$' is not recursive unless $P(X_m)$ is, but tree proof by this rule is recursive provided P(Λ), since every tree is then a proof from every X to every Y. On the other hand tree proof by recursive rules is recursive, since

*Theorem 4.1*† Tree proof is recursive.

> *Proof* The description of an appropriate Turing oracle machine is left to the reader, since it is sufficiently obvious from the definition that the concept of tree proof is an effective one.

A calculus is *axiomatisable* if it admits a recursive notion of proof, i.e. if there is a recursive relation P such that X ⊢ Y iff (∃j)P(j,X,Y), i.e. if ⊢ is *recursively enumerable*. Since there are indenumerably many sets 'recursively enumerable' is strictly a misnomer in connection with conditions on sets in general (cf. Rogers, 1967, Section 15.1), but it is defensible in the present case in view of

*Theorem 4.2* A consequence relation ⊢ is recursively enumerable iff it is compact and $X_m \vdash Y_n$ is recursively enumerable.

> *Proof* Suppose that ⊢ is recursively enumerable, so that there is a recursive P such that X ⊢ Y iff (∃j)P(j,X,Y). Then $P(j,X_m,Y_n)$ is recursive too, so that $X_m \vdash Y_n$ is recursively enumerable. To see that ⊢ is compact, suppose that X ⊢ Y, i.e. that P(j,X,Y) for some j. Let X′ be the set of formulae A about which the question 'is A in X?' is asked and answered affirmatively in the course of the computation to decide P(j,X,Y), and let Y′ be similarly defined. Then X′ and Y′ are finite, and the computation that establishes P(j,X,Y) also establishes P(j,X′,Y′), so that X′ ⊢ Y′ as required. For the converse, if ⊢ is compact then, since any consequence relation is closed under dilution, we have X ⊢ Y iff (∃m,n)($X_m \subset X$ and $Y_n \subset Y$ and $X_m \vdash Y_n$). Hence if $X_m \vdash Y_n$ is recursively enumerable we

have for some recursive P that $X \vdash Y$ iff $(\exists j,m,n)(X_m \subset X$ and $Y_n \subset Y$ and $P(j,m,n))$. Since the condition within the quantifiers is recursive, $\vdash$ is recursively enumerable.

By definition a calculus is axiomatisable iff $\vdash$ can be equated with deducibility based on some recursive proof relation P or other; but there are equivalent conditions which mention proofs of a particular kind, or which do not mention proof at all:

*Theorem 4.3*† A calculus is axiomatisable iff it is characterised by recursive rules of inference.

*Theorem 4.4*† A calculus is axiomatisable iff there exists a recursive notion of tree proof for it.

*Proofs* If L is characterised by recursive rules R then tree proof by R is recursive by 4.1 and so there exists a recursive notion of tree proof for L; and if there is a recursive notion of tree proof for L then for some rules R that characterise L the relation '$\tau_j$ is a tree proof from X to Y by R' is recursive, so that $\vdash_R$ is recursively enumerable and L is axiomatisable. To complete the cycle of implications, and so prove both theorems, it remains to show that any axiomatisable calculus L is characterised by recursive rules. By 4.2 there exists a recursive relation P such that $X_m \vdash Y_n$ iff $(\exists j)P(j,m,n)$. Let R be the rules R1 'from $X_m,A_j$ infer $Y_n$ if $P(j,m,n)$' and R2 'from $X_m$ infer $A_j,Y_n$ if $P(j,m,n)$'. Evidently R is recursive, and to show that R characterises L it is sufficient to show that $R \subset \vdash \subset \vdash_R$. If X R Y then for some j, m and n, $X_m \subset X$ and $Y_n \subset Y$ and $P(j,m,n)$, so that $X_m \vdash Y_n$, and by dilution $X \vdash Y$; thus $R \subset \vdash$. If on the other hand $X \vdash Y$ then by 4.2 there exist finite subsets $X_m$ and $Y_n$ of X and Y such that $X_m \vdash Y_n$. So for some j, $P(j,m,n)$ and consequently

$X_m, A_j \vdash_R Y_n$ and $X_m \vdash_R A_j, Y_n$. By cut for $A_j$ it follows that $X_m \vdash_R Y_n$, and hence $X \vdash_R Y$ by dilution; thus $\vdash \subset \vdash_R$ as required.

Although Theorem 4.4 refers specifically to tree proofs, it ultimately depends only on the fact that they are adequate and recursive. Similar definitions of recursiveness may be formulated for other notions of proof, and a similar result will therefore hold for any that can be shown to be both adequate and recursive, e.g. the universal proofs of Section 3.3 or the varieties of graph proof introduced in Chapters 9 and 10.

## 4.2 *Single-conclusion calculi*

The definitions of Section 4.1 are readily adapted to the single-conclusion case. For example we say that a *single-conclusion rule of inference* R *is recursive* if $X_m$ R B is recursive; that *sequence proof is recursive* if the relation '$\sigma_j$ is a sequence proof of B from X by R', holding between j, X, B and R, is recursive; and that a single-conclusion calculus is *axiomatisable* if its consequence relation is recursively enumerable, i.e. if there is a recursive relation P such that $X \vdash B$ iff $(\exists j)P(j,X,B)$. The first two results carry over straightforwardly to the single-conclusion case, giving

*Theorem 4.5*[†] Sequence proof is recursive.

*Theorem 4.6* A single-conclusion consequence relation $\vdash$ is recursively enumerable iff it is compact and $X_m \vdash B$ is recursively enumerable.

The two conditions given as equivalent to axiomatisability in Theorems 4.3 and 4.4, and hence equivalent to one another, are likewise equivalent to one another in the single-conclusion case:

*Theorem 4.7*† There exists a recursive notion of sequence proof for a single-conclusion calculus iff it is characterised by recursive rules of inference.

*Proof* If L is characterised by recursive rules R then sequence proof by R is recursive by 4.5. Conversely, suppose that sequence proof by R is recursive for some rules R characterising L. It does not necessarily follow that R is recursive, but let R′ be the rule 'from $X_m$ infer B if B is an immediate consequence of $X_m$ by R', where by *immediate consequence by* R we mean, as in the multiple-conclusion case, the closure of R under overlap and dilution. A sequence obtained by arranging the members of $X_m$ in arbitrary order, followed by B, is a proof of B from $X_m$ by R iff $X_m$ R′ B. Since sequence proof by R is recursive it follows that R′ is recursive, and since $R \subset R' \subset \vdash_R$ it follows too that R′ characterises L.

A calculus satisfying either of these conditions will be axiomatisable, just as in the multiple-conclusion case, but at this point the parallels cease, for the converse proposition does *not* carry over from the multiple-conclusion case:

*Theorem 4.8* There exists an axiomatisable single-conclusion calculus which cannot be characterised by recursive rules of inference and for which there exists no recursive notion of sequence proof.

*Proof* Let L have the same formulae as the classical predicate calculus but let it be characterised by the rule 'from A&A infer A∨A if A is a theorem of the predicate calculus'. This rule is *non-iterable* in the sense that no conclusion of any instance of the rule can ever be a premiss of that or any other instance of it. For such rules we have

*Lemma 4.9* Consequence and immediate consequence by a non-iterable (multiple- or single-conclusion) rule coincide.

To prove the lemma, suppose that Y is not an immediate consequence of X by a non-iterable rule R. There is a partition $\langle Z_1,Z_2\rangle$ of V-(X∪Y) such that no formula in $Z_1$ ever figures as a premiss to R, and none in $Z_2$ as a conclusion. So $Z_2$,Y is not an immediate consequence of X,$Z_1$, and by 2.17 $X,Z_1 \nvdash_R Z_2,Y$ and hence $X \nvdash_R Y$. The result for a single-conclusion rule R is a corollary, since the corresponding singleton-conclusion rule $R_1$ is also non-iterable and it is immediate from the definitions of $\vdash_R$ and $\vdash_{R_1}$ that $X \vdash_{R_1} \{B\}$ iff $X \vdash_R B$.

Applying the lemma to the rule given for L we have (1) $X_m \vdash B$ iff either $B \in X_m$ or, for some theorem A of the predicate calculus, B = A∨A and $A\&A \in X_m$. It follows that $X_m \vdash B$ is recursively enumerable and so by 1.12 and 4.6 L is axiomatisable. Now let R be any rules that characterise L. Obviously if A&A R A∨A then A&A ⊢ A∨A. Conversely, let A&A ⊢ A∨A and so let $C_1,\ldots,C_k$ be a proof of A∨A from A&A by R. Not every $C_i$ is A&A, for $C_k$ is A∨A, and so let $C_j$ be the first of the $C_i$ that is not A&A. Then $C_j$ must be the conclusion of an instance of R whose premisses all precede it in the proof: hence either $\Lambda$ R $C_j$ or A&A R $C_j$. But $\Lambda$ R $C_j$ is excluded by (1), so A&A R $C_j$; and so by (1) $C_j$ is A∨A, i.e. A&A R A∨A. We have thus shown that A&A R A∨A iff A&A ⊢ A∨A, i.e. by (1) that A&A R A∨A iff A is a theorem of predicate calculus. Since this latter is not a recursive property it follows that R is not a recursive rule. So L is not characterised by recursive rules, and by 4.7 there is no recursive notion of sequence proof for it.

The situation exposed here is concealed if one thinks of calculi merely as sets of theorems, since every recursively enumerable set can be generated by recursive axioms and rules. For if it is finite one can take all its members as axioms; if infinite and enumerated by the recursive function f one can take rules of the form 'from {f(k): k<n} infer f(n)' for every n. See also Hermes, 1951.

A natural reaction would be to take this result as showing that our definition of axiomatisability simply fails to capture the informal idea: that we ought to have defined axiomatisability in terms of the existence of a recursive notion of sequence proof. Theorem 4.8 would then merely state that not every calculus with a recursively enumerable consequence relation is axiomatisable. But to react in this way is to give unwarranted pride of place to sequence proofs in preference to other eligible methods of proof, such as natural deduction, truth-table evaluation, or some suitably programmed Turing machine. There is no a priori reason for thinking that just because proof for a calculus can be effectively realised by one method it can be effectively realised by another. The calculus of Theorem 4.8 is a case in point. For when Kneale introduced proofs using multiple-conclusion rules he saw them primarily as providing a new method of *single-conclusion* proof, namely by treating a multiple-conclusion proof from X to {B} as constituting a proof of B from X, and his rules were chosen to suit a particular single-conclusion calculus (the classical predicate calculus) rather than any particular multiple-conclusion counterpart of it. In this spirit it is easy to devise multiple-conclusion rules which provide a recursive notion of proof for the single-conclusion calculus of Theorem 4.8 (e.g. by applying the argument of Theorems 4.3 and 4.4 to its axiomatisable multiple-conclusion counterpart described in Theorem 5.18), and we are thus justified in calling it axiomatisable after all.

Theorem 4.4 showed that if there is an effective notion of proof of any sort for a multiple-conclusion calculus, there is an effective notion of tree proof for it. Theorem 4.8 shows that sequence proofs do not have the same privileged position vis à vis single-conclusion calculi. One remedy is to adapt the definition of a sequence proof to suit rules of inference in which the multiplicity of occurrence of premisses is taken into account, i.e. in which the premisses of each instance of the rule are treated as forming an array rather than a set of formulae. A rule of this kind is essentially a binary relation R between finite arrays and individual formulae; if $\chi$ R B we say as in Section 1.2 that $\langle \chi, B \rangle$ is an instance of R, with B as its conclusion and the formulae occurring in $\chi$ as its premisses. To adapt the definition of sequence proof we only need to tack a proviso about multiplicity onto the end of the definition of proof given in Section 1.3:
A finite sequence of formulae $C_1, \ldots, C_n$ is a proof of B from X by R if $C_n$ is B and each $C_i$ is either a member of X or the conclusion of an instance of R each of whose premisses occurs in $C_1, \ldots, C_{i-1}$ at least as often as it occurs in the relevant instance of R.

This treatment of rules is not without support from actual practice. For example, there is at least as much inclination to say that the rule of adjunction 'from A,B infer A&B' always has two premisses as to say that it has sometimes one and sometimes two, depending on whether A = B or not. Nor does this treatment of rules affect their strength. If we take a partition $\langle T, U \rangle$ as satisfying an instance $\langle \chi, B \rangle$ when B $\in$ T or some member of U occurs in $\chi$, we can define $\vdash_R$ as a relation between a set and a formula (rather than between an array and a formula) exactly as in Section 1.2. It is clear that satisfaction - and so consequence by R - is unaffected by the multiplicity of occurrence of the premisses in R, and that each rule in which multiplicity is taken into account is equivalent to (characterises the same consequence relation as) the corresponding rule in which multiplicity is not taken into account. Recursiveness is a different matter. For example, the rule of Theorem 4.8 is not recursive, but is equivalent to a rule in which multiplicity is taken into account and which is recursive, namely 'for each n, from n occurrences of A&A infer A∨A if A is the nth theorem of the predicate calculus'.

Deducibility, as determined by the revised definition of proof given above, is similarly a relation between a set and a formula (rather than between an array and a formula), and is similarly unaffected by the actual multiplicity of occurrence of the premisses of the rules, for it is easy to adapt the argument of Theorem 1.13 to show that deducibility by R is the same as $\vdash_R$, i.e. that

*Theorem 4.10* Sequence proofs are adequate for single-conclusion rules in which multiplicity of occurrence of premisses is taken into account.

This result would have been forthcoming even if we had left the definition of proof unaltered; the point of the alteration is solely to ensure that proof by R is recursive whenever R is recursive. Without the added proviso about multiplicity there would be no effective way of deciding, for example, whether or not an arbitrary sequence of the form A&A,A∨A was a proof of A∨A from A&A by the recursive rule cited in the paragraph above; given the proviso we need only examine the instance of the rule for which n = 1. We thus have

*Theorem 4.11*[†] Sequence proof, using rules in which multiplicity of occurrence of premisses is taken into account, is recursive.

With these modifications to the definitions we are able to show that if there is an effective notion of proof of any sort for a single-conclusion calculus there is an effective notion of sequence proof for it, and hence supply the desired single-conclusion analogues of Theorems 4.3 and 4.4.

*Theorem 4.12*[†] A single-conclusion calculus is axiomatisable iff it is characterised by recursive rules in which the multiplicity of occurrence of premisses is taken into account.

*Theorem 4.13*[†] A single-conclusion calculus is axiomatisable iff there exists a recursive notion of sequence proof for it, using rules in which multiplicity of occurrence of premisses is taken into account.

*Proofs* By 4.10 and 4.11, if L is characterised by recursive rules there exists a recursive notion of sequence proof for it. If there exists a recursive notion of sequence proof for L we have $X \vdash B$ iff $(\exists j,m)(X_m \subset X$ and $\sigma_j$ is a sequence proof of B from $X_m)$, so that L is axiomatisable. To complete the cycle of implications we need to show that L is characterised by recursive rules on the supposition that it is axiomatisable, i.e. on the supposition that $\vdash$ is compact and there exists a recursive relation P such that $X_m \vdash B$ iff $(\exists j)P(j,m,B)$. If L has no theorems we posit the rule 'from j+1 occurrences of each member of $X_m$ infer B if $P(j,m,B)$'. This rule evidently characterises L, and is recursive since in the case of any given nonempty array of putative premisses there is an upper bound to the range of j and m that need to be tested, while an empty array can be rejected out of hand since P(j,m,B) is never true for empty $X_m$, there being no theorems. If on the other hand L has some theorems we posit any one of them, A, as an axiom and posit the rule 'from j occurrences of A and one occurrence of each member of $X_m$ infer B if P(j,m,B)'. By cut for A these evidently characterise L, and the rule is recursive since there is a bound to the range of j and m that need to be tested when assessing any given array of putative premisses, empty or not.

## 4.3 *Decidability*

We model our definition of decidability on the necessary and sufficient conditions for axiomatisability established in Theorems 4.2 and 4.6, and say that a calculus of either kind is *decidable* if (i) it is compact and (ii) $X_m \vdash Y_n$ (or $X_m \vdash B$) is recursive. The connection between decidability and axiomatisability is then immediate:

*Theorem 4.14* A decidable (multiple- or single-conclusion) calculus is axiomatisable.

The use of (ii) on its own (as in Shoesmith and Smiley, 1971) would produce a weaker notion of decidability, for which Theorem 4.14 would only hold for compact calculi. Alternatively a definition of decidability might have been modelled directly on that of axiomatisability, i.e. we might have said that a calculus was decidable iff its consequence relation was recursive. This would produce a stronger notion, for if X ⊢ Y or X ⊢ B is recursive then (arguing as in Theorem 4.2 or 4.6) ⊢ must be compact and $X_m \vdash Y_n$ or $X_m \vdash B$ recursive, but the converse implications do not hold, as can be seen from Theorems

4.15 and 4.16 below. The first of these results shows that decidability in this strong sense is not a serious possibility for multiple-conclusion calculi, and is a reason for proposing the intermediate definition.

*Theorem 4.15* A consequence relation ⊢ is recursive iff V is finite or X ⊢ Y for every X and Y.

*Proof* Every relation defined over subsets of a finite V is recursive, while there is a trivially affirmative decision procedure for ⊢ if X ⊢ Y for every X and Y. For the converse suppose that X ⊬ Y for some X and Y. If there is a decision procedure for ⊢ then for each A either 'Is A in X?' or 'Is A in Y?' must be asked in the process of deciding that X ⊬ Y, since otherwise the same process would decide that X,A ⊬ A,Y. Hence V is finite.

By contrast there are non-trivial single-conclusion calculi with recursive consequence relations, notably those with rules in which every premiss is a subformula of the conclusion, but even these are exceptions. The most striking example of a single-conclusion calculus whose consequence relation is not recursive is the classical propositional calculus. It is obviously decidable, for the relation of tautological consequence is compact and the truth-tables provide a decision procedure for $X_m$ ⊢ B; but there is no effective method of deciding in general (i.e. given an oracle for X) whether or not X ⊢ B.

*Theorem 4.16* Consequence in the classical propositional calculus is not recursive.

*Proof* Let L be the classical propositional calculus with propositional variables $p_1, p_2, \ldots$, and suppose that ⊢ is recursive. If X is empty then X ⊬ $p_1$, and so let A be a formula about which the question 'Is $A\&p_1$ in X?' is not asked in the process of deciding that $X \nvdash p_1$. Then the same process will decide that $A\&p_1 \nvdash p_1$, which is absurd.

Theorem 4.16 may also be proved as a corollary of the stronger result that

*Theorem 4.17* There exists a recursive set X such that the set of tautological consequences of X is not recursive.

*Proof* Let $L_1$ be the predicate calculus (so that the condition $\vdash_1 A$ is not recursive), and for any recursive set of axioms and rules characterising $L_1$ let X consist of (i) $p_n$ for each axiom $A_n$ and (ii) $p_{i_1} \supset . \ldots \supset . p_{i_m} \supset p_n$ for each rule instance 'from $A_{i_1}, \ldots, A_{i_m}$ infer $A_n$'. Then any proof of $A_n$ in $L_1$ can be matched by a proof of $p_n$ from X in L using modus ponens; and conversely if $\nvdash_1 A_n$ each member of X takes value t and $p_n$ takes value f when we assign to each $p_i$ the value t or f according as $\vdash_1 A_i$ or not. Thus $X \vdash p_n$ iff $\vdash_1 A_n$, whence the result follows. This idea of translating a recursively unsolvable problem into the language of a propositional calculus was used by Linial and Post, 1949, to establish a number of undecidability results concerning sets of tautologies, the one closest in form to 4.17 being equivalent to the existence of an undecidable, finitely axiomatisable subcalculus of the classical one.

# 5 · Counterparts

## 5.1 *The range of counterparts*

We have so far treated multiple- and single-conclusion calculi separately, though on parallel lines. We now try to establish some direct connections between them. For this purpose let L stand for an arbitrary single-conclusion calculus and L′ for an arbitrary multiple-conclusion one. We have seen how a set of partitions can be used to characterise a calculus of either kind, and when L and L′ are characterised by the same set of partitions we shall say that they are *counterparts* of each other. Since any partition satisfies $\langle X,B\rangle$ iff it satisfies $\langle X,\{B\}\rangle$ an alternative criterion for two calculi to be counterparts can be formulated directly in terms of their consequence relations:

*Theorem 5.1* A necessary and sufficient condition for a single-conclusion calculus L and a multiple-conclusion calculus L′ to be counterparts is that $X \vdash B$ iff $X \vdash' \{B\}$.

Theorem 5.1 shows that there is only one counterpart L of each L′ and that even though $\vdash$ is not literally a subrelation of $\vdash'$ it is the image of one, namely the subrelation comprising just those instances of $\vdash'$ whose conclusions are singletons. (This subrelation is not itself a consequence relation at all since it does not permit dilution.) We therefore call L *the single-conclusion part* of L′. But although a multiple-conclusion calculus has a unique single-conclusion counterpart the converse is not true. For whereas each multiple-conclusion calculus is characterised by a unique set of partitions, the presence or absence of each pair $\langle T,U\rangle$ being dictated by the falsity or truth of $T \vdash U$, several sets of partitions can characterise the same single-conclusion calculus, since unless U is a unit set there may be no way of ensuring the presence of $\langle T,U\rangle$ by means of a condition on a single-conclusion consequence relation. Indeed the single-conclusion part of a calculus is never sufficient to determine the remainder of it, for it is an immediate corollary of Theorem 5.25 below that

*Theorem 5.2* Every single-conclusion calculus has at least two counterparts.

Our first task is therefore to describe the range of counterparts possessed by any given single-conclusion calculus. Let $\vdash_\cap$ and $\vdash_\cup$ stand for the intersection and union of the consequence relations of the various counterparts of L. Then we have

*Theorem 5.3* L and L′ are counterparts iff $\vdash_\cap \subset \vdash' \subset \vdash_\cup$.

*Proof* Clearly if L′ is a counterpart of L its consequence relation lies between $\vdash_\cap$ and $\vdash_\cup$. For the converse, suppose that $\vdash_\cap \subset \vdash' \subset \vdash_\cup$. If $X \vdash B$ then by 5.1 $X \vdash_\cap \{B\}$ and so $X \vdash' \{B\}$; while if $X \vdash' \{B\}$ then $X \vdash_\cup \{B\}$, so that by 5.1 $X \vdash B$. Hence L and L′ are counterparts by 5.1.

The problem thus reduces to that of describing $\vdash_\cap$ and $\vdash_\cup$. In particular we wish to know whether they are consequence relations themselves, for if so they will determine *minimum* and *maximum* counterparts of L.

*Theorem 5.4* $X \vdash_{\cap} Y$ iff $X \vdash B$ for some $B$ in $Y$.

*Theorem 5.5* $X \vdash_{\cup} Y$ iff, for all $Z$ and $B$, if $Z,A \vdash B$ for every $A$ in $Y$ then $Z,X \vdash B$.

*Proofs* For 5.4, if $X \vdash B$ for some $B$ in $Y$ then for every counterpart $L'$ of $L$ we have $X \vdash' \{B\}$ by 5.1, and so $X \vdash' Y$ by dilution; and hence $X \vdash_{\cap} Y$. For the converse let $X \vdash' Y$ iff $X \vdash B$ for some $B$ in $Y$. Then $\vdash'$ is closed under overlap and dilution since $\vdash$ is, while for cut we let $\langle Z_1,Z_2 \rangle$ be the partition of $Z$ in which $Z_1 = Z \cap \{A: X \vdash A\}$ and use cut for $Z_1$ for $\vdash$ to show that if $X,Z_1 \vdash' Z_2,Y$ then $X \vdash' Y$. By 5.1, therefore, $\vdash'$ determines a counterpart of $L$, so that $\vdash_{\cap} \subset \vdash'$, as required.

For 5.5, suppose that $X \vdash_{\cup} Y$, so that $X \vdash_I Y$ for some set of partitions $I$ which characterises $L$. If $Z,X \nvdash B$ then for some $\langle T,U \rangle$ in $I$ we have $Z \subset T$ and $X \subset T$ and $B \in U$; but since $\langle T,U \rangle$ satisfies $\langle X,Y \rangle$ it follows that $A \in T$ for some $A$ in $Y$ and so $Z,A \nvdash B$. Conversely, suppose that $X$ and $Y$ are such that $Z,X \vdash B$ whenever $Z,A \vdash B$ for every $A$ in $Y$. Let $I$ be the set of partitions that satisfy both $\vdash$ and $\langle X,Y \rangle$. Then if $Z \vdash B$ each partition in $I$ satisfies $\langle Z,B \rangle$, so $Z \vdash_I \{B\}$. On the other hand if $Z \nvdash B$ then $Z \subset T$ and $B \in U$ for some partition $\langle T,U \rangle$ that satisfies $\vdash$. If $\langle T,U \rangle \in I$ then $Z \nvdash_I \{B\}$. If not then $X \subset T$ and $Y \subset U$ and so $Z,X \nvdash B$. By hypothesis it follows that $Z,A \nvdash B$ for some $A$ in $Y$, so that some partition $\langle T',U' \rangle$ satisfies $\vdash$ but not $\langle Z \cup \{A\}, B \rangle$. A fortiori $\langle T',U' \rangle$ invalidates $\langle Z,B \rangle$, and since $A \in Y$ and $A \in T'$ it follows that $\langle T',U' \rangle \in I$. Thus in either case $Z \nvdash_I \{B\}$, so $I$ characterises a counterpart of $L$. But $X \vdash_I Y$ and hence $X \vdash_{\cup} Y$, as required.

*Theorem 5.6* Every single-conclusion calculus has a minimum counterpart.

*Theorem 5.7* Not every single-conclusion calculus has a maximum counterpart.

*Proofs* For 5.6 we observe that given any family of consequence relations characterised by sets $I_i$ of partitions, their intersection is also a consequence relation, namely $\bigcap \vdash_{I_i} = \vdash_{\bigcup I_i}$. In particular $\vdash_\cap$ is a consequence relation as required. Alternatively it is easy to show as in 5.4 that $\vdash_\cap$ is closed under overlap, dilution and cut for sets.

For 5.7, let V be infinite and let X ⊢ B iff B ∈ X or X is infinite. It is easily seen that ⊢ is closed under overlap and dilution. For cut, if X ⊢ A for every A in Z then either X is infinite or Z ⊂ X, and in either case if X,Z ⊢ B then X ⊢ B. Hence ⊢ is a consequence relation. By 5.5 X $\nvdash_\cup$ Y iff there exist Z and B such that Z,X ⊬ B although Z,A ⊢ B for every A in Y; i.e. iff there exist Z and B such that Z and X are finite and B ∉ X∪Z and B = A for every A in Y; i.e. iff X is finite and Y ⊂ {B} for some B not in X. It follows that Λ $\nvdash_\cup$ B; but T $\vdash_\cup$ U,B for every partition ⟨T,U⟩, since either T or U is infinite. Hence $\vdash_\cup$ is not closed under cut for V and so is not a consequence relation.

It may be noted that for every L, $\vdash_\cup$ is closed under cut for formulae. For if X,C $\vdash_\cup$ Y then whenever Z,A ⊢ B for every A in Y, we have Z,X,C ⊢ B, so that by dilution Z,X,A ⊢ B for every A in Y∪{C}. If also X $\vdash_\cup$ C,Y then Z,X ⊢ B; and hence X $\vdash_\cup$ Y as required.

## 5.2 *Compactness*

The single-conclusion part of a compact calculus is always compact, as we shall see. Conversely, a compact single-conclusion calculus always has at least one compact counterpart but in general it will have non-compact ones as well, though some do have compact counterparts exclusively. Unlike single-conclusion calculi in general (Theorem 5.7), a compact one always has a maximum counterpart (and hence a maximum compact one, though the proof of Theorem 5.9 shows that the two need not coincide); and although by Theorem 5.2 it never has a unique counterpart it may well have a unique compact one. The range of compact counterparts is discussed in effect by Scott (1974a) who, though working a la Gentzen with finite sets of premisses and conclusions, proves results tantamount to Theorems 5.8, 5.10, 5.12 and a 'compact' version of 5.3-5.

*Theorem 5.8* The single-conclusion part of a compact calculus is compact.

*Theorem 5.9* There exists a compact single-conclusion calculus which has a non-compact counterpart.

*Theorem 5.10* Every compact single-conclusion calculus has a minimum compact counterpart, namely its minimum counterpart.

*Theorem 5.11* Every compact single-conclusion calculus has a maximum counterpart.

*Theorem 5.12* Every compact single-conclusion calculus has a maximum compact counterpart.

*Theorem 5.13* There exists a compact single-conclusion calculus which has a unique compact counterpart.

*Proofs* 5.8 is immediate from 5.1. For 5.9 let L be the intuitionist propositional calculus. It is sufficient by 5.11 to show that $\vdash_U$ is not compact. If Z,A ⊢ B for every A in the set Y of non-theorems of L then if Z includes a non-theorem we have at once that $Z \vdash B$; while if Z consists entirely of theorems we have $A \vdash B$ for every A in Y, so that in particular $p \vdash B$ for some variable p not appearing in B, whence by substitution $p \supset p \vdash B$ and again $Z \vdash B$. Hence $\Lambda \vdash_{\cup} Y$ by 5.5. On the other hand when Z is empty and B is the disjunction of the members of any finite subset Y′ of Y we have $Z,A \vdash B$ for each A in Y′, but $Z \nvdash B$ by the well-known result of Gödel, 1932. Hence $\Lambda \nvdash_{\cup} Y'$ by 5.5.

Another example for 5.9 is the implicative fragment of the intuitionist calculus, by 5.2 and 5.13; and so indeed (by 5.11, 5.25 and 5.28) is any compact calculus which, like it, is positive in the sense we define in Section 5.4. In all these cases it is the maximum counterpart that is shown not to be compact, but the classical propositional calculus has non-compact counterparts (Section 18.3) even though its maximum counterpart is compact by 18.4. An example of a calculus which has compact counterparts exclusively is one in which $X \vdash B$ for all X and B; this has just two counterparts, both compact (Section 13.2).

For 5.10, if L is compact then $\vdash_{\cap}$ is compact by 5.4, whence by 5.6 the minimum counterpart of L is compact.

For 5.11 we show that when L is compact $\vdash_{\cup}$ is closed under overlap, dilution and cut for sets. This is obvious for overlap and dilution. We noted in the proof of 5.7 that $\vdash_{\cup}$ is always closed under cut for formulae, but this is of no help here since $\vdash_{\cup}$ need not be compact even though $\vdash$ is. Instead we show that $\vdash_{\cup}$ is closed under cut for V and use 2.2. Suppose then that $X \nvdash_{\cup} Y$, so that there exist Z and B such that $Z,A \vdash B$ for each A in Y but $Z,X \nvdash B$. Let S be $\{W: W,X,Z \nvdash B\}$; then S is nonempty since $\Lambda$ is in it, and is of finite character since L is compact (cf. 2.9). By Tukey's lemma S therefore has a maximal member T, and evidently $X \subset T$ and $Z \subset T$. Let U be the complement of T;

then T,A $\vdash$ B for each A in U by the maximality of T, and also for each A in Y by dilution. But X,T $\nvdash$ B, so that X,T $\nvdash_{\cup}$ U,Y as required.

For 5.12, if L is compact then by 2.11 and 5.11 the maximum counterpart of L has a maximum compact subcalculus L′. By 5.10 $\vdash_{\cap}$ determines a compact subcalculus of the maximum counterpart, so that $\vdash_{\cap} \subset \vdash' \subset \vdash_{\cup}$. Hence L′ is a counterpart of L by 5.3 and is therefore its maximum compact counterpart.

For 5.13 we show that the implicational fragment of the intuitionist propositional calculus - Hilbert's positive implicational calculus - has a unique compact counterpart. Let L be any single-conclusion calculus in which modus ponens is valid, i.e. in which A,A⊃B $\vdash$ B; then the *deduction theorem* for ⊃ states that X $\vdash$ A⊃B if (and hence iff) X,A $\vdash$ B, and we have

*Lemma 5.14* If the deduction theorem holds for ⊃ then X $\vdash_{\cup}$ Y iff X,Y⊃B $\vdash$ B for every B.

To prove the lemma we note that by modus ponens Y⊃B,A $\vdash$ B for every A in Y, and so by 5.5 if X $\vdash_{\cup}$ Y then X,Y⊃B $\vdash$ B. Conversely if X $\nvdash_{\cup}$ Y there exist Z and B such that Z,A $\vdash$ B for each A in Y but Z,X $\nvdash$ B. Hence by dilution and the deduction theorem Z,X $\vdash$ A⊃B for each member A⊃B of Y⊃B, so that Z,X,Y⊃B $\nvdash$ B by cut for Y⊃B, and hence by dilution X,Y⊃B $\nvdash$ B.

To prove the theorem, let $L_1$ be the intuitionist calculus and L its implicational fragment. Let L′ be the maximum compact counterpart of L. We show that L′ is also its minimum counterpart, i.e. that if X $\vdash'$ Y then X $\vdash_{\cap}$ Y. Suppose then that X $\vdash'$ Y, so that there exist finite subsets X′ and Y′ of X and Y such that X′ $\vdash'$ Y′ and hence X′ $\vdash_{\cup}$ Y′. The deduction theorem holds for ⊃ in L, so by 5.14 we have X′,Y′⊃B $\vdash$ B for every B. Taking B to be any propositional variable p not appearing in any member of X′

or $Y'$, we thus have that $X', Y' \supset p \vdash_1 p$. Now $Y'$ cannot be empty, since $\langle X', p \rangle$ can be invalidated in the classical two-valued matrix, and so let $Y' = \{B_1, \ldots, B_n\}$ where $n \geq 1$. Substituting $B_1 \vee \ldots \vee B_n$ for p we have that $X', Y' \supset (B_1 \vee \ldots \vee B_n) \vdash_1 B_1 \vee \ldots \vee B_n$ and hence $X' \vdash_1 B_1 \vee \ldots \vee B_n$. But since $\vee$ does not occur in the formulae of L it follows by a result of Harrop, 1960, that $X' \vdash_1 B_i$ for some i. Hence $X' \vdash B_i$ and so $X \vdash_\cap Y$ as required.

Some results about compact counterparts can usefully be expressed in terms of rules of inference. We noted earlier that to every single-conclusion rule R there corresponds a singleton-conclusion rule $R_1$ such that $X\ R_1\ \{B\}$ iff $X\ R\ B$.

*Theorem 5.15* If a single-conclusion calculus is characterised by rules of inference, its minimum counterpart is characterised by the corresponding singleton-conclusion rules.

*Theorem 5.16*† Rules R (single-conclusion) and R′ characterise counterpart calculi iff (a) if $X\ R\ B$ then $X \vdash_{R'} \{B\}$ and (b) if $X\ R'\ Y$ and $Z, A \vdash_R B$ for every A in Y, then $Z, X \vdash_R B$.

*Proofs* 5.15 follows at once from 3.1 and 5.4. For 5.16, let R characterise L; then by 5.3 R′ characterises a counterpart of L iff $\vdash_\cap \subset \vdash_{R'} \subset \vdash_\cup$. By 5.11 $\vdash_\cup$ is closed under overlap, dilution and cut for sets, and so by 2.14 $\vdash_{R'} \subset \vdash_\cup$ iff $R' \subset \vdash_\cup$, i.e. by 5.5 iff (b) holds. Using 5.15 in a similar argument, $\vdash_\cap \subset \vdash_{R'}$ iff the singleton-conclusion rule corresponding to R is contained in $\vdash_{R'}$, i.e. iff (a) holds.

We note that when Y is a unit set (b) takes a simpler form analogous to (a), namely 'if $X\ R'\ \{B\}$ then $X \vdash_R B$', and when Y is empty (b) reduces to 'if $X\ R'\ \Lambda$ then $X \vdash_R B$ for every B'.

## 5.3 *Axiomatisability*

For single-conclusion consequence relations the property of being characterised by recursive rules need not be the same as recursive enumerability (Theorem 4.8). We chose to call the latter 'axiomatisability' and the following results about the axiomatisability of counterparts are to be understood accordingly. If the other choice were made Theorems 5.18 and 5.19 would still hold but Theorem 5.17 would fail, for taken together with Theorem 4.3 and the present version of Theorem 5.18 it would contradict Theorem 4.8.

*Theorem 5.17* The single-conclusion part of an axiomatisable calculus is axiomatisable.

*Theorem 5.18* Every axiomatisable single-conclusion calculus has a minimum axiomatisable counterpart, namely its minimum counterpart.

*Theorem 5.19* Not every axiomatisable single-conclusion calculus has a maximum axiomatisable counterpart.

*Proofs* For 5.17, let L′ be axiomatisable, so that $X \vdash' Y$ iff $(\exists j)P'(j,X,Y)$ where P′ is recursive; and let L be the single-conclusion part of L′. Then P too is recursive, where $P(j,X,B)$ iff $P'(j,X,\{B\})$, and so L is axiomatisable by 5.1.

For 5.18 we are unable to appeal to 5.15 because of the gap between axiomatisability and recursive rules mentioned above. Instead let L be an axiomatisable single-conclusion calculus, so that $X \vdash B$ iff $(\exists j)P(j,X,B)$, where P is recursive. Then by 5.4 we have $X \vdash_{\cap} Y$ iff $(\exists j,B)(B \in Y$ and $P(j,X,B))$, whence $\vdash_{\cap}$ is axiomatisable and the result follows.

For 5.19 we could use the calculus of 4.8, but we prefer an example that will serve whichever way axiomatisability is construed. Let L, then, be the axiomatisable single-conclusion calculus characterised by the non-iterable rules R: 'from A&A infer B⊃B if P(A,B)' and 'from A∨A infer B⊃B if P(A,B)', where P is any recursive relation such that (∃B)P(A,B) is not recursive, i.e. such that 'not (∃B)P(A,B)' is not recursively enumerable. Suppose that L has a maximum axiomatisable counterpart L′. If Λ ⊢′ A&A,A∨A there is no B such that P(A,B), for otherwise by cut for {A&A,A∨A} and by 5.1 it would follow that ⊢ B⊃B, contradicting 4.9. Conversely, if Λ ⊬′ A&A,A∨A consider the calculus $L_1$ characterised by the relevant instance of the rule 'from Λ infer A&A,A∨A' together with the singleton-conclusion rules corresponding to R. By 4.3 $L_1$ is axiomatisable, but by the maximality of L′ it is not an axiomatisable counterpart of L. Hence it is not a counterpart of L and so by 5.16 there exist Z and C such that Z,A&A ⊢ C and Z,A∨A ⊢ C but Z ⊬ C, whence by 4.9 it follows that C is of the form B⊃B where P(A,B). Thus Λ ⊢′ A&A,A∨A iff not (∃B)P(A,B), contradicting the hypothesis that ⊢′ is recursively enumerable.

*Theorem 5.20* The single-conclusion part of a decidable calculus is decidable.

*Theorem 5.21* Every decidable single-conclusion calculus has a minimum decidable counterpart, namely its minimum counterpart.

*Theorem 5.22* Not every decidable single-conclusion calculus has a maximum decidable counterpart.

*Proofs* For 5.20, let L be the single-conclusion counterpart of L′. If ⊢′ is compact and $X_m$ ⊢′ $Y_n$ is recursive it follows by 5.1 that $\vdash$ is compact and $X_m \vdash B$ is recursive.

For 5.21, if $\vdash$ is compact and $X_m \vdash B$ is recursive then by 5.4 $\vdash_\cap$ is compact and $X_m \vdash_\cap Y_n$ is recursive.

For 5.22 we use the example of 5.19, substituting 'decidable' for 'axiomatisable' throughout. The only points at which expansion is needed are the claims that L and $L_1$ are decidable. Consider the relation which holds between X and Y iff (for the given choice of A) X,A&A $\vdash$ B and X,A∨A $\vdash$ C for some B and C in Y. This relation is obviously compact and contained in $\vdash_1$, and it is easy to show that it contains the rules of $L_1$ and is closed under overlap, dilution and cut for formulae. By 2.16 it coincides with $\vdash_1$, and the decidability of $L_1$ is thus a corollary of the decidability of L, which in turn follows from 4.9.

## 5.4 *Sign*

We define the *sign* of a multiple-conclusion calculus to be *negative* or *positive* according as V $\vdash$ Λ or not, i.e. according as ⟨V,Λ⟩ is excluded from or included in the characterising set of partitions. The calculus obtained by deleting ⟨V,Λ⟩ from the set of partitions that characterises a positive calculus L, or adding it to the set that characterises a negative L, evidently has the opposite sign to that of L, and we call it the negative or positive *variant* of L as the case may be.

For an alternative criterion for sign, we recall that a set X of formulae is customarily said to be *inconsistent* in a single-conclusion calculus if X $\vdash$ B for every B. Analogously we say that X is *inconsistent* in a multiple-conclusion calculus if X $\vdash$ Y for every Y. By dilution it follows that X is inconsistent iff X $\vdash$ Λ, and hence iff X is *unsatisfiable* in the set of partitions characterising L (where we say that $\langle T,U \rangle$ *satisfies* X if $X \subset T$ and that X is *satisfiable* in a set I of partitions if it is satisfied by at least one of them; from which it follows that X is satisfiable in I iff $X \nvdash_I \Lambda$). Since by dilution $V \vdash \Lambda$ iff $X \vdash \Lambda$ for some X, it follows that a calculus is negative iff it contains inconsistent (or unsatisfiable) sets. The inconsistent sets, being those X such that $X \vdash \Lambda$, determine what might be called the zero-conclusion part of a calculus. Given the sign of a calculus, its zero-conclusion part is uniquely determined by its single-conclusion part, for we have seen that a positive calculus has no inconsistent sets while our next theorem predicts those of a negative one.

*Theorem 5.23* Inconsistency in a negative calculus is equivalent to inconsistency in its single-conclusion part.

*Theorem 5.24* The (positive or negative) variant of a calculus L is the only calculus L′ other than L such that $X \vdash' Y$ iff $X \vdash Y$ for all X and all nonempty Y.

*Theorem 5.25* Every single-conclusion calculus has both positive and negative counterparts; in particular the minimum counterpart is positive and the maximum counterpart (where one exists) negative.

*Proofs* For 5.23, let L be the single-conclusion part of a negative calculus L′. If $X \vdash' \Lambda$ then by dilution $X \vdash' \{B\}$ for every B and so by 5.1 $X \vdash B$ for every B. Conversely if $X \nvdash' \Lambda$ then by cut $X,T \nvdash' U$ for some partition $\langle T,U \rangle$, and U must be nonempty since by hypothesis $V \vdash' \Lambda$. Hence by dilution $X \nvdash' \{B\}$ for some B and so by 5.1 $X \nvdash B$ for some B.

For 5.24, the variant of L has the property mentioned

in the theorem since $\langle V,\Lambda\rangle$ invalidates $\langle X,Y\rangle$ only if Y is empty; and L and its variant are the only calculi with their respective signs to have the property since, as we have seen, the zero-conclusion part of a calculus is uniquely determined by its sign and its single-conclusion part.

For 5.25, the minimum counterpart of any single-conclusion calculus is positive by 5.4, and its negative variant is also a counterpart by 5.24 and 5.1. Since this is a subcalculus of the maximum counterpart (if one exists), the latter is negative too.

Although it is easy to verify that the positive variant of a compact negative calculus must also be compact, the negative variant of a compact positive one need not be compact: for example, consider a positive calculus in which V is infinite and $X \vdash Y$ iff X and Y overlap. Subject to this caveat, Theorem 5.27 shows how rules for a calculus can be obtained from rules for its variant.

*Theorem 5.26* A calculus characterised by rules of inference is positive iff no instance of the rules has zero conclusions.

*Theorem 5.27* If a negative calculus L and a positive calculus L′ are variants of one another, rules R′ for L′ may be obtained from rules R for L by replacing every instance $\langle X,\Lambda\rangle$ by instances $\langle X,\{B\}\rangle$ for all B; and if L is compact (and so has a finite inconsistent set $X_o$) rules R for L may be obtained from rules R′ for L′ by adding the single instance $\langle X_o,\Lambda\rangle$.

*Proofs* For 5.26, the calculus characterised by rules R is positive iff $V \nvdash_R \Lambda$, i.e. iff some partition satisfies R but invalidates $\langle V,\Lambda\rangle$, i.e. iff $\langle V,\Lambda\rangle$ satisfies R, i.e. iff no instance of R has zero conclusions.

For 5.27, since L is negative $V \vdash \Lambda$, and so if it is compact $X_o \vdash \Lambda$ for some finite set $X_o$. In both cases covered by the theorem $\langle V,\Lambda\rangle$ satisfies R′ but not R, and every other partition satisfies R′ iff it satisfies R, whence the result. (In the second case, since by 5.24 $X_o \vdash' B$, if $\langle T,U\rangle$ satisfies R′ it satisfies $\langle X_o,B\rangle$ for all B and so either satisfies $\langle X_o,\Lambda\rangle$ or is $\langle V,\Lambda\rangle$.)

The term 'positive' first entered the literature as a way of describing single-conclusion calculi that lacked a negation: thus Hilbert's 'positive logic' is the negationless fragment of the intuitionist propositional calculus. Our own definition of sign for multiple-conclusion calculi seems to fit nicely with this. For instance by applying Theorem 5.26 to the rules for PC listed in Section 2.3 we see that any selection containing the rules for negation characterises a negative calculus while any selection excluding them characterises a positive one. And generally, since a positive calculus admits an interpretation in which every formula is true it cannot contain a connective whose invariable effect is to convert truths into untruths. By this criterion, however, all single-conclusion calculi are positive, even if they contain what is supposed to be a negation; for every such calculus is satisfied by the partition $\langle V,\Lambda\rangle$ representing an interpretation on which any formula and its supposed negation are alike true. This is one of the 'non-normal' interpretations which led Carnap (1943) to reject the classical calculus as an inadequate formalisation of truth-functional logic. To test for the presence of inconsistent sets is no better than to appeal to negation: it makes every single-conclusion calculus negative since V is inconsistent in them all. And the presence of unsatisfiable sets is no test at all, since by Theorem 5.25 every single-conclusion calculus is characterised both by a set of partitions including $\langle V,\Lambda\rangle$, in which every set of formulae is satisfiable, and by one excluding $\langle V, \Lambda \rangle$, in which V is unsatisfiable.

It thus seems that there is no coherent way of drawing the desired distinction for single-conclusion calculi in general. There are however two significant classes of calculi for which a definition of sign can be formulated and defended. First there are the compact calculi. Let a set of formulae be said to be *finitely inconsistent* if it contains an inconsistent finite subset. In a compact multiple-conclusion calculus inconsistency and finite inconsistency are equivalent, since $X \vdash \Lambda$ iff $X' \vdash \Lambda$ for some finite subset $X'$ of X. The equivalence fails in single-conclusion logic (e.g. V is not finitely inconsistent in Hilbert's positive logic), though if any inconsistent set is finitely inconsistent then all are (Theorem 5.29). We therefore circumvent the difficulty about the inevitable inconsistency of V by using finite inconsistency instead of inconsistency as a test for sign, saying that a compact single-conclusion calculus is *negative* or *positive* according as it does or does not contain any finitely inconsistent sets. An equivalent criterion is the finite inconsistency of V, or alternatively the presence or absence of finite inconsistent sets.

*Theorem 5.28* A compact single-conclusion calculus is positive iff every compact counterpart is positive.

*Theorem 5.29* In a negative compact single-conclusion calculus, finite inconsistency is equivalent to inconsistency.

*Theorem 5.30* Finite inconsistency in a compact single-conclusion calculus is equivalent to inconsistency in its maximum compact counterpart.

*Proofs* For 5.28, let L be a compact single-conclusion calculus. If L has a negative compact counterpart $L_1$ then $V \vdash_1 \Lambda$, so that there exists a finite set $V'$ such that $V' \vdash_1 \Lambda$. By dilution $V' \vdash_1 \{B\}$ for every B, so by 5.1 $V'$ is inconsistent in L, whence L is negative. For the converse we define $\vdash'$ so that $X \vdash' Y$ iff $X \vdash B$ for some B in Y or X is finitely inconsistent in L. Then $\vdash'$ is evidently compact and closed under overlap and dilution. For cut, either $X \vdash A$ so that if $X,A \vdash' Y$ then $X \vdash' Y$, or $X \nvdash A$ so that if $X \vdash' A,Y$ then $X \vdash' Y$. By 2.10 $\vdash'$ determines a calculus $L'$ which by 5.1 is a compact counterpart of L. If L is negative it has a finitely inconsistent set, which is also inconsistent in $L'$ by construction, so that $L'$ is negative as required.

For 5.29, any negative compact single-conclusion calculus L must contain a finite inconsistent set Y. If X is inconsistent in L then for each $A_i$ in Y we have $X \vdash A_i$ and by compactness there is a finite subset $X_i$ of X such that $X_i \vdash A_i$. By cut, $\bigcup X_i$ is a finite inconsistent subset of X, which is therefore finitely inconsistent, as required.

For 5.30 every finitely inconsistent set in a compact single-conclusion calculus L, being inconsistent in the compact counterpart $L'$ defined in 5.28, is therefore inconsistent in the maximum compact counterpart. Conversely, if a set is inconsistent in the maximum compact counterpart then that counterpart is negative and so the set is finitely inconsistent in L by 5.23, 5.28 and 5.29.

The other single-conclusion calculi between which a distinction of sign can be drawn are the propositional calculi (to be defined formally in Section 13.1). Let a set of formulae of a propositional calculus be said to be *formally inconsistent* if every substitution instance of it is inconsistent. In a multiple-conclusion propositional calculus formal inconsistency is equivalent to inconsistency, for if $X \vdash \Lambda$ then $s(X) \vdash \Lambda$ for every substitution s. This equivalence fails in the single-conclusion case (e.g. V is not formally inconsistent in Hilbert's positive logic), though if any inconsistent set is formally inconsistent then all are (Theorem 5.32). As before, this suggests the use of formal inconsistency instead of inconsistency as a test for sign, a single-conclusion propositional calculus being *negative* or *positive* according as it does or does not contain any formally inconsistent sets. An equivalent criterion is the formal inconsistency of V.

*Theorem 5.31* A single-conclusion propositional calculus is positive iff every propositional counterpart is positive.

*Theorem 5.32* In a negative single-conclusion propositional calculus formal inconsistency is equivalent to inconsistency.

*Proofs* For 5.31, let L be a single-conclusion propositional calculus. If L has a negative propositional calculus $L_1$ as a counterpart then $V \vdash_1 \Lambda$, whence by substitution and dilution $s(V) \vdash_1 \{B\}$ for every B and s. But by 5.1 s(V) is then inconsistent in L for every s, and so L is negative. For the converse we define $\vdash'$ so that $X \vdash' Y$ iff $X \vdash B$ for some B in Y or X is formally inconsistent in L. Then if L is negative $\vdash'$ determines a negative counterpart L′ of L as in 5.28, and L′ is a propositional calculus since $\vdash'$ is preserved under substitution. The proof of 5.32 runs parallel to that of 5.29.

The analogue of Theorem 5.30 fails for propositional calculi, since by Theorem 16.8 below and in contrast to Theorem 5.12, not every single-conclusion propositional calculus has a maximum propositional counterpart. It can however be shown, following the lines of the proof of Theorem 5.30, that formal inconsistency in a single-conclusion propositional calculus is equivalent to inconsistency in its maximum propositional counterpart whenever the latter exists.

These tests succeed in drawing the desired distinction between, for example, the intuitionist propositional calculus, which is negative, and Hilbert's 'positive' fragment of it. In the classical case they distinguish nicely between connectives like conjunction and disjunction and material implication, whose truth-tables have value t when all their arguments are t, and connectives like negation whose truth-table has value f for argument t: the former give rise to a positive calculus, the inclusion of the latter to a negative one. Equally important is the fact that whenever both tests are applicable they produce the same verdict:

*Theorem 5.33* In a compact single-conclusion propositional calculus, finite inconsistency and formal inconsistency are equivalent.

*Proof* By definition, if X is finitely inconsistent it has an inconsistent finite subset X′. Let q be a propositional variable not occurring in any member of X′, and for any B and any substitution s, let s′ be the substitution such that s′(q) = B and s′(p) = s(p) for all other variables p. Then since X′ ⊢ q we have by substitution that s′(X′) ⊢ s′(q), and so by dilution s(X) ⊢ B; i.e. X is formally inconsistent. Conversely if X is formally inconsistent let $p_1, p_2, \ldots$ be any listing of the propositional variables and let $s(p_i) = p_{i+1}$. Since by hypothesis s(X) is inconsistent, s(X) ⊢ $p_1$, whence by the compactness of ⊢ there exists a finite subset X′ of X such that s(X′) ⊢ $p_1$.

For any B let $s'(p_1) = B$ and $s'(p_{i+1}) = p_i$. Then by substitution $s'(s(X')) \vdash s'(p_1)$, i.e. $X' \vdash B$. Thus X is finitely inconsistent.

## 5.5 *Disjunction*

We remarked in the introduction that multiple conclusions might be expected to have the same relation to disjunction as multiple premisses do to conjunction, and the remaining results explore this suggestion. We say first that a single-conclusion calculus contains a *disjunction* if for every pair of formulae A and B there exists a formula A∨B such that $X, A\vee B \vdash C$ iff $X, A \vdash C$ and $X, B \vdash C$. It is not required that ∨ should be a connective. For example, in a calculus with an axiom D but no rules one can take A∨B to be D or A according as A and B are distinct or not; while the calculus of material implication contains a disjunction with A∨B defined as (A⊃B)⊃B (though its intuitionist analogue, the positive implicational calculus, cannot contain a disjunction by Theorems 5.13 and 5.36). We use $A_1\vee\ldots\vee A_m$ to stand indifferently for any of the formulae obtained by disjoining $A_1,\ldots,A_m$, e.g. for $((A_1\vee A_2)\vee\ldots)\vee A_m$ as well as $A_1\vee(A_2\vee(\ldots\vee A_m))$. This systematic ambiguity is justified by

*Lemma 5.34* If each $A_i$ is a $B_j$ then $A_1\vee\ldots\vee A_m \vdash B_1\vee\ldots\vee B_n$.

*Proof* By iterating the defining condition for a disjunction we have that $X, B_1\vee\ldots\vee B_n \vdash C$ iff $X, B_j \vdash C$ for each j, and that $X, A_1\vee\ldots\vee A_m \vdash C$ iff $X, A_i \vdash C$ for each i. If each $A_i$ is a $B_j$ it follows that if $X, B_1\vee\ldots\vee B_n \vdash C$ then $X, A_1\vee\ldots\vee A_m \vdash C$. The result follows by taking $X = \Lambda$ and $C = B_1\vee\ldots\vee B_n$.

The natural multiple-conclusion analogue of the given condition for a single-conclusion calculus to contain a disjunction is that $X,A \vee B \vdash Y$ iff $X,A \vdash Y$ and $X,B \vdash Y$. This is equivalent to postulating that $A \vdash A \vee B$ and $B \vdash A \vee B$ and $A \vee B \vdash A,B$, which are of course the duals of the familiar conditions for a single-conclusion calculus to contain a conjunction. If now L is any single-conclusion calculus containing a disjunction, it follows from Theorem 5.5 that the above conditions are satisfied by $\vdash_U$ and hence by the maximum counterpart of L whenever it exists. In the compact case the result can be carried a stage further, for just as in a compact single-conclusion calculus containing a conjunction we have (for nonempty X) that $X \vdash B$ iff $A_1 \& \ldots \& A_m \vdash B$ for some $A_1,\ldots,A_m$ in X, so we have (for nonempty Y) that in the maximum compact counterpart of one containing a disjunction $X \vdash Y$ iff $X \vdash B_1 \vee \ldots \vee B_n$ for some $B_1,\ldots,B_n$ in Y; or equivalently

*Theorem 5.35* If L′ is the maximum positive compact counterpart of a compact single-conclusion calculus L containing a disjunction, then $X \vdash' Y$ iff $X \vdash' B_1 \vee \ldots \vee B_n$ for some $B_1,\ldots,B_n$ in Y.

*Proof* Since L′ is positive and compact, if $X \vdash' Y$ then $X \vdash' B_1,\ldots,B_n$ for some $B_1,\ldots,B_n$ in Y; and since by 5.1 and 5.34 $B_j \vdash' B_1 \vee \ldots \vee B_n$, it follows by cut for $\{B_1,\ldots,B_n\}$ that $X \vdash' B_1 \vee \ldots \vee B_n$. For the converse we have by 5.5 that $B_1 \vee \ldots \vee B_n \vdash_U B_1,\ldots,B_n$ and so, as in 2.12, $B_1 \vee \ldots \vee B_n \vdash' B_1,\ldots,B_n$. So if $X \vdash' B_1 \vee \ldots \vee B_n$ for some $B_1,\ldots,B_n$ in Y it follows that $X \vdash' Y$.

These parallels, however, only hold for maximum counterparts. At the other end of the spectrum there is no special connection at all between disjunction and multiple conclusions in the minimum counterpart of a single-conclusion calculus that contains a disjunction. And since there is in general a whole range of intermediate counterparts there is room for a corresponding variety of connections of greater or less intimacy between disjunction and multiple conclusions. For example, does the addition of 'from A∨B infer A,B' to the rules for the intuitionist calculus produce a stronger counterpart than the addition of 'from A∨~A infer A,~A'?

We have seen in Theorem 5.15 that there is a simple algorithm for converting rules of inference for a single-conclusion calculus into rules that characterise its minimum counterpart. In general, as Theorem 5.19 shows, there is no corresponding algorithm which produces rules for the maximum compact counterpart, but we can provide one in the case of calculi which contain a disjunction. We give it in Theorem 5.36 as a method of generating rules for the maximum positive compact counterpart; this covers the case of a positive calculus, by Theorem 5.28, and Theorem 5.27 shows how it can be adapted to the negative case by adding the rule 'from $X_o$ infer Λ' for an arbitrary finite inconsistent set $X_o$.

*Theorem 5.36*† If the single-conclusion rules R characterise a calculus L containing a disjunction, its maximum positive compact counterpart L′ is characterised by the corresponding singleton-conclusion rules together with 'from A∨B infer A,B'.

*Theorem 5.37*† If the multiple-conclusion rules R′ characterise a calculus L′ whose single-conclusion counterpart L contains a disjunction, L is characterised by the rules R1 'from X∨C infer C if X R′ Λ', R2 'from X∨C infer $B_1$∨...∨$B_n$∨C if X R′ {$B_1$,...,$B_n$} and n≥1', and R3 'from $A_1$∨...∨$A_m$ infer $B_1$∨...∨$B_n$ if each $A_i$ is a $B_j$'.

*Proofs* For 5.36, let R′ be the multiple-conclusion rules

mentioned in the theorem. Then $R' \subset \vdash'$ by 5.35, and hence $\vdash_{R'} \subset \vdash'$. Conversely if $X \vdash' Y$ we have by 5.35 that $X \vdash' B_1 \vee \ldots \vee B_n$ for some $B_1, \ldots, B_n$ in Y, so that since L and L′ are counterparts $X \vdash_R B_1 \vee \ldots \vee B_n$. Since every partition that satisfies R′ satisfies R, it follows that $X \vdash_{R'} \{B_1 \vee \ldots \vee B_n\}$, and by repeated applications of the rule 'from $A \vee B$ infer A,B' we have $X \vdash_{R'} B_1, \ldots, B_n$ and hence $X \vdash_{R'} Y$.

For 5.37 it is sufficient by 5.16 to show that (a) if X R B then $X \vdash_{R'} \{B\}$ and (b) if X R′ Y and $Z,A \vdash_R B$ for every A in Y, then $Z,X \vdash_R B$. We use 5.1 throughout and observe first that by 5.34 and the definition of disjunction, $Z,A \vee C \vdash_{R'} B \vee C$ whenever $Z,A \vdash_{R'} B$. If $\langle X \vee C, C \rangle$ is an instance of R1 then $X \vdash_{R'} C$, and applying our observation to each member of X in turn, $X \vee C \vdash_{R'} C \vee C \vee \ldots \vee C$, whence by 5.34 $X \vee C \vdash_{R'} C$. If $\langle X \vee C, B_1 \vee \ldots \vee B_n \vee C \rangle$ is an instance of R2 then $X \vdash_{R'} B_1, \ldots, B_n$; and by 5.34 and cut for $\{B_1, \ldots, B_n\}$ we have $X \vdash_{R'} B_1 \vee \ldots \vee B_n$; and arguing as before, $X \vee C \vdash_{R'} B_1 \vee \ldots \vee B_n \vee C$. Finally if $\langle X,B \rangle$ is an instance of R3 then $X \vdash_{R'} B$ by 5.34. This establishes (a). For (b) we need

*Lemma 5.38* If $X \vdash_R B$ then $X \vee C \vdash_R B \vee C$.

To prove the lemma we use induction on the length of a sequence proof of B from X, and for this it is sufficient to show that $X \vee C \vdash_R B \vee C$ whenever X R B. If $\langle X,B \rangle$ is an instance of R1 then $X = X' \vee B$, say, and as $\langle X' \vee (B \vee C), B \vee C \rangle$ is also an instance of R1 it follows by R3 that $X \vee C \vdash_R B \vee C$. If $\langle X,B \rangle$ is an instance of R2 then $X = X' \vee C'$, say, and $B = B_1 \vee \ldots \vee B_n \vee C'$; but $\langle X' \vee (C' \vee C), B_1 \vee \ldots \vee B_n \vee (C' \vee C) \rangle$ is an instance of R2 and again we have $X \vee C \vdash_R B \vee C$ by R3. Finally if $\langle X,B \rangle$ is an instance of R3 so too is $\langle X \vee C, B \vee C \rangle$.

For the proof of (b) we note that R characterises a calculus containing a disjunction, for if $X, A \vee B \vdash_R C$ then, since $A \vdash_R A \vee B$ and $B \vdash_R A \vee B$ by R3, we have $X,A \vdash_R C$ and $X,B \vdash_R C$; and conversely if $X,A \vdash_R C$ and $X,B \vdash_R C$ then $X \vee B, A \vee B \vdash_R C \vee B$ and $X \vee C, B \vee C \vdash_R C \vee C$ by 5.38, so that by R3 $X, A \vee B$

$\vdash_R$ B∨C and X,B∨C $\vdash_R$ C, whence by cut we have X,A∨B $\vdash_R$ C. Now if X R′ Λ then X∨B $\vdash_R$ B by R1 and so Z,X $\vdash_R$ B by R3. Moreover if X R′ $\{A_1,\ldots,A_n\}$ and n≥1 then X∨$A_1$ $\vdash_R$ $A_1\vee\ldots\vee A_n\vee A_1$ by R2 and so X $\vdash_R$ $A_1\vee\ldots\vee A_n$ by R3. If also Z,$A_i$ $\vdash_R$ B for each i, then Z,$A_1\vee\ldots\vee A_n$ $\vdash_R$ B by the definition of disjunction. Hence by cut for $A_1\vee\ldots\vee A_n$ we again have Z,X $\vdash_R$ B, establishing (b) and hence the theorem.

Note that the dummy formula C in R1 and R2 cannot in general be suppressed without weakening the rules: we show in Section 19.7 that even R1 cannot be replaced by 'from X infer C if X R′ Λ'. When R2 contains axioms, however, (i.e. if X is ever empty), the dummy formula can be suppressed in them without loss of strength. For each axiom is of the form $B_1\vee\ldots\vee B_n\vee C$, where Λ R′ $\{B_1,\ldots,B_n\}$; and every instance of this form follows from $B_1\vee\ldots\vee B_n$ by R3, and conversely $B_1\vee\ldots\vee B_n$ follows by R3 from the instance in which C = $B_1$.

# 6 · Infinite rules

## 6.1 *Infinite rules and proofs*

To develop a multiple-conclusion predicate calculus, Carnap and Kneale introduce the rule 'from $(\exists x)Fx$ infer $Fx_1, Fx_2, Fx_3, \ldots$', whose conclusions are meant to represent all values of the propositional function expressed by Fx (cf. Section 20.2). Instances of this rule may obviously have infinitely many conclusions, and the converse rule for the universal quantifier, 'from $Fx_1, Fx_2, \ldots$ infer $(x)Fx$', raises a similar point for single- as well as multiple-conclusion calculi. We shall briefly consider how the admission of such rules affects what we have said so far. We therefore relax our previous requirement that a rule of inference must be *finite* in the sense that X R Y only if X and Y are finite, and allow rules of the form 'from X infer Y if X R Y' without any restriction as to finiteness; and similarly for single-conclusion rules.

An immediate effect of the change is that in general $\vdash_R$ is no longer compact. Hence every calculus, whether compact or not, can be characterised by rules of inference, if necessary by taking the consequence relation of the calculus to be the rule. Calling a calculus *conclusion-compact* if whenever $X \vdash Y$ there is a finite subset $Y'$ of Y such that $X \vdash Y'$, one might well suppose that when every instance of a set of rules has a finite set of conclusions the resulting calculus will at least be conclusion-compact. Curiously enough, this is not so. Consider, for example, the rules R1 'from $\Lambda$ infer $A, \sim A$' and R2 'from $\sim\Diamond V$ infer $\Lambda$', where

V is infinite and $\sim$ and $\Diamond$ are any singulary connectives. For every partition $\langle T,U \rangle$ we have $T \vdash U, \Diamond V$, by R1 if $\sim\Diamond V$ overlaps U and by R2 otherwise. Hence $\Lambda \vdash \Diamond V$ by cut for V. On the other hand whenever U is finite (and so T nonempty) we have $V-(\sim\Diamond T \cup \Diamond U) \nvdash \sim\Diamond T, \Diamond U$ by Lemma 2.17, and so by dilution $\Lambda \nvdash \Diamond U$.

Finite proofs are obviously inadequate to deal with rules with infinitely many premisses or conclusions. We therefore need to make a similar relaxation in the various definitions of proof, by dropping the requirement that proofs must be finite and making transfinite sequences eligible either as single-conclusion proofs or as branches of multiple-conclusion ones. This is sufficient to ensure the continued adequacy of sequence proofs and the various kinds of tree proof. In the case of sequence proofs this can be shown by a straightforward generalisation of the method of Theorem 1.13. In particular, to show that deducibility is closed under cut for sets we suppose that there exist proofs $\sigma$ of B from X,Z and $\sigma_i$ of $A_i$ from X for each $A_i$ in Z, and concatenate them to form the required sequence proof $\sigma_1,\sigma_2,\ldots,\sigma$ of B from X. It is possible similarly to generalise the method of Theorem 3.5 and demonstrate closure under cut in the multiple-conclusion case by grafting tree proofs onto one another to form a proof of the required kind, but it is a surprisingly intricate business and we can avoid it by tackling the results of Section 3.3 in a different order.

We note first that a universal tree $\tau_Z$ may be defined for any Z exactly as in the finite case, and that it has the same properties as expressed by Lemma 3.8 and Theorem 3.9. The definition of universal proof carries over, and we show that each single tree $\tau_V$ is a universal proof from any X to any Y by any R, provided only that $X \vdash_R Y$:

*Theorem 6.1* Any universal proof $\tau_V$ is adequate.

*Proof* We prove a cycle of implications which establishes not only the adequacy of $\tau_V$ but the adequacy of tree proofs, extended proofs and universal proofs in general, i.e. the analogues of 3.5-7 and 3.10. If $X \vdash_R Y$ then, arguing as in 3.9, $\tau_V$ is a universal proof from X to Y by R. If $\tau_V$ is a universal proof from X to Y by R then obviously a universal proof from X to Y by R exists; and by definition if there is a universal proof there is an extended proof. If there is an extended proof from X to Y by R then, arguing as for 3.3, it contains a tree proof from X to Y by R. Finally, if there is a tree proof from X to Y by R then by the argument of 3.5 it follows that $X \vdash_R Y$.

This last step requires the existence of a maximal file of formulae in T, i.e. a branch of the subtree $\{A: A \in T$ and $\hat{A} \subset T\}$. The proposition that every tree has a branch is equivalent to the axiom of choice, without which the argument therefore fails and indeed the equation between $\vdash_R$ and deducibility by R no longer holds. For if a tree without branches were possible the corresponding tree consisting entirely of occurrences of A would constitute a proof from A to $\Lambda$ by arbitrary rules R; but we do not have $A \vdash_R \Lambda$ in general.

The removal of the finiteness restriction on rules means a change in the sense in which it is appropriate to predicate recursiveness of them. We shall now have to say that a rule R is recursive if X R Y (rather than $X_m$ R $Y_n$) is recursive, and we shall have to treat proofs similarly. It is scarcely surprising that the theory developed in Chapter 4 for the finite case collapses in the infinite one. No comment is needed on the results about counterparts in Chapter 5, except to note that although Theorem 5.16, which states conditions for multiple and single-conclusion rules to characterise counterpart calculi, fails in general, it continues to hold when the relevant single-conclusion calculus possesses a maximum counterpart.

## 6.2 *Marking of theorems*

We have already marked with a † those results in Chapters 1-5 that fail to carry over to the case of infinite rules and proofs. We now supply the justification for this by sketching proofs of the unmarked theorems in the infinite case and disproofs of the marked ones. The examples assume that V is infinite.

1.1-7 These results do not involve rules of inference.

1.8-9 Proofs are as in the finite case.

1.10†-12† If R is 'from X infer B if X is infinite' then $X \vdash_R B$ iff $B \in X$ or X is infinite, so that $\vdash_R$ and the calculus characterised by R are not compact. If R is 'from X infer B if X is infinite or $B \neq C$' for a given C, the proof of 1.3 shows that $\vdash_R$ is not the closure of R under overlap, dilution and cut for formulae.

1.13-14 Proof is as in the finite case, except that without 1.11 we need to establish that deducibility is closed under cut for sets in general, as is done in Section 6.1.

2.1-12 These results do not involve rules of inference.

2.13-14 Proofs are as in the finite case.

2.15†-16† If R is 'from X infer Y if X is infinite' then $X \vdash_R Y$ iff X and Y overlap or X is infinite, so that $\vdash_R$ is not compact. If R is 'from X infer Y if X or Y is infinite' the proof of 2.4 shows that $\vdash_R$ is not the closure of R under overlap, dilution and cut for formulae.

2.17 Proof is as in the finite case.

2.18† The counterexample for 2.15 provides a disproof here too.

3.1-4 Proofs are as in the finite case.

3.5-7 See proof of Theorem 6.1.

3.8-9 Proofs are as in the finite case.

3.10 See proof of Theorem 6.1.

4.1† The empty rule R is recursive, so if tree proof is recursive tree proof by R is too. But an empty tree is not a proof from Λ to Λ by R, and since V is infinite there is a formula A about which no questions are asked in deciding this. Then the same process decides that a tree consisting of a single occurrence of A is not a proof from A to A, which is absurd.

4.2 This result does not involve rules of inference.

4.3† Let L be the axiomatisable calculus in which $X \vdash Y$ iff Y is nonempty. Suppose that L is characterised by recursive rules R. Consider the set S of pairs $\langle W_1, W_2 \rangle$ such that for every n there is an instance $\langle X, Y \rangle$ of R such that for each $m \leq n$, (i) $A_m \in X$ if $A_m \in W_1$ but not if $A_m \in W_2$, and (ii) $A_m \notin Y$. As in 2.3 we partially order S so that $\langle W_1, W_2 \rangle \subset \langle W_1', W_2' \rangle$ iff $W_1 \subset W_1'$ and $W_2 \subset W_2'$. For each n there is an instance $\langle X, Y \rangle$ of R such that $A_m \notin Y$ for all $m \leq n$; for otherwise we would have $A_1, \ldots, A_n \nvdash_R A_{n+1}, A_{n+2}, \ldots$ by the infinite analogue of 2.17. Hence $\langle \Lambda, \Lambda \rangle \in S$ and so S is nonempty. Moreover any totally ordered subset $S_o$ of S has an upper bound in S, namely $\langle \cup W_1, \cup W_2 \rangle$ where the unions are taken over all $\langle W_1, W_2 \rangle$ in $S_o$, since for any n there is a member $\langle W_1', W_2' \rangle$ of $S_o$ such that for each $m \leq n$, $A_m \in W_1'$ iff $A_m \in \bigcup W_1$ and $A_m \in W_2'$ iff $A_m \in \bigcup W_2$. So by Zorn's lemma S has a maximal member $\langle T,U \rangle$. Evidently T and U are disjoint, and either $A \in T$ or $A \in U$ for every A. (For since $\langle T,U \rangle \in S$, there is for each n an instance of R, call it $\langle X_n, Y_n \rangle$, such that for each $m \leq n$, $A_m \in X_n$ if $A_m \in T$ but not if $A_m \in U$, and $A_m \notin Y_n$. Either the values of n for which $A \in X_n$, or those for which $A \notin X_n$, are unbounded. In the first case $\langle T \cup \{A\}, U \rangle \in S$ and in the second $\langle T, \{A\} \cup U \rangle \in S$; whence the result by the maximality of $\langle T,U \rangle$.) Now $\langle T, \Lambda \rangle$ is not an instance of R since $T \nvdash \Lambda$, and in the process of deciding this questions about $A_m$ may only be asked for finitely many m, say only for $m \leq n$. Since $\langle T,U \rangle \in S$, there is an instance $\langle X,Y \rangle$ of R such that for each $m \leq n$, $A_m \in X$ iff $A_m \in T$, and $A_m \notin Y$; but the process that decides that $\langle T, \Lambda \rangle$ is not an instance of R will do the same for $\langle X,Y \rangle$, and this provides the required contradiction.

4.4† The empty rule characterises an axiomatisable calculus L. But given any rules R for L, tree proof by R is not recursive by the argument for 4.1† above.

4.5† The empty rule R is recursive, so sequence proof by R is recursive if sequence proof is recursive. But the sequence $A_1, A_2, A_3, \ldots; A_1$ is a proof of $A_1$ from V by R, and, V being infinite, there is a formula $A_n$ (other than $A_1$) about which no questions are asked in deciding this. Then the same process decides that the same sequence is a proof of $A_1$ from $V-A_n$, which it is not.

4.6 This result does not involve rules of inference.

4.7† The empty rule is recursive, but by the argument for 4.5† above the resulting calculus is not characterised by any rules R for which sequence proof by R is recursive.

4.8 We consider the same calculus as in the finite case, and show as there that it is axiomatisable but not characterised by recursive rules. To show that it is not characterised by any rules R for which sequence proof by R is recursive, we argue as for 4.5† above, choosing $A_n$ not to be of the form A∨A.

4.9 The proof is as in the finite case.

4.10 The proof is as for 1.14 above.

4.11† The disproof is as for 4.5† above.

4.12† Consider the axiomatisable calculus with the vocabulary of the classical one but characterised by the non-iterable finite rule 'from A&A infer B∨B', where B is fixed. Suppose it is characterised by recursive rules R. In deciding that ⟨Λ,B∨B⟩ is not an instance of R there is a formula A about which the question 'does A&A occur as a premiss?' is never asked. Then the same process decides that ⟨χ,B∨B⟩ is not an instance of R, where all the members of χ are occurrences of A&A. Hence the partition ⟨A&A, V-A&A⟩ satisfies R but not $\vdash_R$, which is absurd.

4.13† The empty rule characterises an axiomatisable calculus, but by the argument for 4.5† above there are no rules R (whether or not multiplicity of premisses is taken into account) which characterise the calculus and for which sequence proof by R is recursive.

4.14-17 These results do not involve rules of inference.

5.1-14 These results do not involve rules.

5.15 Proof is as in the finite case.

5.16† The result holds good when the calculus characterised by R possesses a maximum counterpart, and the proof is as in the finite case. But it fails in general:

for example, take R to be the rule 'from X infer B if X is infinite', and let R′ be the corresponding singleton-conclusion rule plus 'from Λ infer A,B if A ≠ B'. Then R and R′ satisfy the conditions of the theorem but do not characterise counterparts, since by cut for V we have $\Lambda \vdash_{R'} B$ for every B.

5.17-25 These results do not involve rules.

5.26-7 Proofs are as in the finite case. In 5.27 the proviso about compactness becomes redundant: rules R for L may in every case be obtained from rules R′ for L′ by adding 'from V infer Λ'.

5.28-35 These results do not involve rules.

5.36† Where L is not compact it cannot by 5.8 have any compact counterpart.

5.37† Let R′ be the singleton-conclusion rule corresponding to R3, plus 'from Λ infer A,A∨A,A∨A∨A,...'. Then $X \vdash_{R'} Y$ iff Y is nonempty. But the additional rule cannot be reconstructed in R, which consists of R3 only.

5.38 Proof is as in the finite case.

# Part II · Graph proofs

# 7 · Graph arguments

## 7.1 *Arguments*

Textbooks say that an argument is valid if the conclusion follows from the premisses, but this cannot be right. It would make Figure 7.1 into a valid scheme of argument even though both its steps (i.e. from 2 to 3, and from 1 and 3 to 4) are classic fallacies. The textbook criterion is all right for arguments consisting of a single step but it overlooks the fact that arguments are generally made up of a number of steps. Perhaps then an argument is valid if all its component steps are valid? No, for this would make Figure 7.2 a valid scheme of argument from any A to any B.

| | | |
|---|---|---|
| 1 | All a are b | premiss |
| 2 | All b are c | premiss |
| 3 | All c are b | from 2 |
| 4 | All a are c | from 1,3 |

Figure 7.1

| | | |
|---|---|---|
| 1 | A | premiss |
| 2 | $A \supset B$ | from 3 |
| 3 | $A \& (A \supset B)$ | from 1,2 |
| 4 | B | from 3 |

Figure 7.2

Clearly the validity of an argument is the product of two factors: (i) the individual steps must be valid and (ii) they must be correctly arranged and articulated. Thus Figure 7.1 is invalid because it fails to satisfy (i) and Figure 7.2, with its vicious circle, is invalid because it fails to satisfy (ii). The following chapters are our attempt to elucidate and explore the idea behind (ii). If at the same time we were going to discuss the other factor we should need to coin a distinctive name for this one: 'potential validity', say, or 'structural validity'. The two operate, however, at quite different levels of generality. Validity of steps is relative to a given calculus - a step that is classically valid may be intuitionistically invalid, and so on - but the question of their correct arrangement is independent of the choice of calculus and even of the choice of formulae, as we shall see. It is therefore possible to discuss either condition for validity in a self-contained way without bringing in the other. This being so, we shall speak of arguments as being simply 'valid' or 'invalid' with reference to (ii) alone.

A prerequisite of a theory of validity is a definition of arguments that takes due account of their structure. The concept of a sequence proof is of no use here. First, it is too narrow: it rules out invalid arguments altogether and also as it happens many impeccably valid ones, as we note in Chapter 11. Second, it is inherently indeterminate with regard to argumentative structure. The definition requires that each formula in the sequence be a premiss or an axiom or derived from an (unspecified) selection of preceding formulae, but none of these alternative justifications is put forward as the one actually invoked by the proof. Calculi can easily be devised in which several or even all the possible alternatives are equally admissible, so that the same sequence ambiguously represents a number of different proofs. For example, in a calculus with the rule 'from $C_{i_1},\dots,C_{i_m}$ infer $C_j$ if $0 \le m$ and $i_1 < \dots < i_m < j$' the sequence $C_1,\dots,C_7$ represents nearly ten million proofs of $C_7$ from $C_1,\dots,C_6$, each with its own distinctive structure. Similar objections apply to tree proofs in the context of multiple conclusions.

The answer seems to be straightforward enough: take an argument to be an array of formulae structured by a 'step' relation, so that a step from $A_1,\ldots,A_m$ to B (or to $B_1,\ldots,B_n$ as the case may be) is represented by this relation holding between their occurrences. This can be indicated informally by annotations as in Figures 7.1 and 7.2, or by a diagram as in Figure 7.3 (which shows the same argument as Figure 7.2). A difficulty arises in handling the step relation systematically, however, since the number of premisses varies from step to step. Thus the step relation does not have a fixed valency (binary, ternary, etc.) but relates varying numbers of objects (doubly variable in the multiple-conclusion case); and there exists no developed theory of relations of this sort. It certainly cannot be treated as a binary relation holding between each $A_i$ and B or each $B_j$, since this fails to do justice to the way in which the premisses (and the conclusions in the multiple-conclusion case) function together. There is a vast difference between (a) inferring B from $A_1$ and $A_2$ and (b) inferring B from $A_1$ and from $A_2$; but a relation holding between $A_1$ and B and between $A_2$ and B fails to mark the difference.

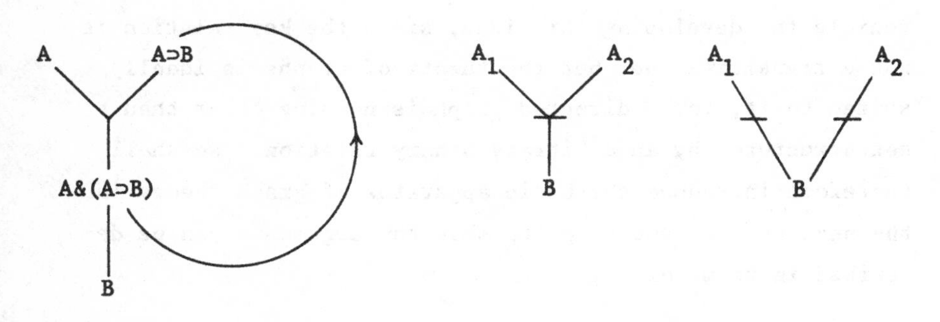


Figure 7.3 Figure 7.4(a) Figure 7.4(b)

When discussing rules of inference and consequence relations we anticipated a similar difficulty by taking sets instead of individual formulae as the objects of a (binary) relation, but here we propose a different solution. This is to interpose a fresh object between the premisses and conclusion(s) of the step, and to represent a step from $A_1,\ldots,A_m$ to $B_1,\ldots,B_n$ by a binary relation holding between each $A_i$ and the mediating object and between it and each $B_j$. The new object can well be thought of as an occurrence of one of those words like 'so' or 'therefore' which help to articulate our daily arguments. (They also mark a distinctive kind of linguistic activity, namely arguing, but this we do not pretend to discuss: an argument for us is an inanimate thing and not a live performance.) We call the object an (*inference*) *stroke* and in picturing arguments we symbolise it by a short horizontal line. In this way Figure 7.4 shows how the ambiguity mentioned above is resolved, for 7.4(a) represents a single step '$A_1$, $A_2$ so B' while 7.4(b) represents a pair of steps '$A_1$ so B' and '$A_2$ so B'.

We thus arrive at the idea of an argument as an array of formulae and inference strokes, structured by a relation between formulae and strokes and vice versa. The theory of orderings (e.g. of sequences or trees) is not an appropriate vehicle for developing this idea, since the key relation is not a transitive one, but the theory of graphs is ideally suited to it, for a directed graph is nothing other than a set structured by an arbitrary binary relation. We shall therefore introduce the basic apparatus of graph theory in the next section and go on to show how arguments can be described in terms of it.

## 7.2 *Graphs*

A (directed) *graph* is a set of *vertices* together with a binary relation between them. An instance ⟨a,b⟩ of this relation is called an *edge* joining a and b; more exactly, it is the edge that goes *from* a *to* b. A vertex is *initial* if no edge goes to it and *final* if no edge goes from it, and a vertex which is neither initial nor final is *intermediate*. A *subgraph* of a graph is any graph contained in it, and we may denote it (somewhat loosely) by listing its vertices and edges between braces. A *redirection* of a graph is obtained by changing the direction (i.e. replacing ⟨a,b⟩ by ⟨b,a⟩) of some or none or all of the edges; if all are reversed the redirection is the *converse* graph, whose edge relation is the converse of that of the original. A *bipartite* graph is one whose vertices are of two sorts, vertices joined by an edge always being of opposite sorts. In particular no edge in a bipartite graph goes from a vertex to itself (i.e. there are no loops).

Let $a_1,E_1,a_2,E_2,\ldots,E_{n-1},a_n$ be a finite sequence of alternate vertices and edges in which each $E_i$ joins $a_i$ and $a_{i+1}$ and no edge occurs more than once. If $n>1$ and no vertex occurs more than once in the sequence, the corresponding subgraph $\{a_1,E_1,\ldots,a_n\}$ is a *path* connecting its *endpoints* $a_1$ and $a_n$. (The endpoints are uniquely determined as the vertices adjoining just one edge in the path.) If $n>2$ and $a_1=a_n$ but otherwise no vertex occurs more than once, the corresponding subgraph is a *circuit*. These definitions take no account of the directions of the edges; if in addition each $E_i$ goes from $a_i$ to $a_{i+1}$ the corresponding subgraph is a *directed path* from $a_1$ to $a_n$ or a *directed circuit* as the case may be. A graph is *connected* if there exist paths connecting every pair of distinct vertices. Two graphs are *disjoint* if they have no vertex, and hence no edge, in common. Any graph is thus the union of zero or more disjoint nonempty connected graphs called its *components*.

In these definitions we have made our own selection from the variety of available terminology (e.g. Ore, 1962). Paths and circuits are usually defined as sequences, but it suits us better to take the corresponding subgraphs (and incidentally avoid the arbitrariness of stipulating a particular starting-point for a circuit). For us a finite graph is one with a finite number of vertices, though some writers require merely that the number of edges be finite. We use a, b, c as notation for vertices, E for edges, $\pi$ for graphs and graph arguments in general, $\gamma$ for circuits and $\alpha$ and $\beta$ for paths. We also write (a,b) to denote any path connecting a and b, and observe that in a circuit-free graph there is at most one such path for each pair of vertices.

If the ancestral of the edge relation that determines a graph is a partial ordering we can dispense with an explicit notation for the direction of edges by adopting the convention that they run down the page. By virtue of the next result this covers all cases except the few which involve directed circuits; in these we will have to mark by an arrow any edges that run up the page. Even a partial ordering of the vertices does not mean that we can dispense with an explicit notation for the edges themselves, but we mention in Section 8.1 the special circumstances in which this too is possible.

*Lemma 7.1* The ancestral of the edge relation is a partial ordering iff there are no directed circuits.

*Lemma 7.2* A nonempty finite bipartite graph with no directed circuits has at least one initial and at least one final vertex.

*Proofs* We write $\geq$ for the ancestral of the edge relation. Like all ancestrals it is reflexive and transitive, so for

7.1 it is sufficient to show that it is antisymmetric iff there are no directed circuits. If there is a directed circuit $\{a_1,E_1,a_2,\ldots,a_n\}$ we have $a_1 \geq a_2$ and $a_2 \geq a_n = a_1$ although $a_1 \neq a_2$, i.e. $\geq$ is not antisymmetric. Conversely, if there exist distinct a and b such that $a \geq b$ and $b \geq a$, there are sequences $a_1,E_1,\ldots,a_m$ and $a_m,E_m,\ldots,a_n$ such that $1<m<n$ and each $E_i$ goes from $a_i$ to $a_{i+1}$ and $a_1 = a_n = a$ and $a_m = b$. By choosing the shortest such sequences we may assume that $a_1,\ldots,a_m$ are all distinct and so are $a_m,\ldots,a_n$. There exist j and k (e.g. 1 and n) such that $j<m<k$ and $a_j = a_k$, and choosing them so that $k-j$ is a minimum we have that $\{a_j,E_j,\ldots,a_k\}$ is a directed circuit.

For 7.2 we note that any finite nonempty partially ordered set has both a maximal and a minimal member, that a vertex b which is maximal with respect to $\geq$ must be initial (since no edge goes from b to itself in a bipartite graph), and similarly that a minimal vertex must be final. The result is then immediate from 7.1.

## 7.3 *Premisses and conclusions*

We suggested that arguments should be envisaged as arrays of formulae and inference strokes, structured by a relation between formulae and strokes. It is assumed that the inference stroke is never a formula, so in graphical terms the idea becomes that of a bipartite graph of formulae and strokes, i.e. one in which each vertex is an occurrence of a formula or of the inference stroke, and each edge goes from a formula vertex to a stroke vertex or vice versa. The description of an argument is however incomplete until we have identified its premisses and conclusions. We therefore define a *(graph) argument* to be a finite bipartite graph of formulae and strokes in which subsets of the initial and final formula vertices are specified as the *premisses* and *conclusions* respectively. In picturing graph arguments we represent the conclusions by triangles, premisses by inverted triangles, and other formula vertices by circles, as in Figures 7.5 and 7.6. These notation are adapted from the sport of orienteering, the physical equivalent of proof-tracing. An isolated vertex, being both initial and final, can in principle be both a premiss and a conclusion, and in such a case we superimpose the two notations as in Figure 7.5.

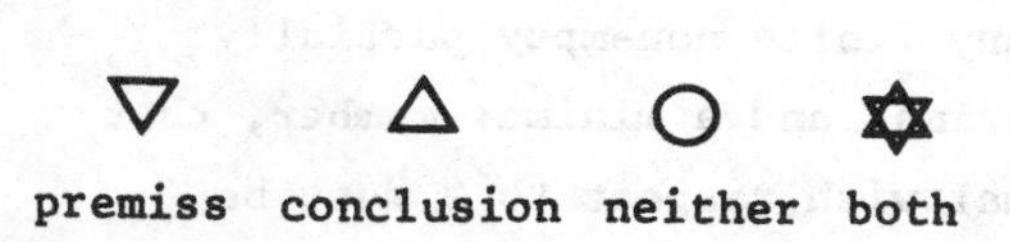


Figure 7.5

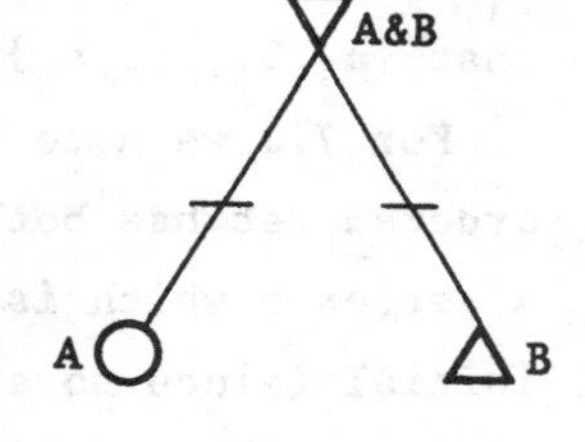


Figure 7.6

Our stipulation as to which vertices are eligible to be the premisses and conclusions of a graph argument is a compromise. We could have made every formula vertex eligible, be it initial or final or intermediate, but by doing so we would not obtain any significantly wider class of graph proofs, as we shall show in Theorem 7.9. Taking the opposite line we might have required that all the initial formula vertices should be regarded as premisses and all the final ones as conclusions, and so incidentally dispense with the need for a special notation like our triangles. As it is, we call an argument satisfying these requirements *standard*. Figure 7.6 is an example of a non-standard argument, but since it is the graphical equivalent of an ordinary sequence proof (viz. the sequence A&B,A,B construed as a proof of B from A&B by the rules 'from A&B infer A' and 'from A&B infer B'), there is a prima facie case for including it. One might propose its exclusion on the ground that what Figure 7.6 really displays is a standard proof (the leg from A&B to B) with some irrelevant trimmings (the leg from A&B to A), and that we lose nothing by ignoring the latter. But we shall see in Chapter 9 that there are proofs which are irreducibly non-standard, i.e. which have no standard subproof, and so do constitute a significantly wider class of proof than the standard ones. Any initial formula vertex that is not a premiss of the argument, and any final one that is not a conclusion, will be called *non-standard*. All other vertices (including all strokes and all intermediate formulae) are *standard*. A standard argument is thus one in which every vertex is standard.

The premisses and conclusions of a graph argument were defined in the first instance as formula vertices, i.e. occurrences of formulae rather than formulae themselves, and we need a notation for the corresponding formulae. Let $X_\pi$ therefore stand for the set of formulae which occur as premisses of the argument $\pi$, and let $Y_\pi$ stand for the set of formulae which occur as conclusions of $\pi$. Similarly if b is any stroke vertex we write $X_b$ for the set of formulae from whose occurrences edges go to b, and $Y_b$ for the set of formulae to whose occurrences edges go from b. We write $R_\pi$ for the rule 'from $X_b$ infer $Y_b$ if b is a stroke in $\pi$', i.e. the rule which exactly covers the steps of $\pi$. Obviously $\pi$ is an argument from $X_\pi$ to $Y_\pi$ by $R_\pi$, and making the customary allowance for dilution we say that $\pi$ is an *argument from* X *to* Y *by* R iff $X_\pi \subset X$ and $Y_\pi \subset Y$ and $R_\pi \subset R$.

Because we ignored questions of multiplicity in defining rules of inference, there is no restriction (other than finiteness) on the multiplicity with which the premisses and conclusions of an instance of a rule can occur as vertices in a corresponding graph argument. For example, each of the three graphs in Figure 7.7 would be admissible as part of an argument by the rule of adjunction 'from A,B infer A&B'. We mentioned in Section 4.2 the alternative possibility of assigning a definite multiplicity to the premisses and conclusions of rules by defining rules in terms of arrays rather than sets, and it is easy to adapt the definition of 'argument by R' to match, since the formula vertices adjacent to the various inference strokes already form appropriate arrays. As it happens none of our present results are affected by the decision to construe rules one way rather than the other, but we shall see in Chapter 12 that some theorems become affected when extended to cover the infinite case.

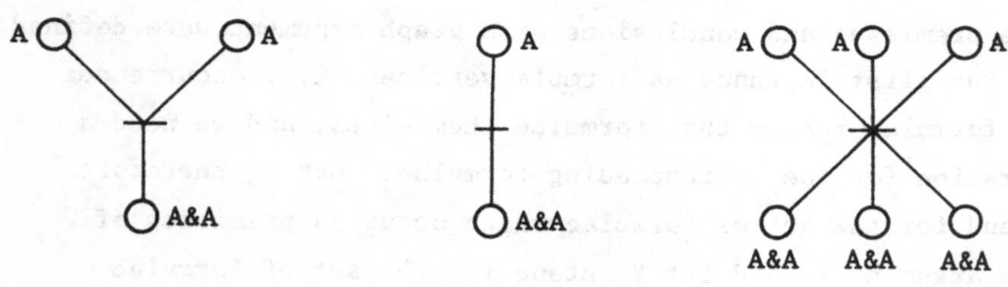


Figure 7.7

## 7.4 *Validity and form*

Our concern is with arguments which are potentially or structurally valid in the sense that they can never be made to lead from true premisses to false conclusions by truth-preserving steps, however the formulae are interpreted. If b is an inference stroke the step represented by b is truth-preserving on a given interpretation iff $\langle X_b, Y_b \rangle$ is satisfied by the appropriate partition, whence all the steps in $\pi$ are truth-preserving iff $R_\pi$ is satisfied. We therefore say that $\pi$ is *valid* if every partition that satisfies $R_\pi$ satisfies $\langle X_\pi, Y_\pi \rangle$, i.e. if $X_\pi \vdash_{R_\pi} Y_\pi$. Equivalently, by 2.13, $\pi$ is valid iff $\langle X_\pi, Y_\pi \rangle$ is valid whenever $R_\pi$ is valid. It is also convenient to say that a partition *satisfies* $\pi$ unless it *invalidates* it in the sense that it both satisfies $R_\pi$ and invalidates $\langle X_\pi, Y_\pi \rangle$, and in these terms $\pi$ is valid iff every partition satisfies it.

A (*graph*) *proof* is a valid graph argument, and in particular a *proof from* X *to* Y *by* R is a valid argument from X to Y by R. For example, the arguments shown below in Figures 7.9-10 are proofs: if one disregards any clues to the intended meanings of the formulae and considers all the possible ways of labelling them as true or false, one can easily verify that it is impossible to make the premisses true and the conclusions false without similarly invalidating at least one of the steps. For expository reasons we give examples whose individual steps are sound when the formulae are construed in the obvious way, but as we stressed in Section 7.1 this is strictly beside the point. The same considerations that show that the standard one-step arguments of Figure 7.10 are valid show that any standard one-step argument by R is a proof by R, regardless of the acceptability of R itself. For example, R could equally well be the fallacious syllogistic rule used in Figure 7.1.

*Theorem 7.3* An argument $\pi$ is valid iff $X \vdash_R Y$ whenever $\pi$ is an argument from X to Y by R.

*Theorem 7.4* Every graph proof is nonempty.

*Proofs* For 7.3, $\pi$ is valid iff $X_\pi \vdash_{R_\pi} Y_\pi$, i.e. iff $X \vdash_R Y$ whenever $X_\pi \subset X$ and $Y_\pi \subset Y$ and $R_\pi \subset R$, i.e. iff $X \vdash_R Y$ whenever $\pi$ is an argument from X to Y by R. For 7.4, every partition invalidates the empty argument. Note that although tree proofs can be empty - the empty tree being a proof from any X to any Y by 'from Λ infer Λ' - the corresponding graph proof is not empty but consists of an isolated inference stroke.

In the subsequent chapters we enquire how the validity of arguments is related to their form. The form of a graph argument is the resultant of two components. One is its broadly graphical structure, i.e. the distribution of the edges and of the various kinds of vertex (strokes, formulae, premisses, conclusions). The other is the pattern of recurrence of its formulae, i.e. the distribution of the vertices that are occupied by the same formulae. Analogous components of form can be discerned in other contexts. For example, the validity of the tautology $A \vee \sim A$ is due partly to the relevant combination of connectives and partly to the fact that the same formula occurs at the end as at the beginning. Recurrence is normally a crucial factor in such cases: what tautologies are there in which no formula occurs more than once? By contrast there is a large and diverse class of graph proofs whose validity can be attributed solely to their graphical structure and owes nothing to the recurrence of formulae. In such a case we can abstract altogether from the formulae and exhibit a skeleton form of graph such that no matter how its vacant vertices are filled with formulae the resulting argument will be a proof. For example, Figure 7.8 shows the abstract form of proof exemplified in Figure 7.9 by (i) a proof from $\sim B, A \supset B$ to $\sim A$ by the rules for PC, and (ii) a proof, cast in the syllogistic mood *Festino*, by the rules of the square of opposition and *Celarent*. Similarly Figure 7.10 shows that proofs by rules as different as modus ponens and adjunction may nevertheless have the same abstract form.

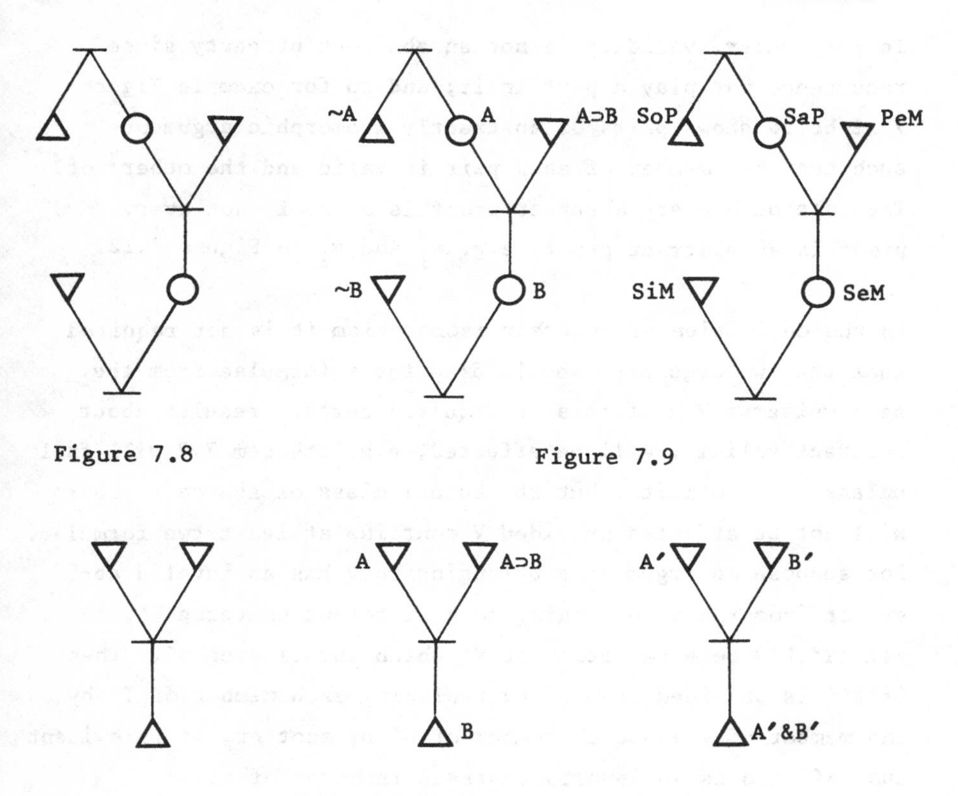


Figure 7.8

Figure 7.9

Figure 7.10

We can best approach the idea of abstract form by saying that two arguments are *abstractly isomorphic* if there is a one-one correspondence between them in which strokes correspond to strokes, formula vertices to formula vertices, premisses to premisses, conclusions to conclusions, and which preserves the edge relation (i.e. edges correspond to edges going to and from corresponding vertices). An argument is *abstractly valid* if every argument abstractly isomorphic to it is valid; and an abstractly valid argument is an *abstract proof*. Examples of abstract proofs are provided by Figures 7.9-10.

A property is *abstract* if it is preserved under abstract isomorphism. Not all the properties we consider are abstract.

In particular, validity is not an abstract property since recurrence may play a part in it; and so for example Figure 7.12 below shows pairs of abstractly isomorphic arguments such that one member of each pair is valid and the other not. Thus although every abstract proof is a proof, not every proof is an abstract proof, e.g. $\pi_1$ and $\pi_2$ in Figure 7.12.

In the definition of abstract isomorphism it is not required that the two arguments should draw their formulae from the same universe V. If this is required certain results about abstract validity will be affected, e.g. Theorem 7.7 will fail unless V is infinite; but the actual class of abstract proofs will not be affected provided V contains at least two formulae. For suppose an argument $\pi$ belonging to V has an invalid abstract isomorph $\pi'$ belonging to a different universe V′, and let $\langle T', U' \rangle$ be a partition of V′ which invalidates $\pi'$. Then if $\pi''$ is obtained from $\pi'$ by replacing each member of T′ by one member of V and each member of U′ by another, it is evident that $\pi''$ too is an invalid abstract isomorph of $\pi$.

When both components of argumentative form are taken into account - recurrence of formulae as well as purely graphical structure - there is a choice of relations on which to base the idea of form. We call them ‘isomorphism’ and ‘conformity’. We say that two arguments are *isomorphic* if (1) there is a one-one correspondence between them in which strokes correspond to strokes etc. and which preserves the edge relation (i.e. they are abstractly isomorphic), and (2) there is a one-one correspondence between their respective formulae such that corresponding vertices are occupied by corresponding formulae. For example the arguments shown in Figure 7.10 are generally isomorphic, using for (1) the obvious correspondence between their vertices, and for (2) making A, B and A⊃B correspond to A′, A′&B′ and B′ respectively. Note however that the second correspondence ceases to be one-one if A = B or A′ = B′, and in such cases the arguments are not isomorphic. Note too that the correspondence postulated under (2) does not have to be in any sense a translation and does not have to take any account of the composition of the formulae. The fact that A⊃B contains A and B as proper parts is only significant (if at all) in the context of a particular calculus, and our definition is designed to abstract from just this sort of thing. We therefore do not require that the formula chosen to correspond to A⊃B should be related in any way to those corresponding to A and B.

We say that one argument *conforms* to another if (1) there is a one-one correspondence between them in which strokes correspond to strokes etc. and which preserves the edge relation (i.e. they are abstractly isomorphic), and (2) there is a one-many correspondence between their respective formulae such that corresponding vertices are occupied by corresponding formulae. Conformity is intermediate in strength between isomorphism and abstract isomorphism, and like them it is transitive, but unlike them it is not necessarily symmetric. In Figure 7.11, for example, assuming that A, B, C are distinct, it can be seen that $\pi_1$ to $\pi_6$ are all abstractly isomorphic, but only $\pi_1$ and $\pi_2$ are isomorphic; that both $\pi_1$ and $\pi_2$ conform to each of $\pi_3$, $\pi_4$ and $\pi_5$, but not vice versa; and that each of $\pi_1$ to $\pi_5$ conforms to $\pi_6$ but not vice versa. Isomorphism and abstract isomorphism can each be defined in terms of conformity, for it can be shown that two arguments are abstractly isomorphic iff both conform to some third argument, and isomorphic iff each conforms to the other. (When infinite arguments are allowed, however, as in Chapter 12, two arguments may conform to one another without being isomorphic. For example, consider two sets of $i_n$ isolated occurrences of

$A_n$ for each n, where according as n is even or odd, $i_n$ is 1 or n in one case and n or 1 in the other.)

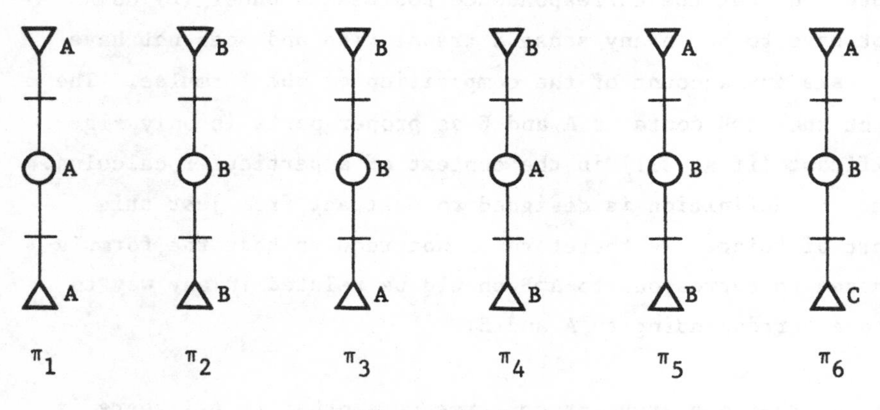


Figure 7.11

An argument form can be associated either with the class of all arguments isomorphic to a given one, or, perhaps more plausibly, with the class of all arguments conforming to a given one. Since each argument belongs to a unique isomorphism class but may belong to several conformity classes, an argument will have a unique form if 'form' is taken in the first sense, but not in the second. For example, $\pi_1$ in Figure 7.11 can plausibly be said to exemplify each of five forms of varying degrees of abstractness, represented by $\pi_2$ to $\pi_6$.

Similarly in calling a property *formal* one could mean that it is preserved under isomorphism or mean that it is preserved under conformity (i.e. whenever π has the property so does any argument conforming to π). The first sense is weaker than the second; for example, invalidity is a formal property in the first sense but not in the second. Validity, however, is formal in both senses, as Theorem 7.5 will show. In particular, the fact that every argument isomorphic to a proof is also a proof means that (unlike the validity attributed to rules or steps, or to formulae) the aspect of validity we are concerned with is independent of the choice of calculus and even of vocabulary. And the fact that every argument which conforms to a proof is also a proof means that we can exhibit semi-abstract patterns of proof like those in Figure 7.12: provided identically numbered vertices are occupied by identical formulae (as they are in $\pi_1$ and $\pi_2$ though not in $\pi_1'$ or $\pi_2'$) these patterns can be filled out with formulae in any way in the knowledge that the resulting arguments will be valid.

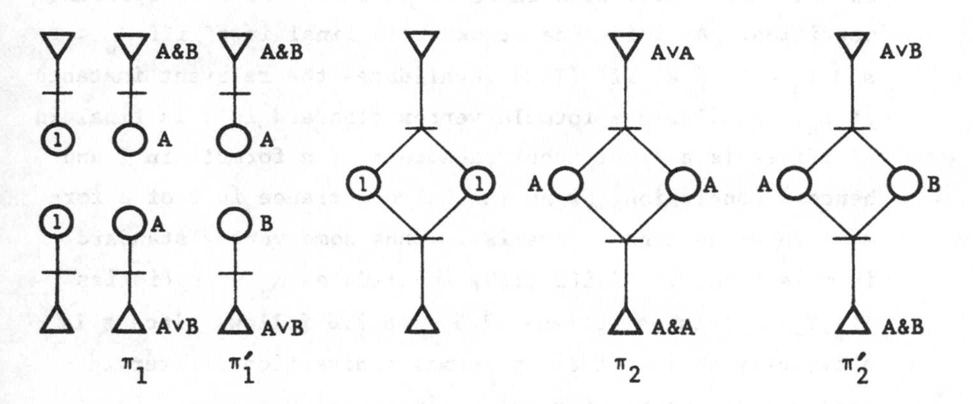


Figure 7.12

A graphical criterion for validity and abstract validity is available through the idea of a redirection of a graph; and the next two theorems show incidentally that validity and abstract validity are recursive properties. Let $\pi'$ be a redirection of the graph of an argument $\pi$. If for each formula A, either all the edges adjoining occurrences of A are reversed or none is, $\pi'$ is said to be a *formal redirection* of $\pi$. If for each formula vertex a either all the edges adjoining a are reversed or none is, $\pi'$ is an *abstract redirection* of $\pi$. Every formal redirection is an abstract redirection but not vice versa.

*Theorem* 7.5 An argument $\pi$ is valid iff every formal redirection has a final vertex standard in $\pi$.

*Theorem* 7.6 An argument $\pi$ is abstractly valid iff every abstract redirection has a final vertex standard in $\pi$.

*Proofs* To each partition $\langle T,U \rangle$ there corresponds a formal redirection $\pi'$ of $\pi$, obtained by reversing all edges adjoining occurrences of formulae of U; and conversely for each formal redirection there is at least one corresponding partition. An inference stroke b is final in $\pi'$ iff $X_b \subset T$ and $Y_b \subset U$, i.e. iff $\langle T,U \rangle$ invalidates the relevant instance of $R_\pi$. Similarly a formula vertex standard in $\pi$ is final in $\pi'$ iff it is a final occurrence in $\pi$ of a formula in T and hence a conclusion, or an initial occurrence in $\pi$ of a formula in U and hence a premiss. Thus some vertex standard in $\pi$ is final in $\pi'$ iff $\langle T,U \rangle$ invalidates $R_\pi$ or satisfies $\langle X_\pi, Y_\pi \rangle$. This establishes 7.5, and 7.6 follows since $\pi$ is abstractly valid iff every formal redirection of every abstract isomorph $\pi'$ of $\pi$ has a final vertex standard in $\pi'$, i.e. iff every abstract redirection of $\pi$ has a final vertex standard in $\pi$.

*Theorem* 7.7 An argument is abstractly valid iff it conforms to a proof in which no formula occurs more than once.

*Theorem* 7.8 If $\pi$ is a proof (or an abstract proof) then so is any argument $\pi'$ obtained by omitting edges from $\pi$.

*Theorem* 7.9 If $\pi$ satisfies the conditions for being a proof (or abstract proof) from X to Y by R, except that there are edges to some premisses or from some conclusions, the argument $\pi'$ obtained from $\pi$ by deleting these edges together with the strokes adjoining them and all edges adjoining these strokes, is a proof (abstract proof) from X to Y by R.

*Proofs* 7.7 follows from 7.5 and 7.6, since if the vertices of an argument are occupied by distinct formulae, the abstract redirections coincide with the formal ones.

For 7.8, although $\pi'$ will not in general be an argument by the same rules as $\pi$, each instance of $R_\pi$ is obtained by dilution from an instance of $R_{\pi'}$. Hence $\vdash_{R_\pi} \subset \vdash_{R_{\pi'}}$, and since $X_{\pi'} = X_\pi$ and $Y_{\pi'} = Y_\pi$ it follows that $X_{\pi'} \vdash_{R_{\pi'}} Y_{\pi'}$ and hence that $\pi'$ is valid. The corresponding result for abstract proofs follows by applying this result to every argument abstractly isomorphic to $\pi$.

For 7.9, it is evident that $\pi'$ is an argument and that $X_{\pi'} = X_\pi$ and $Y_{\pi'} = Y_\pi$ and $R_{\pi'} \subset R_\pi$, and if b is a stroke of $\pi$ not in $\pi'$ either $X_b$ overlaps $Y_\pi$ or $Y_b$ overlaps $X_\pi$. Now every partition $\langle T,U \rangle$ that invalidates $\langle X_{\pi'}, Y_{\pi'} \rangle$ invalidates $\langle X_\pi, Y_\pi \rangle$ and so, since $\pi$ is valid, invalidates $R_\pi$. But $\langle T,U \rangle$ certainly satisfies $\langle X_b, Y_b \rangle$ for each stroke b of $\pi$ not in $\pi'$, and so invalidates $R_{\pi'}$. Hence $\pi'$ is a proof, as required. When $\pi$ is an abstract proof the abstract validity of $\pi'$ follows in the same way as for 7.8.

We are now able to defend the stipulation in Section 7.3 that only initial occurrences of formulae may be premisses and only final ones conclusions. For even if one were to admit arguments - call them quasi-arguments - which violate this stipulation, Theorem 7.9 shows that every valid quasi-argument contains a valid argument in the strict sense.

## 7.5 *Subproofs*

We envisage a subargument as being obtained by omitting some or none or all of the steps in an argument, counting as steps not only (i) the application of a rule of inference but also (ii) the inscription of a formula and (iii) the positing of an inscribed formula as a premiss or conclusion of the argument. A step of either of the first two kinds is represented in graphical terms by the presence of a stroke or a formula vertex, and its omission will require the deletion of the relevant vertex and hence of any adjoining edges. An omission of the third kind does not affect the argument qua graph but only the status of the vertex (symbolised in our notation by the replacement of a triangle by a circle).

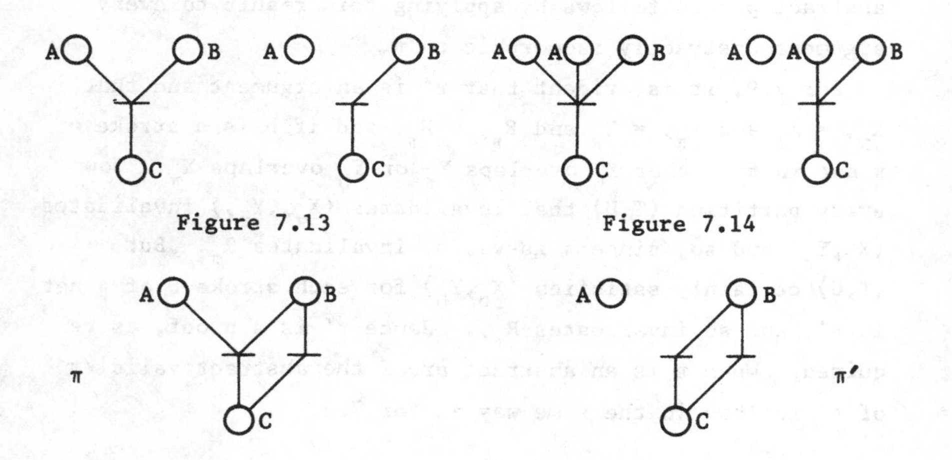


Figure 7.13

Figure 7.14

Figure 7.15

In general we do not allow the removal of edges alone to form a subargument: we do not accept that someone who argues from A and B to C has incidentally argued from B to C (Figure 7.13). On the other hand we should certainly allow the removal of edges where their presence merely reflects the looseness involved in construing rules in terms of sets, there being no difference between a rule ‘from A,A,B infer C’ and ‘from A,B infer C’ (Figure 7.14). The problematic case is illustrated in Figure 7.15, where π′ is obtained from π by an omission which would not be tolerable on its own (cf. Figure 7.13) but which could be defended in the context of the argument as a whole, on the grounds that someone who argues from A,B to C and from B to C *has* incidentally argued from B to C. Allowing such a case would produce an elegant definition of subargument in terms of a requirement that $R_{\pi'} \subset R_{\pi}$, but in our view its defence rests on an equivocation and we choose to exclude it.

We therefore define a *formal subgraph* of a graph $\pi$ to be any subgraph $\pi'$ such that for each stroke b in $\pi'$ the sets $X_b$ and $Y_b$ are the same whether construed with reference to $\pi'$ or to $\pi$. That is to say, if b itself is retained when forming $\pi'$ from $\pi$, an edge to b from an occurrence of A may only be suppressed if some other edge to b from an occurrence of A is retained, as in Figure 7.14; and similarly for edges from b. A *subargument* of an argument $\pi$ is an argument whose graph is a formal subgraph of that of $\pi$ and whose premisses and conclusions are subsets of those of $\pi$. Unlike 'redirection' and 'subgraph', 'subargument' has no prior usage in graph theory and so there is no need to include a distinguishing adjective like 'formal' here. If any were needed 'formal' would be the right one, for the relation of being a subargument - like the relation of being a formal subgraph or a formal redirection - is preserved under the kinds of correspondence envisaged in our definitions both of isomorphism and conformity. If $\pi'$ is a subargument of $\pi$ it follows from the definition that $R_{\pi'} \subset R_{\pi}$ and, applying what it says about premiss and conclusion vertices to the corresponding formulae, that $X_{\pi'} \subset X_{\pi}$ and $Y_{\pi'} \subset Y_{\pi}$. We thus have

*Theorem 7.10* If $\pi'$ is a subargument of $\pi$ then $\pi'$ is an argument from X to Y by R whenever $\pi$ is.

The only aspect of the formation of subarguments that depends on the recurrence of formulae is the omission of edges illustrated in Figure 7.14, and by disallowing this we obtain the appropriate abstract analogue. Thus an *abstract subgraph* of $\pi$ is a subgraph which, for each stroke b that it contains, also contains every edge adjoining b in $\pi$. An *abstract subargument* of an argument $\pi$ is an argument whose graph is an abstract subgraph of that of $\pi$, and whose premisses and conclusions are subsets of those of $\pi$.

These definitions make no mention of validity. A subargument of an invalid argument cannot be valid (Theorems 7.3 and 7.10). A subargument of a proof, on the other hand, may or may not be valid, and we call a valid subargument a *subproof*. Similarly we call an abstractly valid abstract subargument an *abstract subproof*. Only an abstract proof can have an abstract subproof, but not every subproof of an abstract proof is an abstract subproof. In Figure 7.16, for example, $\pi'$ is a valid abstract subargument of $\pi$ but is not abstractly valid, and in Figure 7.17 $\pi'$ is an abstractly valid subargument of $\pi$ but is not an abstract subargument.

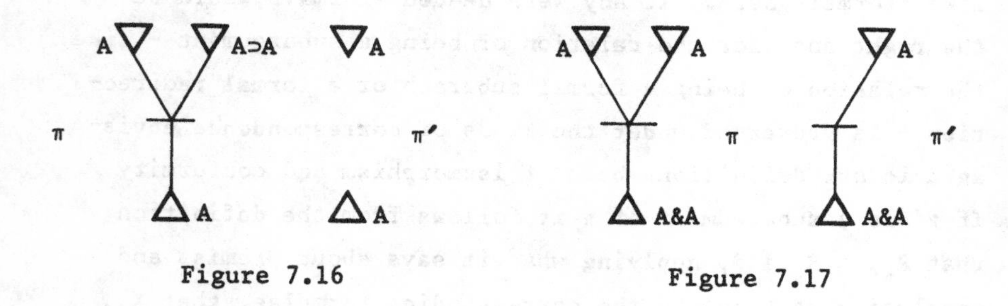


Figure 7.16 Figure 7.17

We say that a subargument $\pi'$ of an argument $\pi$ is *proper* if $\pi' \neq \pi$, even if the difference lies only in the specification of their premisses and conclusions. A proof is *minimal* if it has no proper subproof of any kind: e.g. a minimal standard proof means a standard proof with no proper subproof, standard or otherwise. Although minimality presupposes the idea of subargument, it is not difficult to show that it is unaffected by the decision not to count such cases as Figure 7.15 as subarguments. A proof which is not strictly minimal may nevertheless exemplify a minimal form of proof; and we say that a proof is *concise* if it conforms to a minimal proof. Every minimal proof is concise, but not conversely, as the examples of Figures 7.16 and 7.17 show: in each case $\pi$ is not minimal, since it has the proper subproof $\pi'$, but it is concise since it exemplifies the minimal form shown in Figure 7.10. Since graph arguments are by definition finite, every proof has a minimal subproof and hence a concise one. We also have for the abstract case:

*Theorem 7.11* An abstract proof is concise iff it has no proper abstract subproof.

*Theorem 7.12*† Every abstract proof has a concise abstract subproof.

*Theorem 7.13* Every concise abstract proof is standard and connected.

*Theorem 7.14* Every abstract proof has a standard connected abstract subproof.

*Proofs* Any abstract proof $\pi$ conforms to an abstract proof $\pi'$ in which no formula occurs more than once. Every subproof of $\pi'$ is an abstract subproof of it. So for 7.12, any minimal subproof $\pi'_1$ of $\pi'$ is an abstract subproof by 7.7. The corresponding subargument of $\pi$ is an abstract subargument and is concise since it conforms to $\pi'_1$. Similarly for 7.11, if $\pi$ has no proper abstract subproof it follows that $\pi'$ has no proper subproof of any kind;

i.e. $\pi'$ is minimal and therefore $\pi$ is concise. Conversely if $\pi$ is a concise abstract proof, it conforms to a minimal proof $\pi'$. Since $\pi$ and $\pi'$ are abstractly isomorphic, any proper abstract subproof of $\pi$ would be abstractly isomorphic to a proper subproof of $\pi'$, but there are none by hypothesis.

For 7.13, if an abstract proof $\pi$ has a non-standard vertex b (say an initial formula vertex which is not a premiss), let $\pi'$ be obtained from $\pi$ by specifying b as a (possibly) additional conclusion. Then $\pi'$ satisfies the conditions to be an abstract proof except that there may be edges from b. By deleting strokes and their adjoining edges, we obtain by 7.9 an abstract proof $\pi''$ in which b is an isolated non-standard vertex. By 7.6 the argument obtained from $\pi''$ by deleting b is an abstract proof, and hence a proper abstract subproof of $\pi$. Similarly if $\pi$ is disconnected one of its components must be abstractly valid and so a proper abstract subproof, since otherwise each component independently could be abstractly redirected so as to have no final vertex standard in $\pi$, and the union of these redirections would show that $\pi'$ was not abstractly valid either. In each case, therefore, $\pi$ is not concise by 7.11. 7.14 follows at once from 7.12 and 7.13.

Minimality is a formal property in the weaker but not the stronger sense. Conciseness is a formal property in both senses, and for all practical purposes it can be treated as an abstract one, because by Theorem 7.11 any abstract isomorph of a concise abstract proof is concise. The same theorem shows that conciseness is in practice the abstract analogue of minimality: e.g. Theorem 7.12 states the analogue for abstract proofs of the fact that every proof has a minimal subproof. Note that Theorem 7.12 fails if ‘concise’ is replaced by 'minimal', for Figure 7.16 shows an example of an abstract proof with no minimal abstract subproof.

## 7.6 *Symmetry*

Consequence in a single-conclusion calculus is a heterogeneous relation, holding between a set of formulae on the one side and an individual formula on the other. The converse of such a relation is therefore by definition ineligible to be a consequence relation. The admission of multiple conclusions makes consequence a homogeneous relation between sets of formulae, and so opens up possibilities of symmetry in the following sense.

We say that a property of binary relations is *symmetrical* if whenever a relation possesses the property so does its converse. For example, transitivity, symmetry and reflexivity are symmetrical (and so is asymmetry: whether a property is symmetrical has nothing to do with whether symmetrical relations possess it). The overlap, dilution and cut conditions are all symmetrical in this sense, for a relation is evidently closed under each of them iff its converse is. Hence by Theorem 2.1 the property of being a consequence relation is symmetrical. As we said, this does not imply that a consequence relation is identical to its converse, but simply that the converse is also a consequence relation. In other words, to every calculus L there corresponds a *converse calculus* $L^*$ such that $\vdash^*$ is the converse of $\vdash$. The set I of partitions characterising L may be construed as a relation, its converse $I^*$ comprising the partitions $\langle U,T \rangle$ for which $\langle T,U \rangle \in I$. It is obvious that $\langle T,U \rangle$ satisfies $\langle X,Y \rangle$ iff $\langle U,T \rangle$ satisfies $\langle Y,X \rangle$, and so $L^*$ is characterised by $I^*$: converse calculi are characterised by converse sets of partitions.

Similarly, for each set of rules R there is a converse set determined by the converse relation R*. Thus the converse of 'from X infer Y if X R Y' is 'from X infer Y if X R* Y', or equivalently 'from Y infer X if X R Y'. Since a relation contains R iff its converse contains R*, it follows from Theorem 2.14 that consequence by R* is the converse of consequence by R: converse rules characterise converse calculi.

This observation suggests the possibility of devising *converse proofs*, where the converse of a given proof is to be a proof from Y to X by R* iff the original was a proof from X to Y by R. Let the *converse* $\pi^*$ of a graph argument $\pi$ be obtained by forming the converse graph and taking the conclusions of $\pi$ as premisses of $\pi^*$ and the premisses of $\pi$ as conclusions of $\pi^*$. Given our conventions for drawing diagrams of arguments, the diagram of $\pi^*$ is obtained by simply inverting that of $\pi$. It follows at once from the relevant definitions that $\pi$ is an argument from X to Y by R iff $\pi^*$ is an argument from Y to X by R*, and that $\pi$ is valid iff $\pi^*$ is valid. Hence any valid graph argument and its converse are converse proofs in the required sense. This shows too that the concepts of graph argument and graph proof are symmetrical ones, and likewise all the various species of argument we shall be considering in the next three chapters - standard, circuit-free, concise, abstractly valid, etc - are determined by symmetrical conditions.

Tree proofs, however, are ill suited to the formation of converses. The obvious procedure is to take $\tau^*$ to be the same array as $\tau$ but with the converse ordering; but this fails, since in general $\tau^*$ is not even a tree, let alone a proof from Y to X by R*. The definition of extended proof in Section 3.2 has an appearance of symmetry between rank and file which is entirely spurious in this context, since rank and file are not converse notions with respect to the partial ordering that defines a tree. It is not only this simple recipe which fails to generate a converse proof: there is in general no way of forming one. The reason is the lack of specificity inherent in the idea of tree proof which we noted in Section 7.1. There must exist appropriate relationships between the ranks and files of a tree proof, but they are loose enough for it to be a proof from X to Y by R and at the same time a proof from some other X′ to some other Y′ by some other R′; and when this happens it imposes conditions on a converse proof which generally cannot be met. For example, $\tau$ in Figure 7.18 is a proof from A,~A to $\Lambda$ by the rule 'from A,~A infer $\Lambda$', but is simultaneously a proof from A to ~A by the rule 'from A infer ~A'. Its converse must be a proof (such as $\tau_1$) from $\Lambda$ to A,~A by the rule 'from $\Lambda$ infer A,~A', and so must have a branch ending in ~A; but it must also be a proof (such as $\tau_2$) from ~A to A by 'from ~A infer A' and so can have no branch ending in ~A.

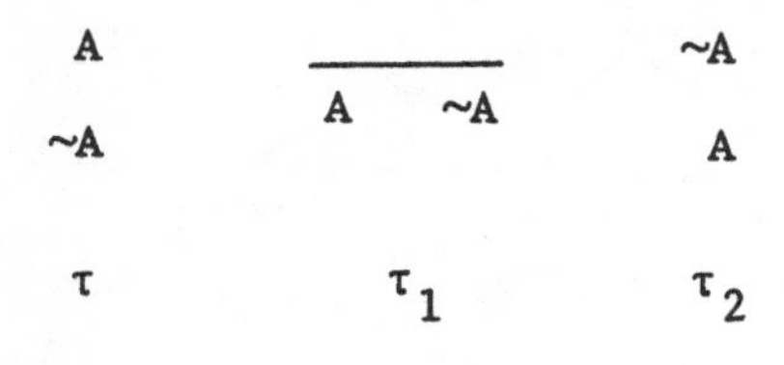


Figure 7.18

The moral we draw is that for a proof to have a converse it must either possess an unambiguous argumentative structure which the converse may be able to represent in mirror fashion, or it must have an even greater argumentative ambiguity than a tree proof, so that the converse may be capable of bearing all the requisite variety of argumentative constructions. We have already explored the first alternative by introducing graph proofs, but we remark that the other is satisfied by the universal proofs of Section 3.3. These are defined as extended tree proofs, and each is a proof from about as many sets of premisses to as many sets of conclusions by as many rules as is possible without obliterating the distinction between what can and cannot be proved; and we have

*Theorem 7.15* Every universal proof is its own converse.

*Proof* From the definition of $\vdash^{Z}_{R}$ in Section 3.3 it is easily shown that $\vdash^{Z}_{R*}$ is the converse of $\vdash^{Z}_{R}$, whence the result follows by 3.9.

# 8 · Kneale proofs

## 8.1 *Developments*

Kneale's 'tables of development' are the pioneer multiple-conclusion proofs. He introduces the idea in these terms: 'The general pattern of a table of development is

$$\frac{A \quad B}{C \quad D}$$

but between the initial formulae which express the premisses and the end formulae which express the limits there may be many intermediate expressions, i.e. expressions that occur both above and below lines' (1956, p.246). Kneale's examples make it clear how this works out in practice; e.g. Figure 8.1 shows a development leading from A⊃B to ~A∨B by the rules for PC.

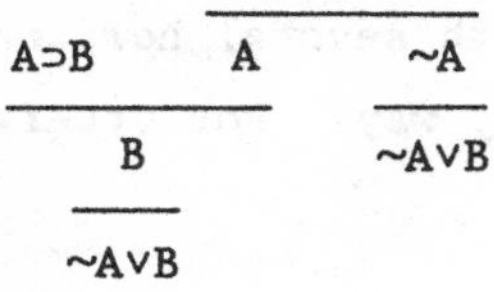

$$\dfrac{\dfrac{A\supset B \qquad A}{B}}{{\sim}A\vee B} \qquad \dfrac{{\sim}A}{{\sim}A\vee B}$$

Figure 8.1

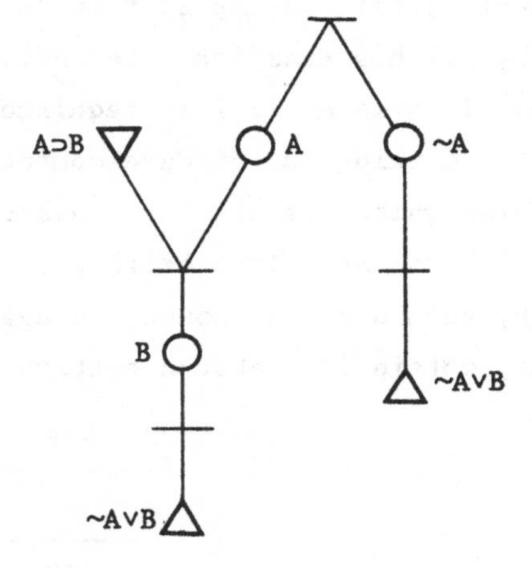


Figure 8.2

The horizontal lines in a table of development have a quite different role from the similar-looking lines used in Chapter 3 in picturing tree proofs. There the lines were introduced not as an integral part of the proofs but merely as a convenience to the reader in following them. As such they were more often used to bring out the branching structure of the proof than its inferential structure (not all rules of inference cause a tree proof to branch, and not all branching in an extended one is caused by rules). Here the lines are invariably associated with the application of a rule and, as our next quotation will show, they must be regarded as being genuine constituents of the development on a par with the formulae. In short, Kneale's lines play the part of inference strokes.

Suppose then that we join each line in a development to the formulae lying immediately above and below it, by means of edges whose directions are taken to run down the page. The result will be a finite bipartite graph of formulae and inference strokes (compare Figure 8.1 with Figure 8.2), with a number of special features. In the first place it will be nonempty and connected; this is implicit in Kneale's account and is borne out by all his examples. Secondly the graph will be circuit-free. This is explicitly required by Kneale (1962, p. 543): 'In the working out of developments it is essential that token formulae which are already connected, either directly by one single horizontal line or indirectly through several horizontal lines, should not be connected again in any way. For otherwise we may obtain such absurd patterns as

$$\frac{\dfrac{A \vee B}{A \qquad B}}{A \& B},$$

This second feature exposes an additional reason, peculiar to multiple-conclusion logic, for incorporating inference strokes into the definition of an argument. For if we do not do so - if we really do regard the connection between the formulae immediately above and below a line in a development as being 'direct' - then it is impossible to satisfy Kneale's requirement when a rule has several conclusions as well as several premisses. Thus in his original example of a development (reproduced as $\pi$ in Figure 8.3), A and C are connected 'directly by one single horizontal line', and so too are A and D, and D and B, and B and C. Hence A and C are connected again via D and B, contrary to Kneale's stipulation. If we represent developments by graphs of formulae alone, $\pi$ becomes $\pi_1$, which is a circuit. However if the horizontal is treated as a genuine part of the development the connections between the formulae above and below it cease to be literally direct, and we obtain instead the circuit-free graph $\pi_2$. On the other hand Kneale's 'absurd patterns' are still excluded: his own example becomes the graph shown in Figure 8.4, which has a forbidden circuit.

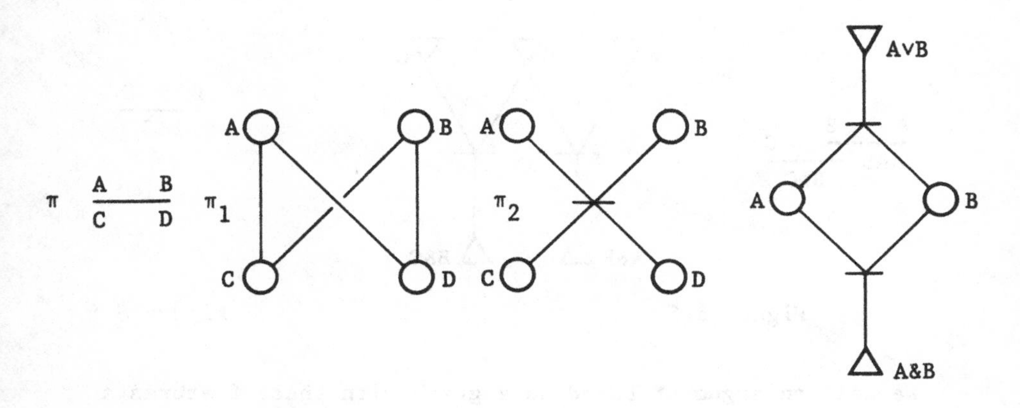


Figure 8.3 Figure 8.4

The third feature of a graph derived from a development is that it will be corner-free, where by a *corner* in a bipartite graph of formulae and strokes we mean a set $\{E_1,b,E_2\}$ in which b is a formula vertex and $E_1$ and $E_2$ are distinct edges both going to b or both from it. For example, the graph in Figure 8.5 has a corner at B; but Figure 8.4 is corner-free, since the only vertices to (or from) which a pair of edges go are inference strokes. There will be a corner in a graph argument whenever the same occurrence of a formula figures as premiss (or as conclusion) in more than one step. A corner-free graph thus corresponds to a Kneale-style table of development in which only one line can occur immediately above (or below) any formula. Tables can certainly be devised in which this condition is broken; Figure 8.5 shows one example and shows too how the corresponding graph has a corner. But it is clear both from Kneale's account and from his examples that he does not envisage tables of this kind, and even if they were allowed they would not constitute a significantly wider class of proofs, for we show in effect in Theorems 8.1 and 8.4 that every such table has an appropriate Kneale-style development as a sub-table (e.g. the table shown in Figure 8.5 contains the development shown in Figure 8.6).

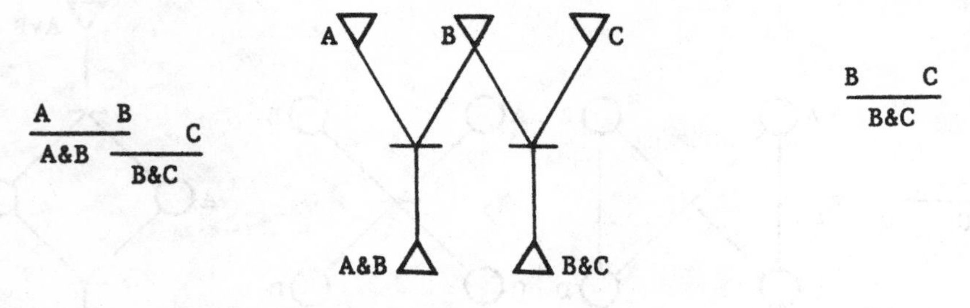


Figure 8.5 Figure 8.6

We call an argument based on a graph with these features a *Kneale argument*. A Kneale argument is thus connected, non-empty, circuit-free and corner-free. It is clear from our first quotation that in representing a development by a Kneale argument we should specify all the initial formula vertices as premisses and all the final ones as conclusions, for what Kneale there calls the 'limits' of a development are just what we call conclusions. Every development thus corresponds to a standard Kneale argument. Conversely, every standard Kneale argument can be represented by a table of development, provided one is allowed to distort the vertical scale to deal with the more convoluted cases. We can therefore safely assert that the developments and the standard Kneale arguments are essentially the same.

## 8.2 *Validity*

The soundness of the development as a form of argument is assured by

*Theorem 8.1*† Every standard nonempty circuit-free argument is abstractly valid.

*Proof* Any redirection of a nonempty circuit-free argument $\pi$ is circuit-free, and so has a final vertex by 7.2. If all vertices in $\pi$ are standard it follows by 7.6 that $\pi$ is abstractly valid.

Standardness is a necessary as well as a sufficient condition for a Kneale argument to be abstractly valid:

*Lemma 8.2* No connected corner-free argument has a proper nonempty standard abstract subargument.

*Theorem 8.3*† A Kneale argument is an abstract proof iff it is standard.

*Proofs* For 8.2, let $\pi'$ be a proper nonempty standard abstract subargument of a connected argument $\pi$. Since $\pi'$

is standard the difference between $\pi$ and $\pi'$ cannot be just a difference in the specification of their premisses and conclusions, but must involve the omission of vertices or edges or both; and since $\pi$ is connected and $\pi'$ nonempty this means that there is a vertex b of $\pi'$ adjoining an edge E of $\pi$ (going from b, say) such that E is not in $\pi'$. Since $\pi'$ is an abstract subargument b is not a stroke, and is not a conclusion of $\pi$ and hence not of $\pi'$. But $\pi'$ is standard, so b is not final in $\pi'$, i.e. some edge E′ of $\pi'$ goes from it. Hence $\pi$ contains the corner {E,b,E′}.

For 8.3, a standard Kneale argument is an abstract proof by 8.1. Conversely a non-standard Kneale argument can have no standard abstract subproof (not an improper one, by hypothesis, and not a proper one by 8.2 and 7.4), and so cannot be an abstract proof by 7.14.

We call a valid Kneale argument a *Kneale proof*, and an abstractly valid Kneale argument an *abstract Kneale proof*. The terms ‘standard Kneale argument’, ‘abstract Kneale proof’, and ‘standard Kneale proof’ are interchangeable by Theorem 8.3, and we generally use the last of the three. These proofs can be characterised in other ways too, as we now show.

*Theorem 8.4*† The standard Kneale proofs are the concise circuit-free abstract proofs.

*Theorem 8.5*† The standard Kneale proofs are the concise corner-free abstract proofs.

*Theorem 8.6*† The standard Kneale proofs are the connected corner-free abstract proofs.

*Proofs* For 8.4, a standard Kneale proof is circuit-free by definition, an abstract proof by 8.3, and concise by

7.11, 7.14, 7.4 and 8.2. Conversely any concise circuit-free abstract proof $\pi$ is nonempty (by 7.4) and standard and connected (by 7.13); and so to show that it is a standard Kneale proof it is sufficient to show that it is corner-free. If $\pi$ has a corner $\{E,b,E'\}$, where E joins b and c, the argument obtained by deleting E is not connected and c is not in that component $\pi'$ which includes b, for otherwise there would be a path (b,c) in $\pi'$ and $(b,c)\cup\{E\}$ would be a circuit in $\pi$. Now $\pi'$ is a proper nonempty abstract subargument of $\pi$, and since $\{E,b,E'\}$ is a corner, b is standard in $\pi'$ and hence $\pi'$ is standard. It follows by 8.1 that $\pi'$ is a proper abstract subproof of $\pi$, contrary to 7.11.

For 8.5 and 8.6, we note that every standard Kneale proof is a concise corner-free abstract proof by 8.4, and every concise corner-free abstract proof is also connected by 7.13. It is therefore sufficient to show that every connected corner-free abstract proof $\pi$ is a standard Kneale proof. By 7.4, $\pi$ is nonempty, and by 7.4 and 8.2 it has no proper standard abstract subproof, and hence by 7.14 it must itself be standard. It remains to show that $\pi$ is circuit-free. If not, let E be an edge joining a formula vertex b and a stroke c in some circuit $\gamma$, and let $\pi'$ be obtained from $\pi$ by deleting E. Since $\pi$ is corner-free b is the unique non-standard vertex of $\pi'$, but $\pi'$ is connected since b and c are still connected by the path $\gamma\cap\pi'$. By 8.2, $\pi'$ has no standard abstract subproof and so cannot be an abstract proof by 7.14. So by 7.6 it has an abstract redirection in which there is no final vertex except perhaps b, and by redirecting E as necessary this can be extended to an abstract redirection of $\pi$ in which b must be final, by 7.6. Hence the edges of $\gamma$ which adjoin b must form a corner, contrary to hypothesis.

The analogue of Theorem 8.6 for circuit-free proofs is false: there are connected circuit-free abstract proofs which are not Kneale arguments, as Figure 8.7 shows.

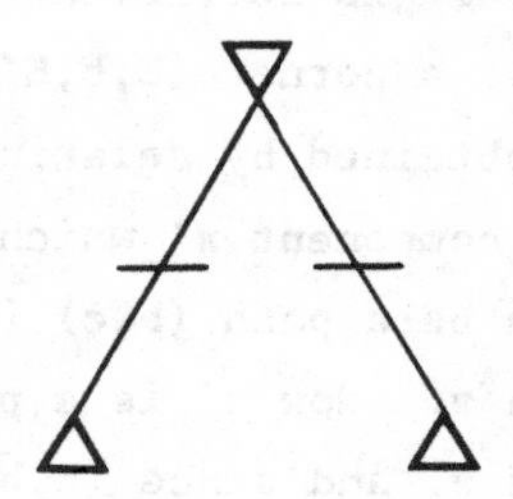

Figure 8.7

The results of this section give us some purely syntactical criteria for abstract validity. A circuit-free argument is an abstract proof iff it has a standard nonempty abstract subargument (Theorems 7.4, 7.14 and 8.1); and a circuit-free or corner-free argument is an abstract proof iff it has a standard Kneale argument as an abstract subargument (Theorems 7.12 and 8.3-5).

## 8.3 *Inadequacy*

Kneale illustrated his rules for PC by constructing developments corresponding to the axioms and rules of the Principia Mathematica version of the propositional calculus. Since every tautology is deducible from the Principia axioms by the Principia rules it ought to follow that there is a corresponding development for every tautology; but in fact it does not, and this failure exposes the inadequacy of the development as an instrument of proof. To establish the fact we need to define two more properties of arguments: simplicity and normality.

A graph of formulae and strokes is *simple* if there are never edges going from different occurrences of the same formula to one and the same stroke, or to different occurrences of the same formula from one and the same stroke. An argument is *simple* if it is based on a simple graph. For example, in each of Figures 7.7 and 7.17 just one of the arguments is simple (and so is $\pi$ of Figure 7.16, since although two edges join occurrences of A and the same stroke, one is to the stroke and one from it). We have seen that in general it is necessary to distinguish between subarguments and abstract subarguments, but in the simple case the two coincide: a graph is simple iff every formal subgraph is an abstract one, and an argument is simple iff every subargument is an abstract one.

*Theorem 8.7* Every proof has a simple subproof, and every abstract proof has a simple, abstractly valid subproof.

*Proof* If $\pi$ is any argument let $\pi'$ be obtained by deleting, for each stroke, all edges in excess of one going to it from occurrences of the same formula and all edges in excess of one from it to occurrences of the same formula. Evidently $\pi'$ is simple and is a subargument of $\pi$ (though not in general an abstract subargument), and by 7.8 $\pi'$ is valid or abstractly valid if $\pi$ is.

It follows from this result that every minimal proof is simple, but a concise proof need not be simple, $\pi$ of Figure 7.17 being a case in point. Simplicity is thus a formal property only in the weaker sense. Moreover a proof of a particular kind need not have any simple subproofs of the same kind: for example Figure 8.8 shows a standard proof with no simple standard subproof, and a Kneale proof with no simple Kneale subproof. Nevertheless every standard Kneale proof has a simple standard Kneale subproof; for the deletion of, say, n edges as prescribed by Theorem 8.7 creates at most n non-standard vertices, but since the proof is circuit-free it creates n+1 components, one of which must therefore be standard and so valid by Theorem 8.3.

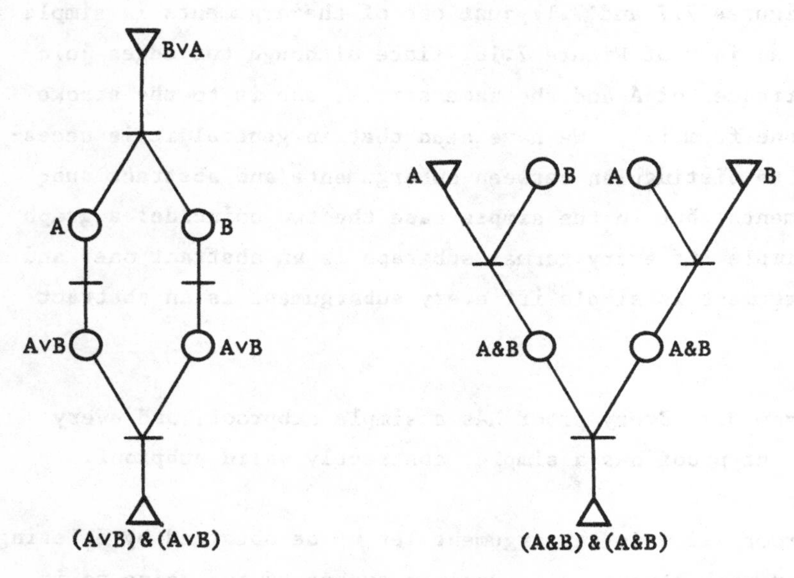


Figure 8.8

The idea of normality depends on that of a subformula, and is therefore not a formal property in either sense. In the case of PC we shall take the subformula relation literally, as the ancestral of the relation which holds between A and each of A⊃B, B⊃A, A&B, B&A, A∨B, B∨A and ~A; but when we come to deal with predicate calculus in Section 20.2 we shall follow Gentzen (1934) in allowing Fb to count as a subformula of (∃x)Fx. Indeed any partial ordering of the formulae can be defined as the subformula relation for the purpose of establishing the 'subformula property' of normal proofs expressed by Theorem 8.8 below. Given such a definition, and the idea of a proper subformula of A as any subformula of A other than A itself, we may first divide the edges of any graph argument into major and minor ones as follows. Let $E_1,\ldots,E_n$ be all the edges to and from a stroke b, and let the other vertex of each $E_i$ be an occurrence of $A_i$; then $E_i$ is *minor* if $A_i$ is a proper subformula of some $A_j$ and *major* otherwise. An argument is *normal* if no formula vertex has both a major edge to it and a major edge from it. For examples of arguments which are abnormal in the context of PC see Figure 9.7, where the occurrences of A&A and (A&B)&B have major edges both to and from them.

*Theorem 8.8*† Every formula that occurs in a standard normal argument from X to Y is a subformula of a member of X or Y.

*Proof* Let π be any argument from X to Y and let Z be the set of formulae which occur in π but are not subformulae of members of X or Y. Z is finite since π is; hence if Z is nonempty it has a maximal member A in the partial ordering determined by the subformula relation. Every edge adjoining an occurrence of A is major, and π therefore fails either to be normal or to be standard, according as the occurrence is or is not intermediate.

*Theorem 8.9* Whenever there is a standard Kneale proof from X to Y by the rules for PC there is one which is both simple and normal.

*Proof* Let π be a standard Kneale proof from X to Y by the rules for PC, with the smallest possible number of edges. Then π is simple, since otherwise as we remarked above it would have a (proper) simple standard Kneale subproof. If π is not normal some formula vertex has a major edge to it and a major edge from it. An inspection of the rules shows that π must be of one of the nine forms sketched in tabular form in Figure 8.9, where in each case $\pi_1$ and $\pi_2$ are Kneale arguments with the indicated occurrences of A as their only non-standard vertices. But the argument obtained from $\pi_1$ and $\pi_2$ by identifying these occurrences (Figure 8.10) is a standard Kneale argument - and so by 8.3 a standard Kneale proof - with fewer edges than $\pi$, contrary to hypothesis.

```
               π1            π1  π3        π1         π1  π3        π3  π1
               :             :   :         :          :   :         :   :
               :             A   :         A          A   B         B   A
   ______      :            _____:        ___        _______       _______
  A  A⊃B  A                 B⊃A  B        A&A          A&B           B&A
  :  _______                _______       ___          ___           ___
  :     B                      A           A            A             A
  :     :                      :           :            :             :
  π2    π3                     π2          π2           π2            π2

  π1           π1             π1                              π1
  :            :              :                               :
  A            A              A                               :
 ___          ___            ___                   _______    :
 A∨A          A∨B            B∨A                   A    ~A    A
 ___        _______        _______                 :   _______
  A          A    B         B    A                 :
  :          :    :         :    :                 :
  π2         π2   π3        π3   π2                π2
```

Figure 8.9

```
  π1
  :
  A
  :
  π2
```

Figure 8.10

*Theorem 8.10*† There exists a tautology B such that there is no standard Kneale proof from $\Lambda$ to B by the rules for PC.

*Proof* Take B to be (A⊃A)&(A∨(A⊃A)), where A is a propositional variable. By 8.9, if there is a standard Kneale proof from $\Lambda$ to B there is such a proof $\pi$ which is also simple and normal, so that by 8.8 every formula occurring in $\pi$ is B or one of its proper subformulae A∨(A⊃A), A⊃A and A. Since $\pi$ is simple the only contexts in which an inference stroke can occur are $\pi_1$ to $\pi_9$, shown in tabular form in Figure 8.11.

$$\pi_1 \ \frac{\quad}{A \quad A\supset A} \qquad \pi_2 \ \frac{A}{A\supset A} \qquad \pi_3 \ \frac{A\supset A}{A\vee(A\supset A)}$$

$$\pi_4 \ \frac{A\supset A \qquad A\vee(A\supset A)}{(A\supset A)\&(A\vee(A\supset A))} \qquad \pi_5 \ \frac{A}{A\vee(A\supset A)} \qquad \pi_6 \ \frac{A \quad A\supset A}{A}$$

$$\pi_7 \ \frac{A\vee(A\supset A)}{A \quad A\supset A} \qquad \pi_8 \ \frac{(A\supset A)\&(A\vee(A\supset A))}{A\supset A} \qquad \pi_9 \ \frac{(A\supset A)\&(A\vee(A\supset A))}{A\vee(A\supset A)}$$

Figure 8.11

Since $\pi$ is a standard proof from $\Lambda$ no formula occurs initially. Since $\pi$ is normal occurrences of B cannot be intermediate, and so all must be final. Hence only the contexts $\pi_1$ to $\pi_7$ can occur. Similarly if there is an occurrence of A∨(A⊃A) in the context $\pi_7$ it cannot be initial and so must also belong to $\pi_3$ or $\pi_5$, contradicting the normality condition as before. Hence only $\pi_1$ to $\pi_6$ occur. The same argument prohibits the occurrence of A⊃A in the context $\pi_6$, so that only $\pi_1$ to $\pi_5$ can occur. But now deleting the occurrences of B (which, as we have seen, must be final) and the edges adjoining them leaves a nonempty graph in which exactly two edges adjoin each formula vertex (at least two since the vertex, being standard, must be intermediate, and at most two since the graph is corner-free), and exactly two edges adjoin each stroke (by inspection of $\pi_1$ to $\pi_5$). Each component of such a graph, being finite, must be a circuit, and hence $\pi$ contains a circuit, contrary to hypothesis.

The proof is designed to go through whether or not the rules are construed having regard to multiplicity of occurrence (some passages in Kneale suggest the one course, some the other). Both proof and result can easily be modified to match Kneale's use of arbitrary formulae instead of $\Lambda$ in rules and developments.

It follows from Theorem 8.10 that either the rules proposed for PC are incomplete or developments (standard Kneale proofs) are inadequate as a means of proof. In fact the rules are complete (Section 18.1), but it is not necessary to establish this in order to show that the trouble is due to the idea of proof and not the rules: it is sufficient to show that the particular tautology in question follows from $\Lambda$ by the rules, and for this we only have to cite the tree proof shown in Figure 8.12.

$$\begin{array}{cc} A & A\supset A \\ A\supset A & A\vee(A\supset A) \\ A\vee(A\supset A) & (A\supset A)\&(A\vee(A\supset A)) \\ (A\supset A)\&(A\vee(A\supset A)) & \end{array}$$

Figure 8.12

We have therefore provided an example of rules R and premisses X and conclusions Y such that $X \vdash_R Y$ but there is no standard Kneale proof from X to Y by R. In other words:

*Theorem 8.11*[†] Standard Kneale proofs are inadequate.

It follows from Theorems 7.12 and 8.4 or 8.5 as the case may be, that

*Theorem 8.12* Circuit-free abstract proofs are inadequate.

*Theorem 8.13* Corner-free abstract proofs are inadequate.

We can similarly use Theorems 8.4 and 8.5 together to establish the inadequacy of the entire class of proofs which are (i) abstractly valid and (ii) circuit-free or corner-free or both. If we are to obtain an adequate concept of proof we must therefore abandon one or other of these features. In the next chapter we consider proofs which are circuit-free or corner-free or both, but which are not in general abstract proofs; and in Chapter 10 abstract proofs which may contain both circuits and corners.

# 9 · Cross-reference

## 9.1 *Junction and election*

We have seen that the standard Kneale proofs are inadequate. If we can discover the cause we may be able to remedy it. For a given set of rules R, therefore, let $K_R$ be the relation which holds between X and Y iff there is a standard Kneale proof from X to Y by R.

*Lemma 9.1* $K_R$ contains R and is closed under overlap and dilution.

*Proof* Closure under dilution follows at once from the definition of a graph proof. The other two properties are demonstrated in Figure 9.1, $\pi_1$ being a standard Kneale proof from X to Y in the case where A is common to X and Y, and $\pi_2$ a similar proof from the premisses $A_1,\ldots,A_m$ of any rule to its conclusions $B_1,\ldots,B_n$.

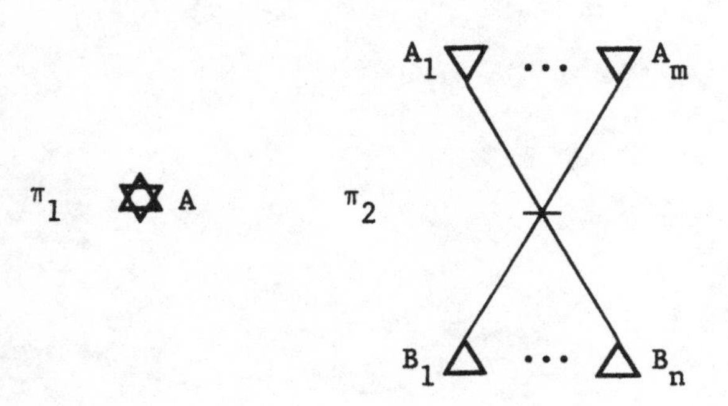


Figure 9.1

Since $K_R \subset \vdash_R$ (Theorem 7.3) but $K_R$ and $\vdash_R$ are not in general identical (Theorem 8.11), $K_R$ must fail to be closed under cut for formulae (Theorem 2.16), let alone cut for sets in general. To see why cut fails, let $\pi_1$ and $\pi_2$ be arguments which have just one vertex b in common, b being a conclusion of $\pi_1$ and a premiss of $\pi_2$. We say that $\pi$ is the *junction* of $\pi_1$ and $\pi_2$ if the graph of $\pi$ is the union of those of $\pi_1$ and $\pi_2$, the premisses of $\pi$ being those of $\pi_2$ (other than b) plus those of $\pi_1$, and the conclusions of $\pi$ being those of $\pi_1$ (other than b) plus those of $\pi_2$. Figure 9.2 illustrates the idea.

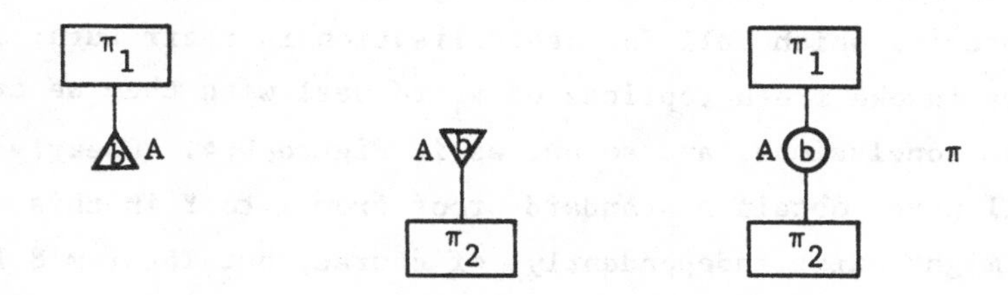


Figure 9.2

*Theorem 9.2* Junction preserves validity and abstract validity.

*Proof* Suppose that $\pi$ is the junction of proofs $\pi_1$ and $\pi_2$ whose common vertex is an occurrence of A, say. Then $Y_{\pi_1}$ is a consequence of $X_{\pi_1}$ by $R_{\pi_1}$ and so by dilution $X_\pi \vdash_{R_\pi} A, Y_\pi$. Similarly $X_\pi, A \vdash_{R_\pi} Y_\pi$, and so by cut for A we have $X_\pi \vdash_{R_\pi} Y_\pi$; i.e. $\pi$ is valid. The preservation of abstract validity follows, since any argument abstractly isomorphic to $\pi$ is the junction of arguments abstractly isomorphic to $\pi_1$ and $\pi_2$.

Now suppose that $\pi_1$ is a standard Kneale proof from X to A,Y and $\pi_2$ is a standard Kneale proof from X,A to Y. If A occurs just once as a conclusion in $\pi_1$ and just once as a premiss in $\pi_2$, the junction of $\pi_1$ and $\pi_2$ (or of suitably chosen replicas of them) will evidently be a standard Kneale proof from X to Y. If A occurs several times in one role (say as conclusion in $\pi_1$) but only once in the other, a single operation of junction will leave final occurrences of A outstanding as conclusions, but these can be dealt with by invoking appropriate replicas of $\pi_2$ and joining them as in Figure 9.3. If these were all the cases that could arise, $K_R$ would be closed under cut. But if A occurs more than once as conclusion in $\pi_1$ and also more than once as premiss in $\pi_2$, then each time we invoke a replica of $\pi_2$ in order to neutralise an unwanted occurrence of A as a conclusion, we add other occurrences of A as premiss which call for neutralisation in their turn; and if we invoke fresh replicas of $\pi_1$ to deal with them we create fresh conclusions, and so on, as in Figure 9.4. Clearly we shall never obtain a standard proof from X to Y in this way. One might exist independently, of course, but Theorem 8.11 excludes this as a general possibility.

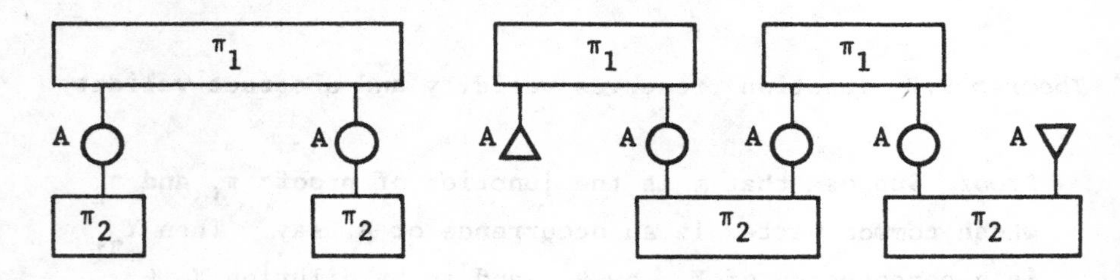


Figure 9.3 Figure 9.4

We can see in this the Achilles' heel of the strategy implicit in Kneale's Principia illustration. It would be natural to suppose that a development for any tautology could be obtained by junction from the developments corresponding to the various steps in its Principia proof, but Figure 9.5 shows why this breaks down in the case we considered in Theorem 8.10. The table $\pi_1$ is a development from $\Lambda$ to $A \supset A$ and $\pi_2$ is a development from $A \supset A$ to $(A \supset A) \& (A \vee (A \supset A))$; but $A \supset A$ occurs twice as premiss in $\pi_2$ and twice as conclusion in $\pi_1$, and consequently their junction $\pi$ fails to supply the expected development from $\Lambda$ to $(A\supset A)\&(A\vee(A\supset A))$.

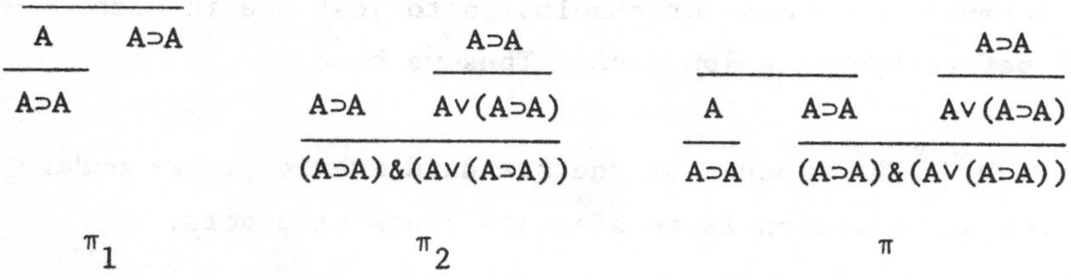

$$\frac{\overline{\;A\qquad A\supset A\;}}{A\supset A}\qquad\qquad \frac{A\supset A\qquad \dfrac{A\supset A}{A\vee(A\supset A)}}{(A\supset A)\&(A\vee(A\supset A))}\qquad\qquad \overline{\;A\qquad A\supset A\;}\ \ \frac{A\supset A}{A\vee(A\supset A)}$$

$$\frac{A}{A\supset A}\qquad \frac{A\supset A\qquad A\vee(A\supset A)}{(A\supset A)\&(A\vee(A\supset A))}$$

$\pi_1$ $\pi_2$ $\pi$

Figure 9.5

Can $\pi$ be regarded as a proof from $\Lambda$ to $(A\supset A)\&(A\vee(A\supset A))$ despite the presence of initial and final occurrences of $A\supset A$? It would have to be what we have called a non-standard proof, and Kneale's tabular notation is inadequate for non-standard arguments without some supplementation to indicate which of the initial and final occurrences of formulae are meant to be premisses or conclusions. We therefore revert to our own explicitly graphical notation.

Let $\pi'$ be a subargument of an argument $\pi$, differing from it only in the specification of its premisses and conclusions, and such that $X_{\pi'} = X_{\pi}$ and $Y_{\pi'} = Y_{\pi}$. In other words the graphs of $\pi$ and $\pi'$ are the same but a premiss of $\pi$ need not be a premiss of $\pi'$ provided some other occurrence of the same formula is a premiss of $\pi'$; and similarly with conclusions. In such a case we say that $\pi'$ is obtained from $\pi$ by the *election* of representative premisses and conclusions. It is immediate from the definition of validity that $\pi'$ is a proof from X to Y by R whenever $\pi$ is, so that

*Theorem 9.3* Election preserves validity.

Unlike junction, election does not preserve abstract validity, since its application exploits the recurrence of the elected formulae. But it overcomes the difficulty about junction and cut, for by election we can reduce the number of occurrences of a formula as premiss or conclusion to just one in each case, before applying junction. Thus we have

*Theorem 9.4* The closure of the standard Kneale proofs under junction and election is an adequate class of proofs.

> *Proof* Let S be the closure of the standard Kneale proofs under junction and election, and for any given R let $X \vdash Y$ mean that there is an argument in S from X to Y by R. By 9.2 and 9.3 any argument in S is a proof, i.e. $\vdash \subset \vdash_R$. Since $K_R \subset \vdash$, by 9.1 $\vdash$ contains R and is closed under overlap; and it is evidently closed under dilution. By 2.16 it only remains to show that $\vdash$ is closed under cut for formulae. Suppose then that $X \vdash A,Y$ and $X,A \vdash Y$, so that there exist proofs $\pi_1$ and $\pi_2$ in S from X to A,Y and from X,A to Y respectively. If either is a proof from X to Y then at once $X \vdash Y$. Otherwise let $\pi_1'$ and $\pi_2'$ be obtained from $\pi_1$ and $\pi_2$ by election, so that A occurs just once as a conclusion in $\pi_1'$ and just once as a premiss in $\pi_2'$. Then $\pi_1'$ and $\pi_2'$ are both in S, and the junction of appropriate replicas of them is the required proof from X to Y.

Figure 9.6 illustrates these various ideas in the case we have been considering. The graph proofs $\pi_1$ and $\pi_2$ correspond to the developments $\pi_1$ and $\pi_2$ of Figure 9.5, and $\pi_1'$ and $\pi_2'$ are obtained from them by electing one of the two occurrences of $A \supset A$ in each case. The desired proof from $\Lambda$ to $(A \supset A) \& (A \vee (A \supset A))$ can now be obtained by joining $\pi_1'$ and $\pi_2'$ to form $\pi$ as in the figure. Whereas $\pi_1$ and $\pi_2$ are standard abstract proofs, $\pi_1'$, $\pi_2'$ and $\pi$ are non-standard and, though proofs, are not abstract ones. Someone arguing his way through $\pi_2'$, say, could justify the inscription of the non-standard occurrence of A⊃A on the ground that he was merely reiterating a formula which he had elsewhere posited as a premiss; but this sort of justification by cross-reference depends entirely on its being the same formula which occurs in both places, and so is not abstract.

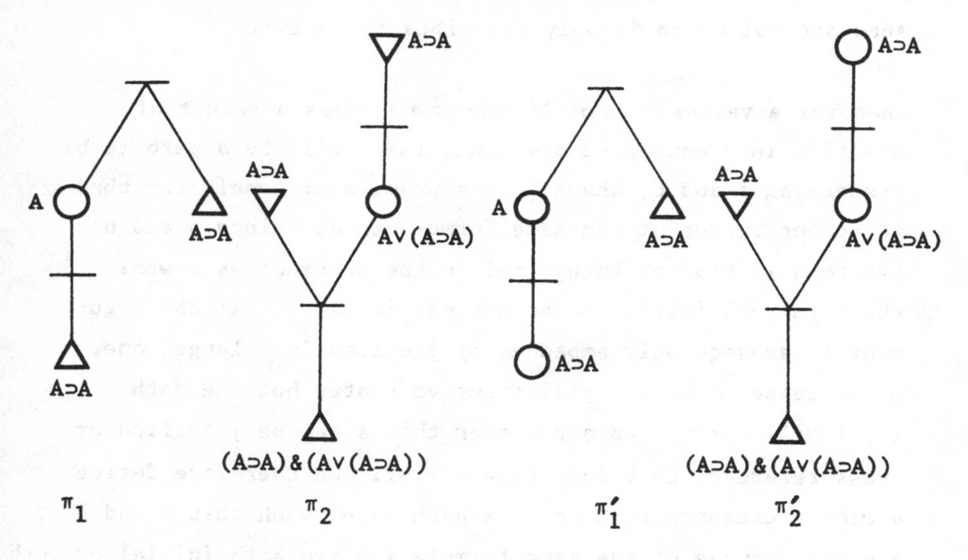


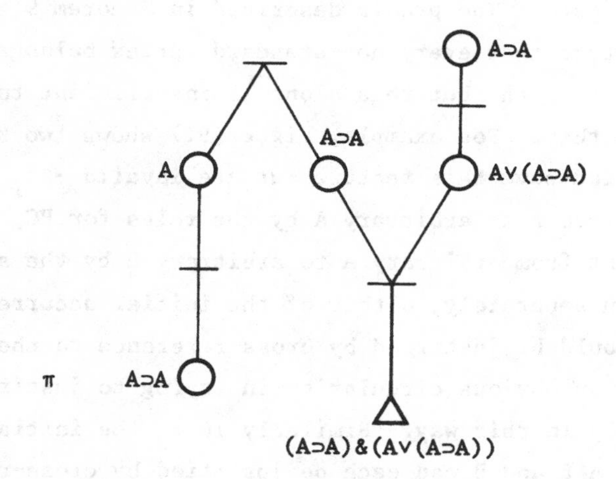


Figure 9.6

## 9.2 *Cross-referenced Kneale proofs*

Theorem 9.4 ensures incidentally that graph proofs as a whole are adequate, but its immediate interest lies in the challenge of finding a more explicit characterisation of the particular variety of proof which the theorem shows to be adequate but which it only describes inductively.

Whenever a vertex a is made non-standard as a result of election in a connected argument, there will be a path (a,b) connecting a and b, where b is a premiss or conclusion that is an occurrence of the same formula as a. Since a and b are both initial or both final in the argument as a whole, they are both initial or both final in (a,b). If the argument is subsequently embedded by junction in a larger one, b may cease to have any distinctive status but the path (a,b) will survive as a reminder that a can be justified by cross-reference to b (cf. Figure 9.6). We therefore define a *cross-reference path* to be a path (a,b) such that a and b are occurrences of the same formula and are both initial or both final in the path. The proofs described in Theorem 9.4 all have the feature that every non-standard vertex belongs to a cross-reference path, but this alone is insufficient to characterise them. For example, Figure 9.7 shows two Kneale arguments which have this feature but are invalid - $\pi_1$ being an argument from $\Lambda$ to arbitrary A by the rules for PC, and $\pi_2$ an argument from arbitrary A to arbitrary B by the same rules. Taken separately, either of the initial occurrences of A in $\pi_1$ could be justified by cross-reference to the other, but there is an obvious circularity in trying to justify them simultaneously in this way. Similarly in $\pi_2$ the initial occurrences of A&B and B can each be justified by cross-reference to the intermediate occurrence of the same formula, but again there is a circularity involved in doing both things at once:

the initial occurrence of A&B is justified by cross-reference to the intermediate one, which in its turn is justified by being deduced from (inter alia) the initial occurrence of B, which is justified by cross-reference to the intermediate occurrence of B, which, completing the circle, is justified by being deduced from the initial occurrence of A&B.

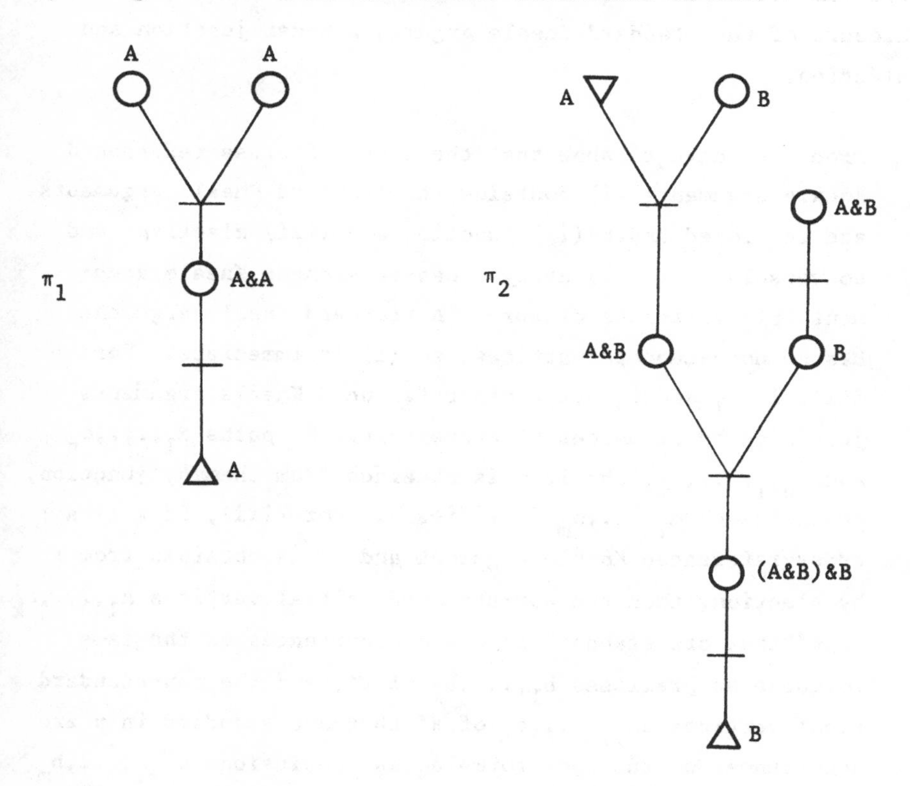


Figure 9.7

We therefore say that a Kneale argument is *cross-referenced* if its non-standard vertices can be arranged in a sequence $a_1,\ldots,a_n$ such that there exist cross-reference paths $(a_1,b_1),\ldots,(a_n,b_n)$, where no $b_i$ appears in $(a_j,b_j)$ for $j>i$ unless $b_i=b_j$. In such a case we say that the sequence of paths *justifies* the argument. (An adequate class of proofs may be obtained by omitting the concession for the case $b_i=b_j$

and requiring that no $b_i$ shall appear in $(a_j,b_j)$ for any $j>i$. But this unnaturally excludes proofs such as $\pi$ in Figure 9.6, in which two non-standard vertices are justified by cross-reference to the same standard one.)

*Theorem 9.5*† The cross-referenced Kneale arguments are the closure of the standard Kneale arguments under junction and election.

*Proof* We have to show that the class of cross-referenced Kneale arguments (i) contains the standard Kneale arguments and is closed under (ii) junction and (iii) election; and conversely that (iv) every cross-referenced Kneale argument belongs to the closure. A standard Kneale argument has no non-standard vertices, so (i) is immediate. For (ii), if $\pi_1$ and $\pi_2$ are cross-referenced Kneale arguments justified by sequences of cross-reference paths $\alpha_1,\dots,\alpha_n$ and $\alpha_{n+1},\dots,\alpha_m$, and if $\pi$ is obtained from them by junction, the sequence $\alpha_1,\dots,\alpha_m$ justifies $\pi$. For (iii), if $\pi$ is a cross-referenced Kneale argument and $\pi'$ is obtained from $\pi$ by election, then the non-standard initial vertices $a_1,\dots,a_k$ of $\pi'$ that are standard in $\pi$ are occurrences of the same formulae as premisses $b_1,\dots,b_k$ of $\pi'$, and the non-standard final vertices $a_{k+1},\dots,a_m$ of $\pi'$ that are standard in $\pi$ are occurrences of the same formulae as conclusions $b_{k+1},\dots,b_m$ of $\pi'$. It follows that if the sequence $(a_{m+1},b_{m+1}),\dots,(a_n,b_n)$ justifies $\pi$ then $(a_1,b_1),\dots,(a_n,b_n)$ justifies $\pi'$; for since $\pi'$ is corner-free and since for each $i$ $(1\le i\le m)$ $b_i$ is standard and either initial or final in $\pi'$, $b_i$ can lie on a path only as an endpoint and $b_i\ne a_j$ for each $j$.

For (iv) we proceed by induction on the number $n$ of non-standard vertices in $\pi$. If $n=0$, $\pi$ is a standard Kneale argument and hence in the closure. If $n\ge1$, let $(a_1,b_1)$, $\dots,(a_n,b_n)$ be a sequence of cross-reference paths which justifies $\pi$. Then $b_1$ is distinct from $a_1$ and by the condition on the sequence it is also distinct from $a_2,\ldots,a_n$ and so is standard. Let $\pi_1$ be the union of $b_1$ and the paths of $\pi$ in which $b_1$ is a final endpoint, and $\pi_2$ the union of $b_1$ and the paths in which $b_1$ is an initial endpoint. Each vertex is to be a premiss or conclusion of $\pi_1$ or $\pi_2$ if it is a premiss or conclusion of $\pi$, and in addition $b_1$ is a conclusion of $\pi_1$ and a premiss of $\pi_2$. Evidently $\pi_1$ and $\pi_2$ are connected and nonempty, and hence they are Kneale arguments. Since $\pi$ is connected they exhaust the vertices and edges of $\pi$, and since $\pi$ is circuit-free they have only $b_1$ in common. It follows that $\pi$ is the junction of $\pi_1$ and $\pi_2$. Since $b_1$ is standard in both $\pi_1$ and $\pi_2$ we may suppose that $a_1,a_{i_2},\ldots,a_{i_k}$ and $a_{i_{k+1}},\ldots,a_{i_n}$ $(k\geq 1)$ are the subsequences of $a_1,\ldots,a_n$ that comprise the non-standard vertices of $\pi_1$ and $\pi_2$ respectively. Since $b_1$ lies on none of $(a_2,b_2),\ldots,(a_n,b_n)$ unless as an endpoint, $\pi_2$ is a cross-referenced Kneale argument justified by the sequence $(a_{i_{k+1}},b_{i_{k+1}}),\ldots,(a_{i_n},b_{i_n})$, and is in the closure by the induction hypothesis. Let $\pi_1'$ differ from $\pi_1$ only in that $a_1$ is a premiss or conclusion according as it is initial or final in $(a_1,b_1)$. Then $\pi_1'$ too is a cross-referenced Kneale argument justified by the sequence $(a_{i_2},b_{i_2}),\ldots,(a_{i_k},b_{i_k})$, and it too is in the closure by the induction hypothesis. But now $\pi_1$ is obtained from $\pi_1'$ by election, and $\pi$ from $\pi_1$ and $\pi_2$ by junction, so that $\pi$ is in the closure as required.

Theorem 9.5 answers our question about the characterisation of the class of proofs introduced by Theorem 9.4, and from the two results (with Theorem 8.3) we have at once

*Theorem 9.6*[†] Every cross-referenced Kneale argument is valid which means that we may use the terms 'cross-referenced Kneale argument' and 'cross-referenced Kneale proof' interchangeably, and

*Theorem 9.7* Cross-referenced Kneale proofs are adequate.

## 9.3 *Cross-referenced circuits*

We know from Theorems 8.1 and 8.12 that standard circuit-free proofs are inadequate, and consequently the cross-referenced Kneale proofs only achieve adequacy by sacrificing standardness. In this section we show contrariwise that an adequate class of standard corner-free proofs can be obtained provided we disregard Kneale's prohibition of circuits. Given that Kneale ignores questions of recurrence there is a good deal of justification for his position, for Theorem 8.5 shows that

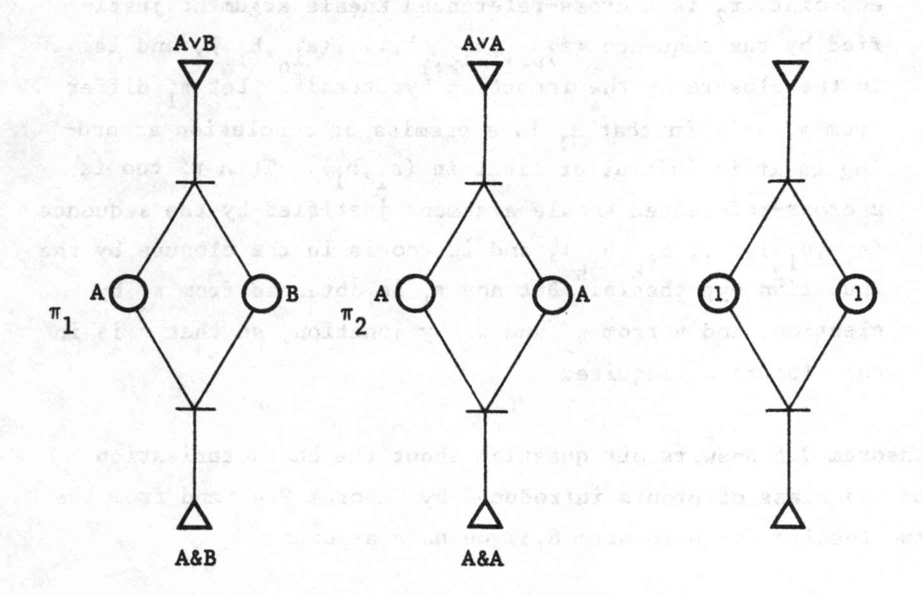


Figure 9.8

a circuit in a corner-free abstract proof is never more than an excrescence. If we are content with mere validity, however, a circuit can make an essential contribution to a proof. Even Kneale's example (reproduced as $\pi_1$ in Figure 9.8) is fallacious only if the formulae A and B are distinct; otherwise we have the proof $\pi_2$ exemplifying the valid form also shown in the figure.

We can see moreover how circuits of this sort can be utilised to solve the problem of junction and cut. Consider our usual example, where A⊃A occurs twice as premiss or conclusion in each of the standard proofs $\pi_1$ and $\pi_2$ shown in Figure 9.6. By joining $\pi_1$ and $\pi_2$ at both places simultaneously we obtain the argument of Figure 9.9, which contains a circuit but is a standard proof from Λ to (A⊃A)&(A∨(A⊃A)) by the rules for PC.

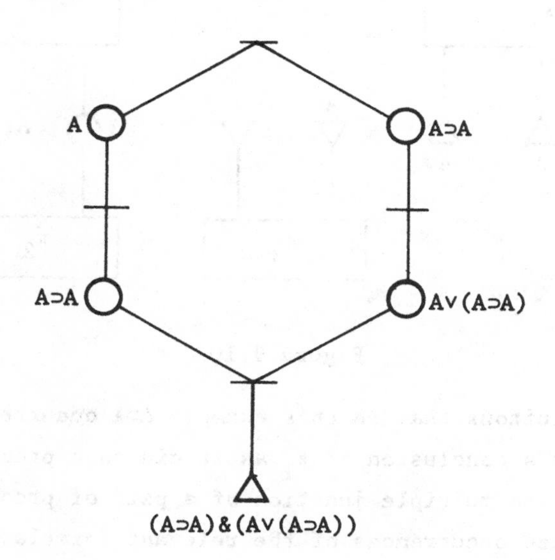


Figure 9.9

In general, let $\pi_1$ and $\pi_2$ be arguments whose common vertices $a_1,\ldots,a_k$ ($k\geq1$) are conclusions of $\pi_1$ and premisses of $\pi_2$, and are all occurrences of the same formula. We say that $\pi$

is the *multiple junction* of $\pi_1$ and $\pi_2$ if the graph of $\pi$ is the union of those of $\pi_1$ and $\pi_2$, the premisses of $\pi$ being those of $\pi_2$ (other than $a_1,\ldots,a_k$) plus those of $\pi_1$, and the conclusions of $\pi$ being those of $\pi_1$ (other than $a_1,\ldots,a_k$) plus those of $\pi_2$. Figure 9.10 illustrates the idea. We may argue exactly as for Theorem 9.2 to show that

*Theorem 9.8* Multiple junction preserves validity

but multiple junction does not in general preserve abstract validity since it exploits the recurrence of the formula occurring at the common vertices. Figure 9.9 is a case in point, where the multiple junction of a pair of abstract proofs is not itself abstractly valid.

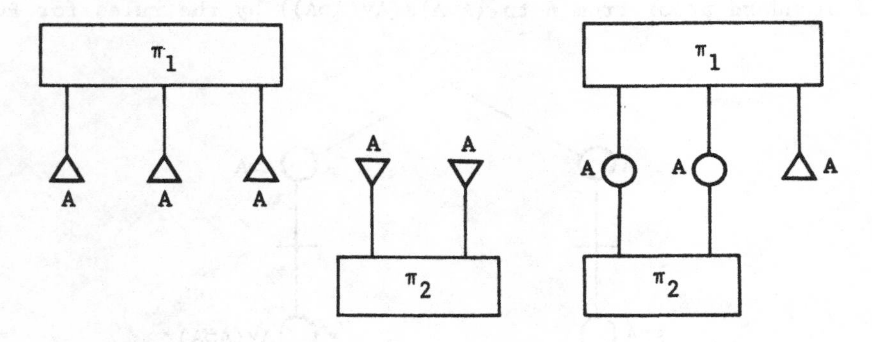


Figure 9.10

It was fortuitous that in this example A⊃A occurred exactly as often as a conclusion of $\pi_1$ as it did as a premiss of $\pi_2$. In general the multiple junction of a pair of proofs will have unwanted occurrences of the relevant formula, as in Figure 9.10. We have already come across a similar situation in Section 9.1 in the case of simple junction. Here however we can deal with it, as we could not do then, by coining sufficiently many replicas of the two proofs. For if the same formula occurs m times as a conclusion in $\pi_1$ and n times as a premiss in $\pi_2$, we can first bring these up to the same number m.n by considering n replicas of $\pi_1$ and m of $\pi_2$, and then use multiple junction, as in Figure 9.11.

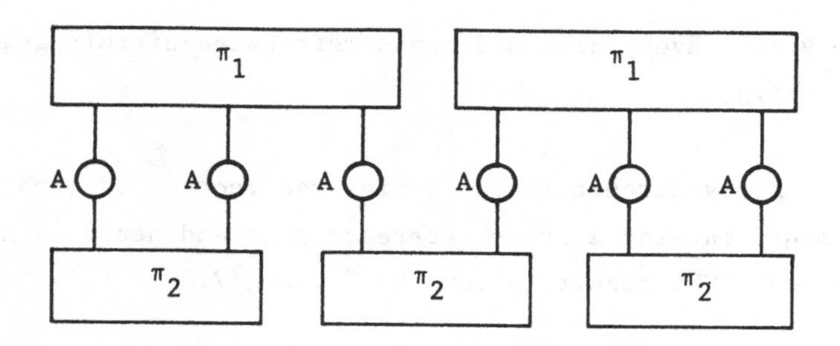


Figure 9.11

*Theorem 9.9* The closure of the standard Kneale proofs under multiple junction is an adequate class of proofs.

*Proof* Let S be the closure of the standard Kneale proofs under multiple junction. It is sufficient to show that, for any given rules, deducibility by means of proofs in S is closed under cut for formulae, the remainder of the argument following precisely the lines of 9.4. So let $\pi_1$ be a proof in S from X to A,Y in which A occurs as conclusion as few times as possible, say $m_1$ times; and let $\pi_2$ be a proof in S from X,A to Y by the same rules in which A occurs as premiss as few times as possible, say $m_2$ times. Then either $m_1$ or $m_2$ is zero, since otherwise the multiple junction of appropriate replicas of $\pi_1$ and $\pi_2$ is a proof in S from X to A,Y with $m_1-m_2$ occurrences of A as conclusion, or else it is a proof from X,A to Y with $m_2-m_1$ occurrences of A as premiss, contradicting the choice of $\pi_1$ or $\pi_2$. Hence either $\pi_1$ or $\pi_2$ is a proof from X to Y as required.

We call an argument a *cross-referenced-circuit argument* if it is connected, nonempty and corner-free, and every circuit contains a cross-reference path. For example, every Kneale argument is one, as is the argument shown in Figure 9.9.

*Theorem 9.10*† Every standard cross-referenced-circuit argument is valid.

> *Proof* Every circuit in any formal redirection of such an argument contains a cross-reference path and hence is not directed. The result follows by 7.2 and 7.5.

Both the standard arguments and the cross-referenced-circuit arguments are closed under multiple junction (in the latter case because any circuit created as a result of multiple junction contains a cross-reference, and the stipulation in the definition that k≥1 ensures that the multiple junction of connected arguments is connected). So the standard cross-referenced-circuit proofs are closed under multiple junction, and by Theorem 9.9 we have

*Theorem 9.11* Standard cross-referenced-circuit proofs are adequate.

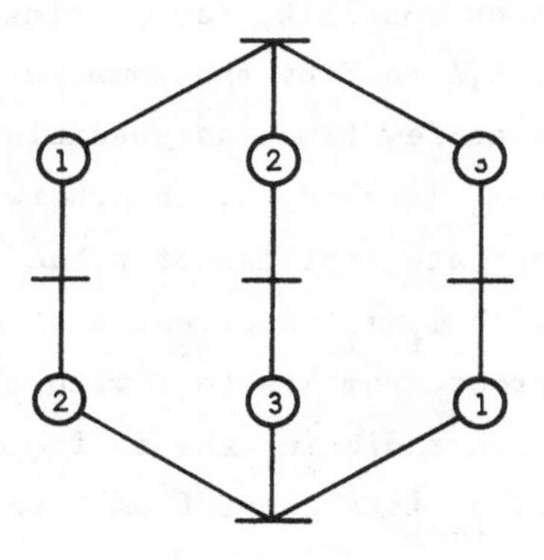


Figure 9.12

Although the class of proofs defined in Theorem 9.9 consists exclusively of standard cross-referenced-circuit proofs it does not exhaust them, as Figure 9.12 shows.

## 9.4 *Other non-abstract proofs*

The proofs we have considered so far have all been corner-free, but they do not exhaust the corner-free proofs. The cross-referenced Kneale proofs do not even exhaust the Kneale proofs, as the example of a concise Kneale proof in Figure 9.13 shows. We leave to the reader the problem of finding syntactical criteria for a Kneale argument to be (1) a proof and (2) a concise proof. The second of these problems may perhaps be solved by equating the concise Kneale proofs with the cross-referenced ones, in a suitably extended sense of 'cross-referenced'. For example, the idea of a cross-refer-

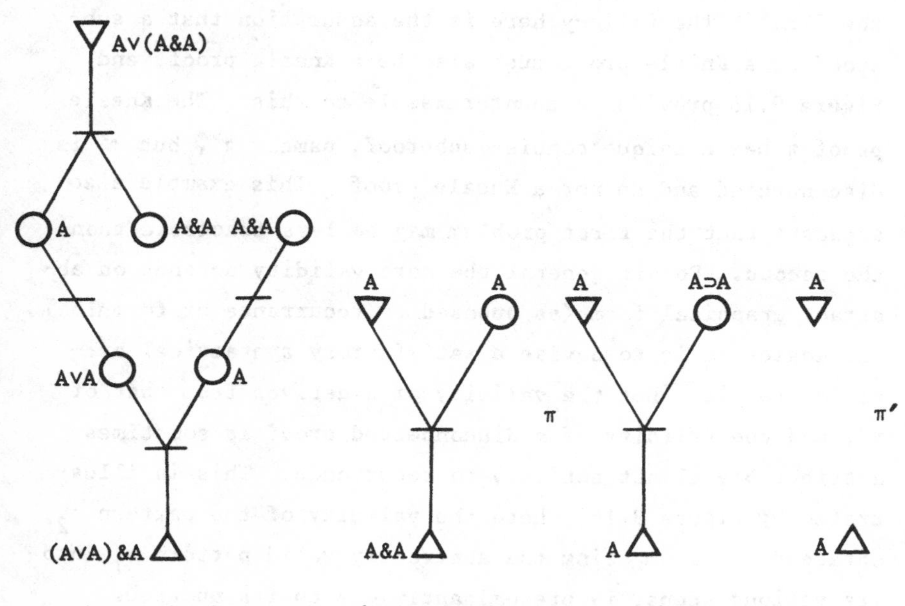


Figure 9.13 Figure 9.14 Figure 9.15

ence path might be extended to include paths joining an initial and a final occurrence of the same formula, provided they contain cross-reference paths of the orthodox kind. The path joining the non-standard vertices in Figure 9.13 is like this, and the validity of arguments containing such paths can be explained by regarding them as cross-referenced-circuit proofs whose circuits have been prised open. But any solution along these lines would require us to take account of multiplicity of occurrence in rules of inference, or equivalently to redefine 'subargument' to mean what we presently call an abstract subargument; for as things stand, even a cross-referenced proof of the orthodox kind need not be concise, as Figure 9.14 shows.

Since every proof has a concise subproof, it might be thought that a Kneale argument will be a proof iff it has a concise Kneale proof as a subargument, and hence that a solution to the second problem would automatically provide a solution to the first. The fallacy here is the assumption that a subproof of a Kneale proof must also be a Kneale proof, and Figure 9.15 provides a counterexample to this. The Kneale proof $\pi$ has a unique concise subproof, namely $\pi'$, but $\pi'$ is disconnected and so not a Kneale proof. This example also suggests that the first problem may be less tractable than the second. For in general the more validity depends on abstract graphical form (as opposed to recurrence of formulae) the easier it is to devise a satisfactory syntactical criterion for it. But the validity of $\pi$ derives from that of $\pi'$, and the validity of a disconnected proof is sometimes attributable almost entirely to recurrence. This is illustrated by Figure 9.16, where the validity of the pattern $\pi_2$, obtained by decomposing the abstractly valid pattern $\pi_1$ into its various steps, is predominantly due to its numerous cross-references.

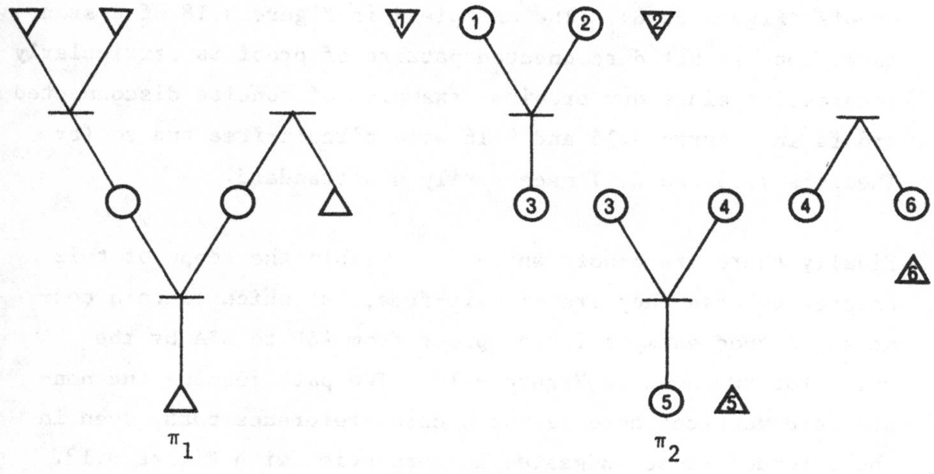


Figure 9.16

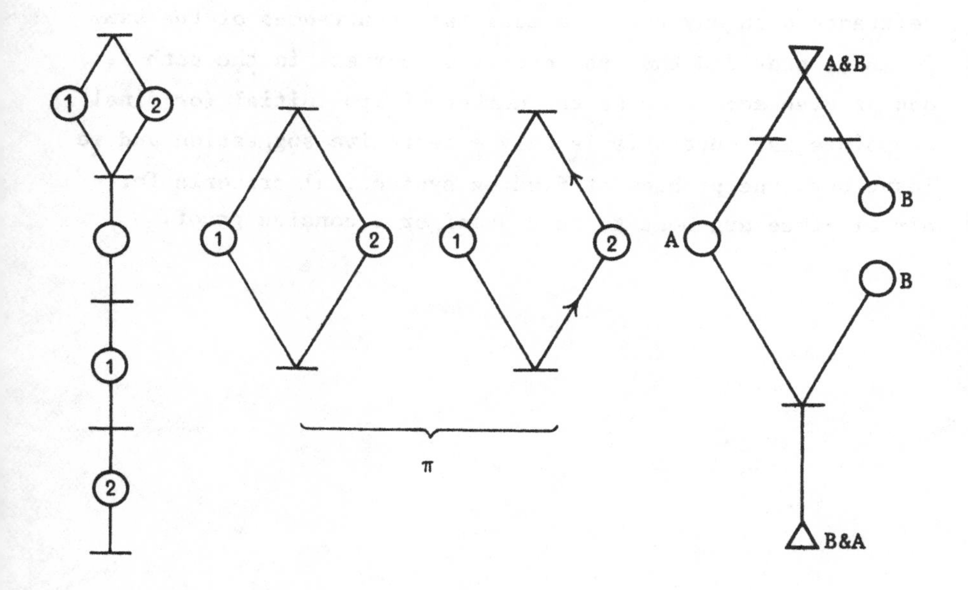


Figure 9.17 Figure 9.18 Figure 9.19

Cross-referenced-circuit proofs pose similar problems in the context of corner-free proofs in general. The cross-referenced-circuit proofs exhaust neither the standard connected corner-free proofs (Figure 9.17), nor the standard concise corner-free proofs (Figure 9.18). The example $\pi$ in Figure 9.18 of a standard, concise but disconnected pattern of proof is particularly interesting since our previous examples of concise disconnected proofs in Figures 9.15 and 9.16 were circuit-free and so (by Theorems 7.13 and 8.1) necessarily non-standard.

Finally there are proofs which fall within the scope of this chapter because they are circuit-free, but which contain corners. A good example is the proof from A&B to B&A by the rules for PC shown in Figure 9.19. The path joining the non-standard vertices here is not a cross-reference path, even in the extended sense suggested in connection with Figure 9.13. It might be feasible to licence it by counting as a cross-reference path any that connects two occurrences of the same formula, provided that the number of corners in the path is odd or even according as the number of its initial (or final) endpoints is. But this is only a tentative suggestion and we leave open the problem of finding syntactical criteria for a circuit-free argument to be a proof or a concise proof.

# 10 · Abstract proofs

## 10.1 *Cornered-circuit proofs*

We know from Theorems 8.12 and 8.13 that an adequate class of abstract proofs must permit the formation of both corners and circuits. With this in mind we offer a third solution to the problem of junction and cut posed in Section 9.1, by invoking the operation of identifying or coalescing different occurrences of the same formula in a proof.

We say that $\pi'$ is obtained from a graph argument $\pi$ by *identification* if it is the result of omitting vertices $a_2,\dots,a_n$ from $\pi$ and replacing every edge to or from any of them by a corresponding edge to or from the vertex $a_1$, where all the $a_i$ are occurrences of the same formula; provided that if any $a_i$ is a premiss (conclusion) then $a_1$ is a premiss (conclusion) and all the $a_i$ are initial (final). For example, the pattern shown in Figure 10.1 is obtained by identifying the same-numbered vertices in the pattern of Figure 9.18. By prohibiting the identification of premisses with non-initial vertices (or conclusions with non-final ones), the proviso in the definition ensures that the result of identification is always a graph argument; and by prohibiting the identification of a premiss $a_i$ with a non-premiss $a_1$ (or a conclusion with a non-conclusion $a_1$) it ensures that $X_{\pi'} = X_\pi$ and $Y_{\pi'} = Y_\pi$ and hence, since evidently $R_{\pi'} = R_\pi$, that identification preserves validity. Note that $\pi'$ may have fewer edges than $\pi$, since if there are edges in $\pi$ from the same stroke to more than one of the $a_i$, they will be replaced by a single edge in $\pi'$.

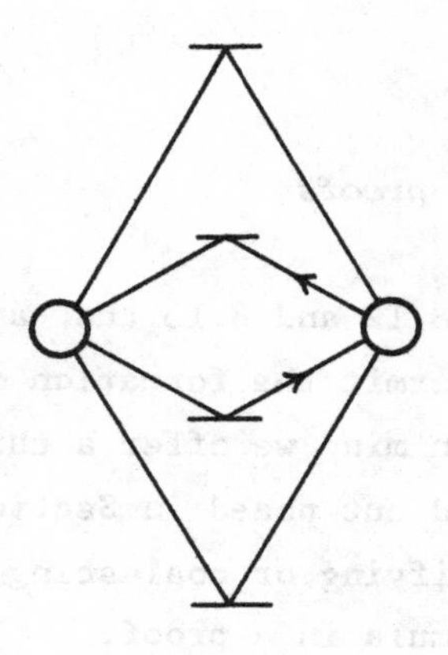

Figure 10.1

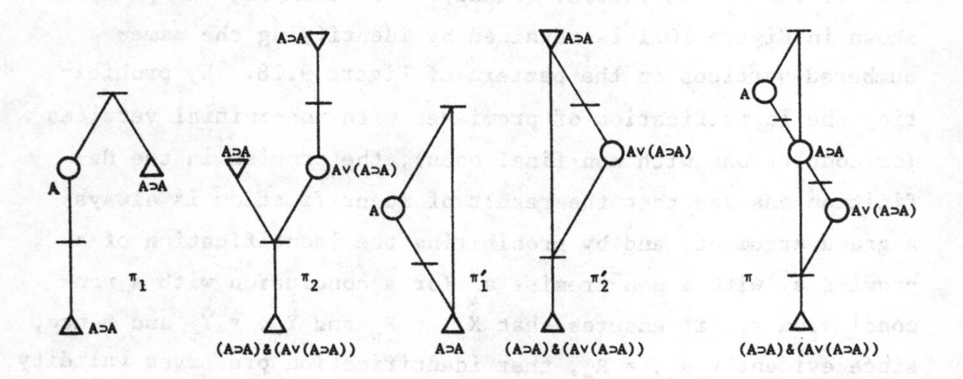


Figure 10.2

It is sufficient for our immediate purpose to consider only the *identification of premisses and conclusions*, where the vertices to be identified are either all premisses or all conclusions. For instance, returning to our stock example from Theorem 8.10, $\pi_1'$ of Figure 10.2 is obtained from $\pi_1$ by identification of conclusions, and $\pi_2'$ from $\pi_2$ by identification of premisses. Evidently $\pi_1'$ is an abstract proof from $\Lambda$ to A⊃A and $\pi_2'$ from A⊃A to (A⊃A)&(A∨(A⊃A)), and as each has only one occurrence of A⊃A we can obtain by junction the desired abstract proof $\pi$ from $\Lambda$ to (A⊃A)&(A∨(A⊃A)).

*Theorem 10.1* Identification preserves validity and abstract validity.

*Theorem 10.2* The closure of the standard Kneale proofs under identification of premisses and conclusions and junction is an adequate class of abstract proofs.

*Proofs* For 10.1, we have shown in the text that identification preserves validity, and since any argument abstractly isomorphic to one obtained by identification from $\pi$ can also be obtained by identification from one abstractly isomorphic to $\pi$, identification preserves abstract validity also.

For 10.2, every argument in the closure is an abstract proof by 8.3, 9.2 and 10.1, and as in the proof of 9.9 it is sufficient to show that if there are in the closure abstract proofs $\pi_1$ from X to A,Y and $\pi_2$ from X,A to Y, then there is also in the closure an abstract proof from X to Y. Let $a_1,\ldots,a_m$ be the occurrences of A as conclusions of $\pi_1$ and let $b_1,\ldots,b_n$ be the occurrences of A as premisses of $\pi_2$. If m=0 or any of the $b_j$ is a conclusion of $\pi_2$ then $\pi_1$ is the required proof from X to Y; and similarly if n=0 or any of the $a_i$ is a premiss of $\pi_1$ then $\pi_2$ is the required proof. Otherwise, choosing suitable replicas so that $a_1=b_1$ but $\pi_1$ and $\pi_2$ have no other vertex in common, we may obtain a proof from X to Y by identifying $a_2,\ldots,a_m$ with $a_1$, identifying $b_2,\ldots,b_n$ with $b_1$, and joining the resulting arguments (cf. Figure 10.3).

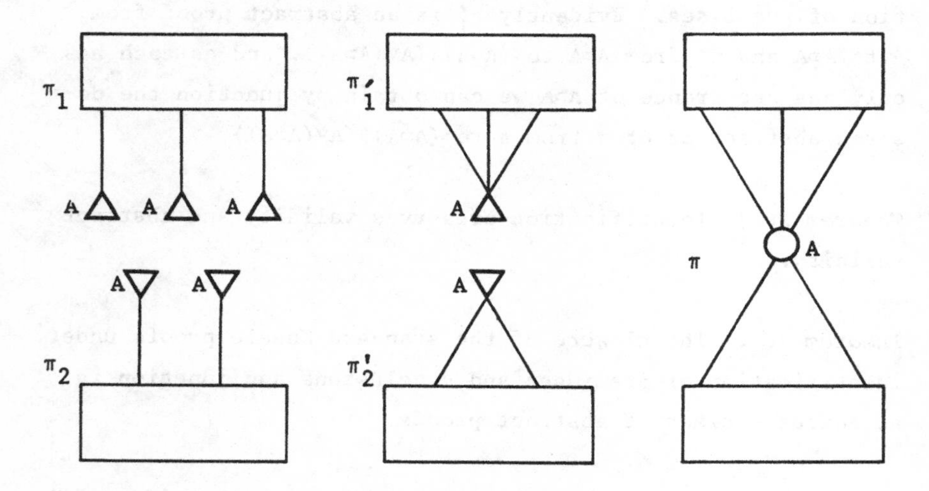


Figure 10.3

When identification is confined to premisses and conclusions, and hence to vertices which are all initial or all final, any circuit formed as a result of identifications will necessarily be *cornered*, i.e. will contain a corner. For example, all the circuits in Figure 10.2 are cornered. A nonempty argument in which every circuit is cornered we call a *cornered-circuit* argument.

*Theorem 10.3*† Every standard cornered-circuit argument is abstractly valid.

*Theorem 10.4* Standard cornered-circuit proofs are adequate.

*Proofs* For 10.3, since abstract redirection converts corners into corners, every circuit in an abstract redirection $\pi'$ of a cornered-circuit argument $\pi$ is cornered and so not directed. By 7.2 $\pi'$ therefore has a final vertex, and if every vertex in $\pi$ is standard it follows by 7.6 that $\pi$ is abstractly valid.

For 10.4, we have noted that the class of cornered-circuit arguments is closed under identification of premisses and conclusions, and it is closed too under junction, which never creates circuits. Since every Kneale proof is a cornered-circuit argument, and since both operations preserve standardness, it follows that every member of the class of proofs defined in 10.2 is a standard cornered-circuit proof, whence the result follows by 10.2.

The proof of Theorem 10.4 has shown that all the proofs that can be obtained from standard Kneale ones by identification of premisses and conclusions and junction are standard cornered-circuit proofs, but the example of Figure 8.7 shows that they do not exhaust the standard cornered-circuit proofs. We return to them in Theorem 10.12; meanwhile we show that the standard cornered-circuit proofs constitute the closure under the same two operations of a wider class than the standard Kneale proofs:

*Theorem 10.5*† The standard cornered-circuit arguments are the closure of the standard nonempty circuit-free arguments under identification of premisses and conclusions and junction.

*Lemma 10.6*† If a cornered-circuit argument has any formula vertices, it has one which is not intermediate in any circuit.

*Proofs* For any graph $\pi$ let $n(b,\pi)$ be 1 or 0 according as $\pi$ does or does not contain a corner at b, and let $n(\pi)$ be the number of corners in $\pi$. For 10.6, let $\pi$ be a cornered-circuit argument containing a formula vertex $a_1$. Suppose that each formula vertex b of $\pi$ is intermediate in some circuit $\gamma(b)$; where there is a choice we take $\gamma(b)$ to have as few corners as possible. Starting at $a_1$ and moving along $\gamma(a_1)$ we trace out a walk $a_1, E_1, a_2, \ldots$ through $\pi$, leaving $\gamma(a_i)$ only when we reach a vertex $a_j$ at which $\gamma(a_i)$ has a corner and transferring to $\gamma(a_j)$ in the direction that avoids the creation of a corner at $a_j$. The walk is to stop as soon as it reaches a vertex $a_k$ on a circuit $\gamma(a_i)$ which the walk joined at $a_i$ and left at $a_j$ $(a_j \neq a_k)$. We have in this way defined three paths connecting $a_j$ and $a_k$ which have only their endpoints in common (Figure 10.4):

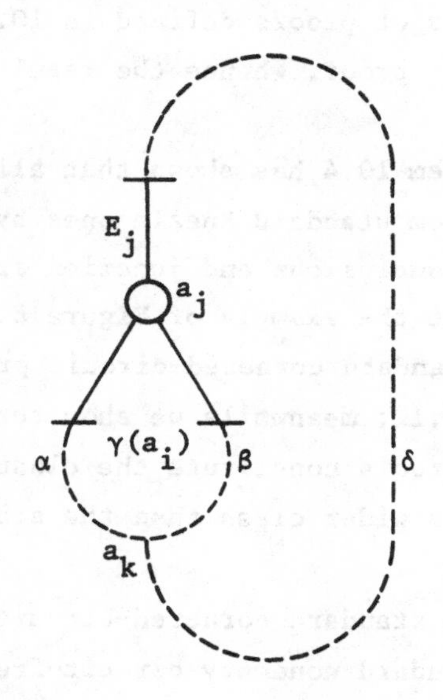


Figure 10.4

the terminal portion $a_j, E_j, \ldots, a_k$ of the walk, which we shall call $\delta$, and the two paths $\alpha$ and $\beta$ which together make up $\gamma(a_i)$. By construction $n(a_j, \alpha\cup\beta) = 1$ and $n(a_j, \alpha\cup\delta) = n(a_j, \beta\cup\delta) = n(\delta) = 0$. Since $\beta\cup\delta$ is cornered, $1 \leq n(\beta\cup\delta) = n(\beta)+n(\delta)+n(a_k, \beta\cup\delta)+n(a_j, \beta\cup\delta) = n(\beta)+n(a_k, \beta\cup\delta)$. Evidently $n(a_k, \alpha\cup\beta)+1 \geq n(a_k, \alpha\cup\delta)+n(a_k, \beta\cup\delta)$, so $n(a_k, \alpha\cup\delta) \leq$

$n(\beta)+n(a_k,\alpha\cup\beta)$. Hence $n(\alpha\cup\delta) = n(\alpha)+n(\delta)+n(a_k,\alpha\cup\delta)+n(a_j,\alpha\cup\delta) < n(\alpha)+n(\beta)+n(a_k,\alpha\cup\beta)+n(a_j,\alpha\cup\beta) = n(\alpha\cup\beta) = n(\gamma(a_i))$. Thus $n(\alpha\cup\delta) < n(\gamma(a_i))$, and similarly $n(\beta\cup\delta) < n(\gamma(a_i))$. It follows that $a_i$ is not intermediate either in $\alpha\cup\delta$ or in $\beta\cup\delta$, since otherwise one of these circuits would have been chosen as $\gamma(a_i)$. On the other hand $a_i$ is intermediate in $\alpha\cup\beta$, which is $\gamma(a_i)$, and so is intermediate in $(\alpha\cup\delta)\cup(\beta\cup\delta)$. It must therefore be initial in $\alpha\cup\delta$ and final in $\beta\cup\delta$ or vice versa, so that there are at least two edges to it and at least two from it in the subgraph $\alpha\cup\beta\cup\delta$. But this is impossible since $a_i$ adjoins at most three edges of the subgraph; hence the result follows by reductio ad absurdum. (This proof is spelt out more fully and set in a more general context in Shoesmith and Smiley, 1979.)

For 10.5, since we know that the standard cornered-circuit arguments contain the standard nonempty circuit-free ones and are closed under identification of premisses and conclusions and junction, it is sufficient to show that every standard cornered-circuit argument $\pi$ belongs to the closure in question. We proceed by induction on the number of circuits in $\pi$. If there are none, $\pi$ is itself a standard nonempty circuit-free argument. If there are circuits in $\pi$, their union $\pi'$ contains by 10.6 a formula vertex b which is intermediate in no circuit of $\pi'$ and hence in no circuit of $\pi$. It follows that the union $\pi_1$ of b and paths of $\pi$ in which b is a final endpoint has only b in common with the union of b and paths in which it is an initial endpoint. Let $\pi_2$ be this latter union together with any components of $\pi$ that do not contain b. Then specifying b to be a conclusion in $\pi_1$ and a premiss in $\pi_2$, but otherwise treating their vertices as in $\pi$, we have that $\pi$ is the junction of $\pi_1$ and $\pi_2$. Every circuit of $\pi$ is either wholly in $\pi_1$ or wholly in $\pi_2$, and since b is in $\pi'$ it follows that some corner {E,b,E′} is in a circuit of

$\pi_1$, say. Let $\pi_1'$ be the argument obtained from $\pi_1$ by replacing E′ by an edge to a new conclusion b′, this being an occurrence of the same formula as b. Then $\pi_1'$ has fewer circuits than $\pi_1$, and so $\pi_1'$ and $\pi_2$ have between them fewer circuits than $\pi$. Hence by the induction hypothesis each of them belongs to the closure. But $\pi_1$ is obtained from $\pi_1'$ by identification of the conclusion b′ with b, and since $\pi$ is the junction of $\pi_1$ and $\pi_2$ it follows that $\pi$ is in the closure, as required.

## 10.2 *Conciseness*

Although we know from Theorem 7.13 that a concise abstract proof is necessarily standard and connected, these conditions are not sufficient for conciseness even for cornered-circuit proofs. For example, the proof shown in Figure 8.7 is standard and connected, but either leg is a proper abstract subproof. We show that the concise cornered-circuit abstract proofs comprise precisely the class of proofs defined inductively and shown to be adequate in Theorem 10.2, but first we establish a number of non-inductive conditions for a cornered-circuit argument to be a concise abstract proof: there are intriguingly many.

Given a graph argument $\pi$, suppose we delete every edge in excess of one to each formula vertex, and every edge in excess of one from each formula vertex. The result is a graph with the same vertices as $\pi$ but without corners, and we call it a corner-free residue or simply a *residue* of $\pi$. A residue is, in other words, a maximal corner-free subgraph. In general a graph has a number of different residues: indeed there is a unique residue iff the original graph is corner-free, when its only residue is itself. The cornered-circuit property is equivalent to the requirement that every residue should be circuit-free. We are concerned only with the residue as a graph and not with its possible status as an argument, but we note that a formula vertex is initial, final or intermediate in a residue according as it is initial, final or intermediate in the original graph.

*Theorem 10.7* An abstract proof in which every residue is connected is concise.

> *Proof* Let $\pi$ be an abstract proof which is not concise. By 7.11 $\pi$ has a proper abstract subproof, which in turn has a standard abstract subproof $\pi'$ by 7.14. Let $\mu'$ be a residue of $\pi'$ and let $\mu$ be a residue of $\pi$ that contains $\mu'$. Since $\pi'$ is standard a formula vertex of $\mu'$ is initial (or final) in $\mu'$ iff it is initial (final) in $\mu$. So any edge of $\mu$ not in $\mu'$ adjoins a formula vertex not in $\pi'$, and must join it and a stroke not in $\pi'$, since $\pi'$ is an abstract subargument of $\pi$. Hence each component of $\mu'$ is a component of $\mu$. But since $\pi'$ is standard and a proper abstract subargument of $\pi$, it must lack some stroke of $\pi$, so that $\mu' \neq \mu$. Since $\mu'$ is nonempty it follows that $\mu$ is not connected.

In general the converse of this result fails: even a standard abstract proof may fail to have any connected residue and still be concise (Figure 10.11). If however we restrict our attention to cornered-circuit arguments we obtain a condition that is both necessary and sufficient. In the statement of this condition in Theorem 10.9 the ambiguity of ‘any’ is deliberate, being justified by Theorem 10.8. Its effect is that we can test for conciseness by drawing a residue at random and seeing whether or not it is connected.

*Theorem 10.8*† Either all residues of a cornered-circuit argument are connected or none is.

*Theorem 10.9*† A cornered-circuit argument is a concise abstract proof iff it is standard and any residue is connected.

*Proofs* For 10.8 it is sufficient to consider the special case in which two residues differ by only one edge, the general result following by iteration. Suppose then that the residues $\mu$ and $\mu'$ of a cornered-circuit argument $\pi$ differ only in that $\mu$ has an edge E joining a formula vertex a and a stroke b, while $\mu'$ has an edge E′ joining a and a stroke b′, where {E,a,E′} is a corner. We show that if $\mu$ is connected then so is $\mu'$. Since all residues of a cornered-circuit argument are circuit-free, at most one path in $\mu'$ (or in $\mu$) connects any two vertices. The path connecting a and b′ in $\mu'$ is {a,E′,b′}, so the path which by hypothesis connects them in $\mu$ cannot be contained in $\mu'$ and so must include E; i.e. it is of the form {a,E,b,...,b′}. Accordingly b and b′ are connected in $\mu$ by a path excluding E, and hence a and b are connected in $\mu'$. The connectedness of $\mu'$ then follows from that of $\mu$.

For 10.9, a standard cornered-circuit argument in which every residue is connected is abstractly valid by 10.3 and concise by 10.7. Conversely a concise cornered-circuit abstract proof $\pi$ is standard by 7.13. Let $\mu$ be a maximal corner-free connected subgraph of $\pi$, and let $\pi'$ be the subgraph of $\pi$ obtained by adding to $\mu$ all edges of $\pi$ that join vertices of $\mu$. If some formula vertex a of $\pi'$ is initial (or final) in $\pi'$ but not in $\pi$, then some edge E of $\pi$ goes to a from a stroke b (or vice versa), where b is not in $\mu$. Also if some edge E, joining a stroke a of $\pi'$ and a formula vertex b, is not in $\pi'$ then b is not in $\mu$. In either case $\mu \cup$ {E,b} is a corner-free connected subgraph of $\pi$, contrary to our choice of $\mu$. So, specifying the initial formula vertices of $\pi'$ as premisses and the final ones as conclusions, $\pi'$ is a standard subargument of $\pi$, and so is an abstract subproof by 10.3. Since $\pi$ is concise it follows by 7.11 that $\pi' = \pi$, and so $\pi$ has $\mu$ as a connected residue.

*Theorem 10.10*† A cornered-circuit argument is a concise abstract proof iff it is standard and connected and every corner is the unique corner in some circuit.

*Proof* If a cornered-circuit abstract proof $\pi$ is concise it is standard by 10.9, and since by 10.9 all its residues are connected $\pi$ itself must be connected. If $\{E,a,E'\}$ is a corner, where E joins a and b, let $\mu$ be a residue including E′. Since $\mu$ is connected, a and b are connected in it by a corner-free path $\alpha$. But then $\alpha\cup\{E\}$ is a circuit which can have a corner only at a, and this corner must be $\{E,a,E'\}$ as required. For the converse let $\pi$ be a connected cornered-circuit argument in which every corner is the unique corner in some circuit. It is sufficient to show that if $\pi$ has any corners then for some edge E in one the graph $\pi$-E obtained from $\pi$ by removing E is connected and retains the property that every corner is the unique corner in some circuit; for by iteration of this result we eventually obtain a connected residue and establish the result by 10.9. The union $\pi'$ of the circuits of $\pi$ is nonempty if $\pi$ has a corner, and so by 10.6 includes a formula vertex a not intermediate in any circuit of $\pi$. So a is in a corner $\{E,a,E'\}$, which in turn is in a circuit, so that the omission of E from $\pi$ leaves the graph connected. Every corner of $\pi$ at a vertex other than a is the unique corner in some circuit of $\pi$, and this circuit does not include a and so does not include E. Similarly, of course, every corner of $\pi$-E at a is the unique corner in a circuit of $\pi$ that excludes E. Hence $\pi$-E has the required property that every corner is the unique corner in some circuit.

*Theorem 10.11*[†] A cornered-circuit argument is a concise abstract proof iff it is standard and every pair of distinct vertices is connected by a corner-free path.

*Proof* If a cornered-circuit abstract proof $\pi$ is concise, by 10.9 it is standard and all its residues are connected. So every pair of distinct vertices is connected in any residue by a path which must be corner-free as required. Conversely, let $\pi$ be a standard cornered-circuit argument in which every pair of distinct vertices is connected by a corner-free path. Let {E,a,E′} be a corner, where E and E′ are, say, from a to b and b′ respectively, and let $\alpha$ be a corner-free path connecting b and b′. If a is in $\alpha$ then a is intermediate in $\alpha$ and so a and b, say, are connected by a corner-free path $\beta$ in which a is final. But then $\beta \cup \{E\}$ is a corner-free circuit, contrary to hypothesis. Hence a is not in $\alpha$ and {E,a,E′} is the unique corner in the circuit $\alpha \cup \{E,a,E'\}$, so that $\pi$ is a concise abstract proof by 10.10.

*Theorem 10.12*[†] The concise cornered-circuit abstract proofs are the closure of the standard Kneale proofs under identification of premisses and conclusions and junction.

*Proof* A standard Kneale proof is nonempty and standard, its circuits are cornered, and every pair of distinct vertices is connected by a corner-free path; and these properties are preserved under identification of premisses and conclusions and under junction. So by 10.11 every argument in the closure is a concise cornered-circuit abstract proof.

For the converse we proceed by induction on the complexity of a concise cornered-circuit abstract proof $\pi$, i.e. the sum of the number of circuits and the number of edges. If there are no circuits, $\pi$ is a standard Kneale proof. Otherwise, applying 10.6 to the union of the circuits of $\pi$, some formula vertex b of a circuit is not intermediate in any circuit. If b is initial (final) in $\pi$ then since by 7.13 $\pi$ is standard b is a premiss (conclusion), so that $\pi$ can be obtained by identification of premisses and conclusions from an argument $\pi'$ in which exactly one edge adjoins each of the premisses (conclusions) identified with b. By 10.10 $\pi'$ is a concise cornered-circuit abstract proof, and since it has fewer circuits than $\pi$ it has a smaller complexity. So by the induction hypothesis $\pi'$ is in the closure and hence $\pi$ is too. If on the other hand b is intermediate in $\pi$ then $\pi$ is obtainable by junction from two concise cornered-circuit abstract proofs of lesser complexity and, arguing as before, $\pi$ is in the closure.

*Theorem 10.13*† An argument is a concise cornered-circuit abstract proof iff it can be obtained from a cross-referenced Kneale proof by identifying the endpoints of each member of a justifying sequence of cross-reference paths.

*Proof* Let $\pi$ and $\pi_o$ be arbitrary arguments obtained from $\pi'$ and $\pi'_o$ by identifying each $a_i$ with $b_i$, where $\pi'$ and $\pi'_o$ are cross-referenced Kneale proofs justified by the respective sequences of cross-reference paths $(a_1,b_1),\ldots,(a_n,b_n)$ and $(a_{n+1},b_{n+1}),\ldots,(a_m,b_m)$; and let S be the set of all such arguments. (The conditions for a cross-reference path ensure that the proviso in the definition of identification is met in each case). It is sufficient by 10.12 to show that S is the closure of the standard Kneale proofs under identification of premisses and conclusions and junction.

Evidently S contains the standard Kneale proofs. If $\pi_1$ can be obtained from $\pi$ by identifying the premisses $c_2,\ldots,c_k$ with the premiss $c_1$, then each $c_j$ is a premiss in $\pi'$. The argument $\pi_1'$, which differs from $\pi'$ only in that $c_2,\ldots,c_k$ are not premisses, is a cross-referenced Kneale proof justified by the sequence of paths $(c_2,c_1),\ldots,(c_k,c_1)$, $(a_1,b_1),\ldots,(a_n,b_n)$; and $\pi_1$ is obtained from $\pi_1'$ by identifying the endpoints of these paths, so that $\pi_1 \in S$. Hence S is closed under identification of premisses, and similarly it is closed under identification of conclusions. If $\pi_1$ is obtained from $\pi$ and $\pi_o$ by junction, their common vertex is, say, a conclusion of $\pi$ (and hence of $\pi'$) and a premiss of $\pi_o$ (and hence of $\pi_o'$). We may assume that none of the $a_i$ is in both $\pi'$ and $\pi_o'$, so that the junction of $\pi'$ and $\pi_o'$ is a cross-referenced Kneale proof justified by the sequence $(a_1,b_1),\ldots,$ $(a_m,b_m)$; and $\pi$ may be obtained from it by identifying the endpoints of these cross-reference paths. Hence the closure is contained in S. For the converse we prove by induction on n that $\pi$ is in the closure. If n=0 then $\pi$ is a standard Kneale proof. If $n\geq 1$ then $b_1$ is standard and does not occur in $(a_2,b_2),\ldots,(a_n,b_n)$ unless as one of the $b_j$. So $\pi'$ is the junction of cross-referenced Kneale proofs $\pi_1'$ and $\pi_2'$ with common vertex $b_1$ and justified by complementary subsequences of $(a_1,b_1),\ldots,(a_n,b_n)$. We may suppose that, say, $a_1$ is in $\pi_1'$, and that $a_1$ and $b_1$ are both final in $\pi_1'$ (see Figure 10.5, noting that $b_1$ may be the sole vertex of $\pi_2'$). Let $\pi_3'$ be obtained from $\pi_1'$ by specifying $a_1$ as an additional conclusion. Now $\pi_2'$ and $\pi_3'$ are cross-referenced Kneale proofs justified by complementary subsequences of $(a_2,b_2),\ldots,(a_n,b_n)$, so that by the induction hypothesis the arguments $\pi_2$ and $\pi_3$ obtained by identifying the endpoints of these cross-reference paths are in S. But $\pi$ is the junction of $\pi_2$ and the argument obtained by identifying the conclusions $a_1$ and $b_1$ of $\pi_3$, and so is in S as required.

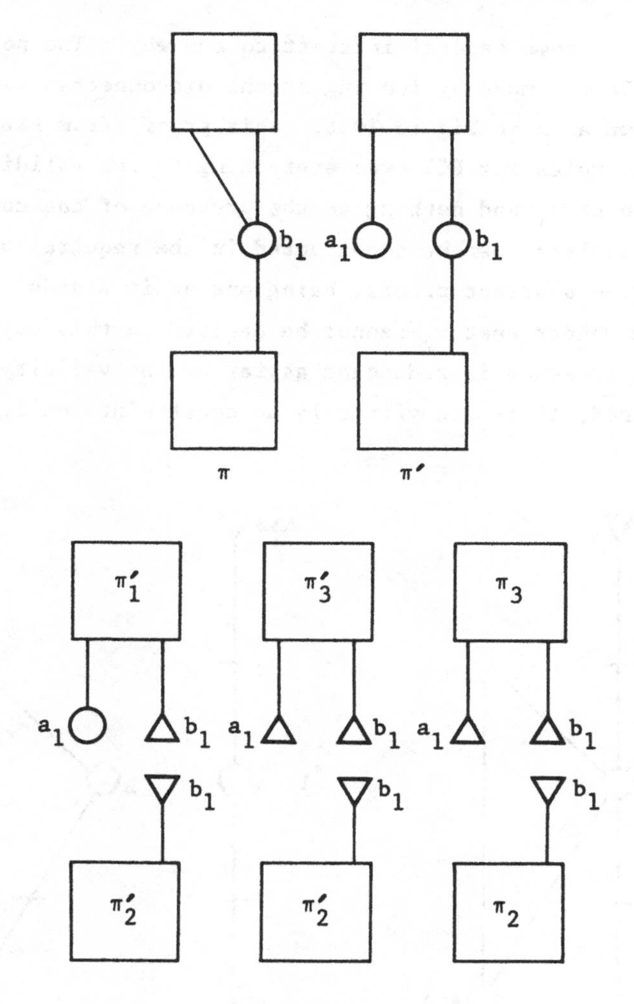


Figure 10.5

## 10.3 *Relevance*

**We have seen how the standard cornered-circuit proofs can be constructed from circuit-free ones by identification of premisses and conclusions and junction (Theorem 10.5); but not all cornered-circuit abstract proofs can be derived in this way from circuit-free ones ($\pi'$ of Figure 10.6 is an example),**

and it is of some general interest to see why. The point can be most clearly made by looking at the disconnected variant of $\pi'$ shown as $\pi$ in Figure 10.6. This proof (from A&B to A∨B by the rules for PC) owes everything to the validity of its component $\pi_1$ and nothing to the presence of the component $\pi_2$. Certainly $\pi_1$ can be constructed in the required way from circuit-free abstract proofs, being one as it stands. But it is little wonder that $\pi_2$ cannot be derived in this way, for since its presence is redundant as far as the validity of $\pi$ is concerned, there are virtually no constraints on its structure.

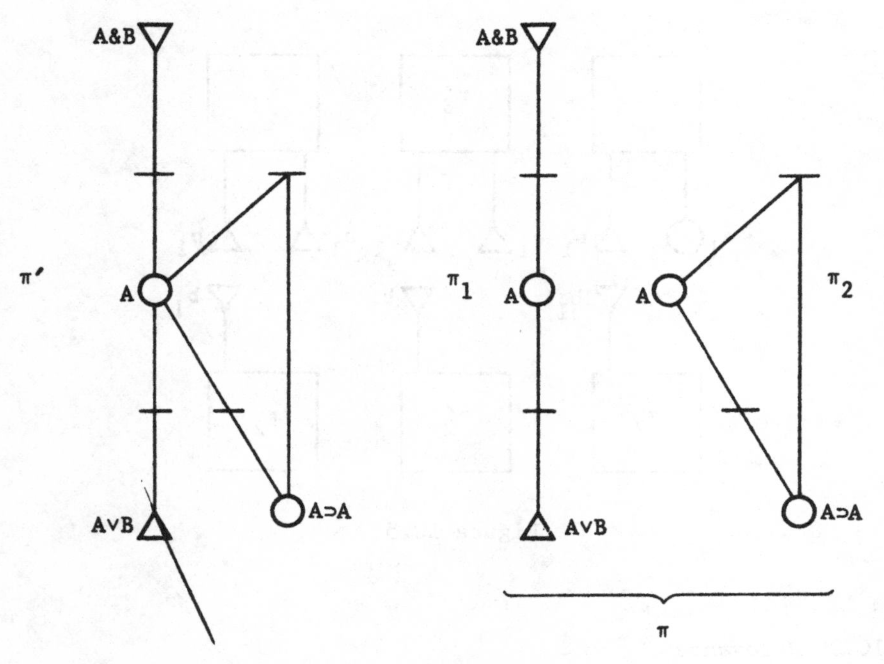


Figure 10.6

In discussing redundancy in proofs we need to distinguish three cases. First there are minimal proofs - proofs which have no proper subproofs and in which consequently every part can be claimed to make an essential contribution to the validity of the whole. Next there are proofs in which some elements are *redundant* in the sense that they belong to no minimal subproof and so their presence is never essential to any defence of the validity of the whole. (We count as elements for this purpose each vertex, each edge, and each specification of a vertex as premiss or as conclusion.) For example, in Figure 10.6 $\pi_2$ is a wholly redundant part of $\pi$, while $\pi'$ shows that even a connected proof may contain redundant elements. Between these extremes are proofs which are not minimal but in which the redundancy cannot be located exclusively in any one part, since each element belongs to some minimal subproof. A good example is Figure 8.7, in which either leg of the proof could be discarded without loss of validity and yet there is no conceivable reason for preferring one to the other.

We can draw parallel distinctions in terms of conciseness instead of minimality, and we say that an element of a proof is *irrelevant* if it belongs to no concise subproof. For example, in Figure 10.6 the entire component $\pi_2$ is irrelevant to $\pi$, and in Figure 7.6 the left leg is irrelevant to the proof. As before three cases are to be distinguished, namely concise proofs, proofs that contain irrelevant elements, and proofs which are not concise but in which no one element can be singled out as irrelevant. Irrelevance evidently implies redundancy, but an interesting example of redundancy without irrelevance is illustrated by Figure 10.7 in connection with the paradoxes of strict implication. One of these, namely the claim that if $\vdash B$ then $A \vdash B$ for arbitrary A, rests solely upon dilution (and from a multiple-conclusion point of view the same is true of the other, since e.g. $A, {\sim}A \vdash B$ comes by dilution from $A, {\sim}A \vdash \Lambda$). Dilution is customarily built into the logician's definition of proof: if $\pi_1$ is a proof of B it is counted as a fortiori a proof of B from A. It can reasonably be objected that this overlooks the distinction between premisses which are and those which are not used in a proof - that $\pi_1$ is not really a proof of B *from* A since (as we may suppose) A does not even occur in it. The construction of $\pi$ from $\pi_1$ seems to take the sting out of this objection by showing how any formula A can be made into a working premiss after all, provided we assume the classical rules for conjunction. In our terms A is not irrelevant in $\pi$; indeed $\pi$ is concise whenever $\pi_1$ is. But A is redundant, since $\pi$ has the proper subproof $\pi_1'$; and it is therefore still moot whether the construction has succeeded in its object.

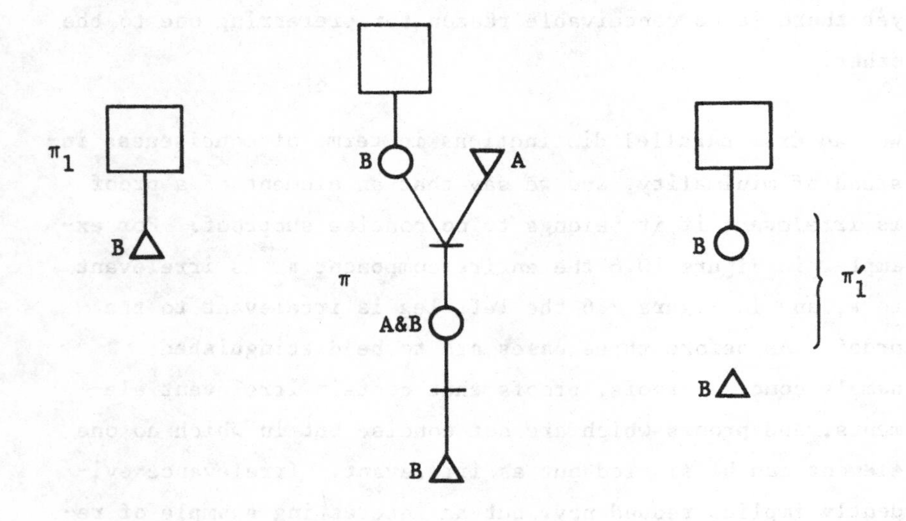


Figure 10.7

Both our previous examples of irrelevance involved nonstandard proofs, but Figure 10.8 shows a standard proof (from A&B to A∨B by the rules for PC) in which the circuit is irrelevant. It is no accident that the circuit here is not cornered, for

*Theorem 10.14*† Nothing in a standard cornered-circuit proof is irrelevant.

*Proof* Let S be the set of those subgraphs of the graph of a standard cornered-circuit proof $\pi$ which include a given edge or vertex and which have the property that every pair of distinct vertices is joined by a corner-free path. Evidently S is not empty, for it contains the graph consisting only of the given vertex, or only of the given edge together with its endpoints, as the case may be. Hence S contains a graph $\pi'$ (not necessarily unique) which is not a subgraph of any other member of S. We take the initial formula vertices of $\pi'$ as premisses and the final ones as conclusions, so that $\pi'$ is standard, and we show that (i) for each premiss (or conclusion) a of $\pi'$, every edge E of $\pi$ to (from) a is also in $\pi'$, and (ii) for each stroke a of $\pi'$, every edge E of $\pi$ adjoining a is also in $\pi'$. In each case let b be the other vertex of E. For any vertex c of $\pi'$ other than a, there is a corner-free path $\alpha$ connecting a and c in $\pi'$. If b is in $\alpha$ then either $\alpha$ contains a corner-free path connecting b and c or else b=c. If b is not in $\alpha$ then $\alpha \cup \{E,b\}$ is a corner-free path connecting b and c, for in neither case can it contain a corner at a: not in case (i) since E goes to (or from) a and $\pi'$ contains no edge to (from) a, and not in case (ii) since a is a stroke. It follows that in any case $\pi' \cup \{E,b\}$ shares with $\pi'$ the property that every pair of distinct vertices is connected by a corner-free path, and hence that $\pi' \cup \{E,b\} \in S$. So $\pi' \cup \{E,b\} = \pi'$ by our choice of $\pi'$; whence E belongs to $\pi'$, establishing (i) and (ii). But these conditions imply that $\pi'$ is an abstract subargument of $\pi$, since every premiss of $\pi'$ is initial in $\pi$ and is therefore ($\pi$ being standard) a premiss of $\pi$, and similarly for conclusions. Hence by 10.11

$\pi'$ is a concise abstract subproof of $\pi$, and a fortiori a concise subproof as required.

The ideas we have discussed in this section have abstract analogues: an element of an abstract proof is *abstractly irrelevant* if it belongs to no concise abstract subproof. The corresponding abstract analogue of redundancy is the property of belonging to no abstract subproof which itself has no proper abstract subproof, but it is evident from Theorem 7.11 that this is the same property as abstract irrelevance. Irrelevance implies abstract irrelevance and, as we have seen, redundancy, but the two latter properties are independent of one another: in Figure 10.7 the vertex A is redundant in $\pi$ but not abstractly irrelevant, while in Figure 10.9 the vertices A are abstractly irrelevant but not redundant.

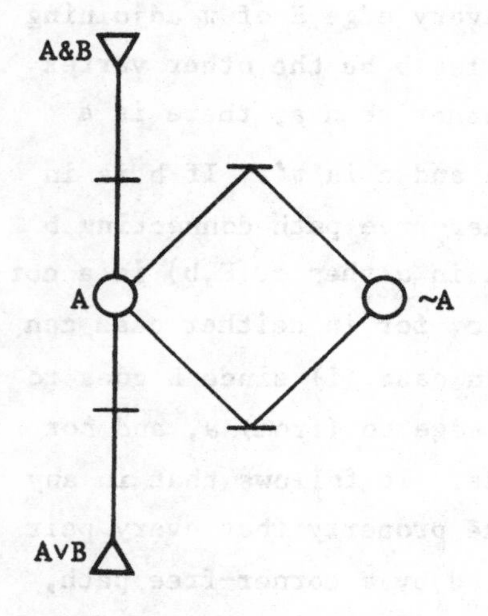


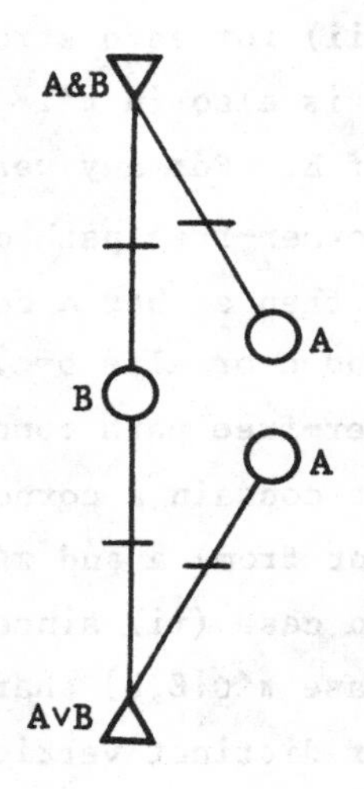


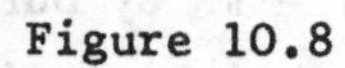

Figure 10.8 Figure 10.9

It will be seen that the proof of Theorem 10.14 establishes the stronger result that nothing is abstractly irrelevant in a standard cornered-circuit proof. On the other hand, since any non-standard vertex in an abstract proof will also be non-standard in any subargument to which it may belong, it cannot by Theorem 7.13 belong to any concise abstract subproof, and so must be abstractly irrelevant. Accordingly

*Theorem 10.15* Every non-standard abstract proof contains an abstractly irrelevant formula vertex.

## 10.4 *Articulated tree proofs*

We have already commented on the lack of specificity in the definition of tree proof, and how it permits the same tree to count as a proof from X to Y by R for quite unrelated choices of X, Y and R. We now look more closely at the relation between a tree $\tau$ of formulae and the graph arguments that embody the various ways of construing $\tau$ as a tree proof.

We say that $\pi$ is an *articulation of* $\tau$ (*qua tree proof*) if $\pi$ can be obtained by performing one of the following operations on each file $\phi$ of $\tau$: (i) specifying as a premiss the occurrence which is the sole member of $\rho(\phi)$, (ii) specifying as a conclusion the occurrence which is the final member of the branch $\phi$, (iii) adding an inference stroke and edges going from it to the members of $\rho(\phi)$ and to it from the members of a subset of $\phi$. An *articulated tree proof* is any such articulation of a tree. (The same tree may also be articulated qua extended proof etc., with different results: e.g. an articulated extended proof will not in general be an abstract proof because of its provision for repetition.) This idea of articulation is a parsimonious one, in which we supply a unique justification for each rank of the tree. By allowing more than one of the three operations to be performed on a file we would obtain a wider variety of articulations, but to be fullblooded about this would mean abandoning the requirement that premisses must be initial and conclusions final. Only the empty tree has a unique articulation but

*Theorem 10.16* If $\pi$ is any articulation of the tree $\tau$ then $\tau$ is a tree proof from $X_\pi$ to $Y_\pi$ by $R_\pi$.

*Theorem 10.17* Every articulated tree proof is a cornered-circuit abstract proof.

*Theorem 10.18* A tree of formulae is a tree proof from X to Y by R iff it has an articulation which is a graph proof from X to Y by R.

*Proofs* 10.16 is immediate from the definition of tree proof. For 10.17, if $\pi$ is an articulation of $\tau$ then by 10.16 $\tau$ is a tree proof from $X_\pi$ to $Y_\pi$ by $R_\pi$, so that $X_\pi \vdash_{R_\pi} Y_\pi$ by 3.5 and hence $\pi$ is valid. Similarly any argument abstractly isomorphic to $\pi$ is an articulation of a tree isomorphic to $\tau$, and so is valid too. Thus $\pi$ is an abstract proof. It follows that $\pi$ is nonempty by 7.4 and so it remains only to show that every circuit $\gamma$ in $\pi$ is cornered. With each vertex b of $\pi$ we associate a file $\phi(b)$: if b is a stroke this is to be the file of $\tau$ with respect to which b was introduced in the articulation, and if b is a formula vertex it is to be the file of $\tau$ that ends in b. If an edge goes from $a_1$ to $a_2$ then (1) $\phi(a_1)$ is a subfile of $\phi(a_2)$ and (2) if moreover $a_1$ is a stroke, $\phi(a_2)$ is the file $\phi(a_1), a_2$. Given any nonempty subgraph $\pi'$ there is a file - call it $\phi(\pi')$ - which is $\phi(b)$ for some vertex b in $\pi'$ although none of its proper subfiles are of this form. By (1), if $\pi'$ is connected then (3) $\phi(\pi')$ is a subfile of $\phi(b)$ for every b in $\pi'$. Suppose that (4) the only vertex of $\gamma$ with which $\phi(\gamma)$ is associated is a stroke b; then by (1) neither of the two edges E and E′ adjoining b in $\gamma$ can go to b, and so they go from b to distinct vertices c and c′; but then by (2) $\phi(c)$ is $\phi(b)$,c and $\phi(c')$ is $\phi(b)$,c′, so that by (3) $\phi(\gamma-\{E,b,E'\})$ is a subfile of $\phi(\gamma)$, which contradicts (4). So $\phi(b) = \phi(\gamma)$ for some formula vertex b of $\gamma$, and by (2) there is no edge to b in $\gamma$. Hence $\gamma$ has a corner at b as required.

For 10.18, if $\tau$ has an articulation $\pi$ which is a proof from X to Y by R then $X_\pi \subset X$ and $Y_\pi \subset Y$ and $R_\pi \subset R$, and so by 10.16 $\tau$ is a proof from X to Y by R. Conversely, if $\tau$ is a tree proof from X to Y by R it is possible to perform at least one of the following operations on each file $\phi$: (i) specifying as a premiss the occurrence of a formula in X which is the sole member of $\rho(\phi)$; (ii) specifying as a conclusion the occurrence of a formula in Y which is the final member of the branch $\phi$; (iii) adding an inference stroke and edges going from it to the members of $\rho(\phi)$ and to it from the members of a subset of $\phi$, these being respectively the conclusions and premisses of an instance of R. Let $\pi$ be the result of performing just one of these operations on each file. Then $\pi$ is an articulation of $\tau$ and so valid by 10.17, and it is easy to verify that $\pi$ is an argument from X to Y by R.

Every initial formula vertex in an articulated tree proof must be a premiss, but such proofs need not be standard or even connected. For example, $\tau$ in Figure 10.10 has the non-standard and disconnected articulation $\pi_2$ as well as the expected classical one $\pi_1$. We leave it to the interested reader to formulate necessary and sufficient conditions for a graph argument to be an articulated tree proof, an articulated extended tree proof, etc.

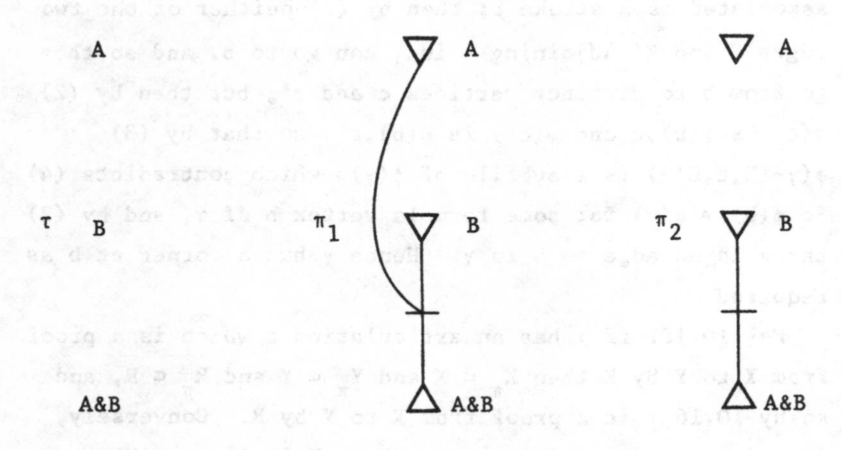


Figure 10.10

## 10.5 *Abstract proofs in general*

Cornered-circuit arguments do not exhaust the abstract proofs. We have already seen an example of this in Figure 10.8, where an irrelevant corner-free circuit was added to a cornered-circuit proof. In contrast Figure 10.11 shows two proofs in which corner-free circuits play an essential part. $\pi_1$ is a concise abstract proof from A∨B,B∨C,C∨A to A&B,B&C,C&A by the rules for PC, and similar proofs from $A_1 \vee A_2, \ldots, A_{n-1} \vee A_n, A_n \vee A_1$ to $A_1 \& A_2, \ldots, A_{n-1} \& A_n, A_n \& A_1$ can be devised for any odd number of formulae; the inference is invalid if n is even. This family of examples is particularly interesting because they are entirely built up of instances of the very paradigm whose absurdity was cited by Kneale as a reason for banning circuits altogether (compare $\pi_1$ with Figure 8.4). The other example, $\pi_2$, is a concise abstract proof from A∨(A&A) to A&(A&A) by the same rules; it actually contains a directed circuit, which in isolation exemplifies the gross fallacy of circularity.

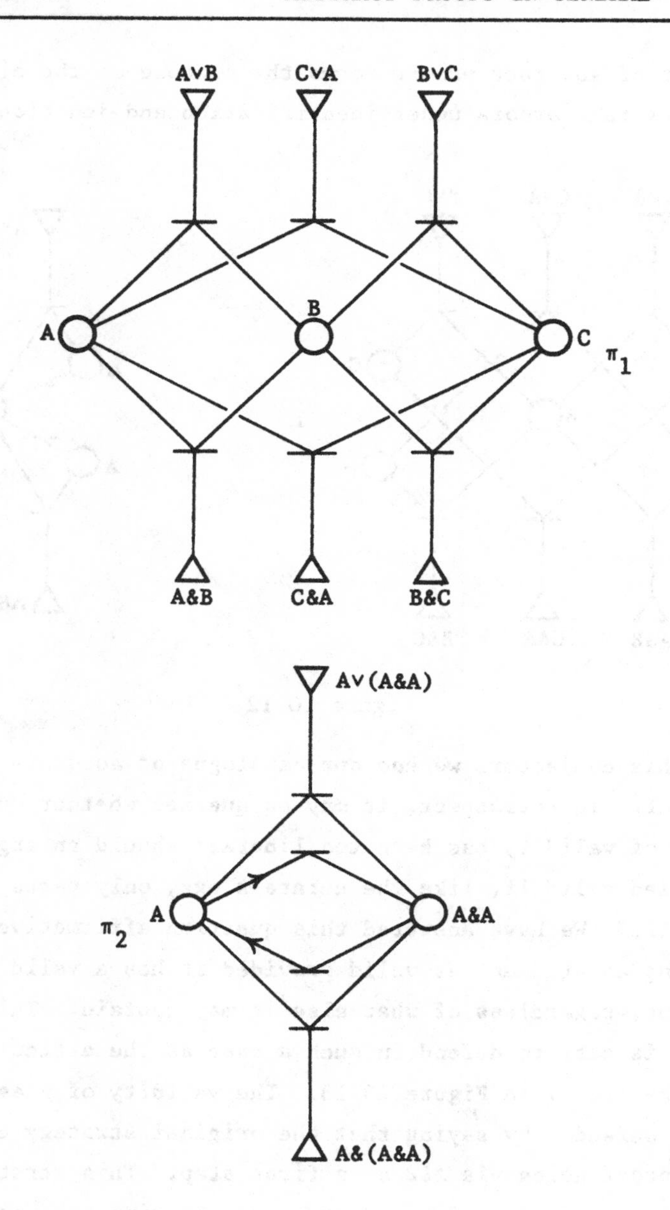


Figure 10.11

Both $\pi_1$ and $\pi_2$ can be obtained from cornered-circuit abstract proofs $\pi_1'$ and $\pi_2'$ by identification, $\pi_1$ in two steps and $\pi_2$ in one, as Figure 10.12 shows; and we conjecture that in general the set of abstract proofs forms the closure of the circuit-free abstract proofs under identification and junction.

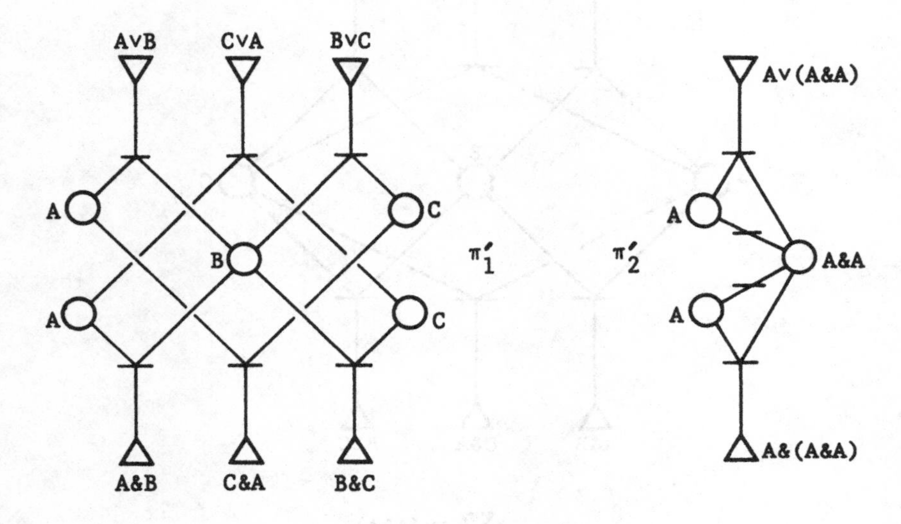


Figure 10.12

With this conjecture we end our catalogue of adequate classes of proof. In retrospect, it may be queried whether our definition of validity has been too liberal: should an argument be called valid if, like the curate's egg, only parts of it are valid? We have answered this question affirmatively, counting an argument as valid provided it has a valid sub-argument, regardless of what else it may contain. This answer is easy to defend in such a case as the articulation $\pi$ of the tree $\tau$ in Figure 10.13. The validity of $\pi$ as a whole can be defended by saying that the original strategy called for a proof going via A&B as a first step. This strategy was abandoned in favour of a direct proof, leaving the uncompleted workings; and what could be the harm in that? Although the leg from (A&B)&A to A&B is invalid in isolation (because it has no conclusion) and irrelevant in $\pi$, its innocuousness is plain from the existence of plenty of proofs in which it plays an operative part: for example, the whole of $\pi$ can be embedded in $\pi'$, in which the leg to A&B is no longer irrelevant or even redundant.

It might be expected that on the other hand directed circuits and Kneale's 'absurd patterns' would be incorrigibly fallacious - invalid as they stand, and incapable of featuring relevantly in any valid argument. Our previous discussion in this section confounds this expectation. The irrelevant circuit in Figure 10.8, for example, could be defended as the abandoned workings of a projected proof on the lines of $\pi_1$ of Figure 10.11, and in general the presence of directed circuits can be defended in a similar way with reference to $\pi_2$ of the same figure. The only candidate we have found for an incorrigibly fallacious pattern of argument is the smallest circuit of all, shown in Figure 10.14.

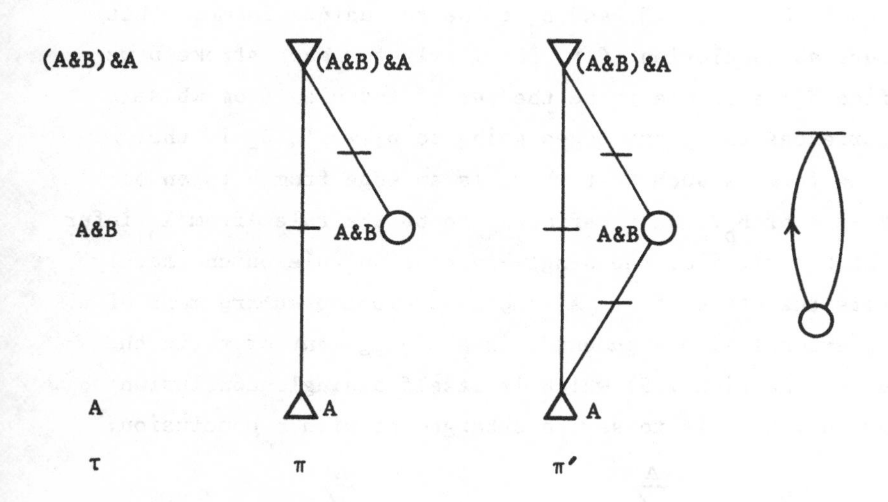


Figure 10.13

Figure 10.14

# 11 · Single-conclusion proofs

## 11.1 *Single-conclusion arguments*

By a *single-conclusion argument* we mean a graph argument in which exactly one edge goes from each inference stroke, and exactly one formula vertex is specified as a conclusion. To develop a self-contained treatment of single-conclusion arguments we need to define appropriate analogues of the ideas of Chapter 7. Given a single-conclusion argument $\pi$ we define $X_\pi$ to be the set of formulae that occur as premisses of $\pi$, as in Section 7.3, and $B_\pi$ to be the unique formula that occurs as conclusion of $\pi$. Similarly for each stroke b we define $X_b$ as before to be the set of formulae from whose occurrences there are edges going to b, while $B_b$ is the unique formula such that there is an edge from b to an occurrence of $B_b$. We redefine $R_\pi$ to be the rule 'from $X_b$ infer $B_b$ if $b \in \pi$', i.e. the single-conclusion rule which exactly covers the steps of $\pi$. A *single-conclusion subargument* of a single-conclusion argument $\pi$ is a subargument of $\pi$ (in the sense of Section 7.5) which is itself a single-conclusion argument; that is to say, a subargument with a conclusion.

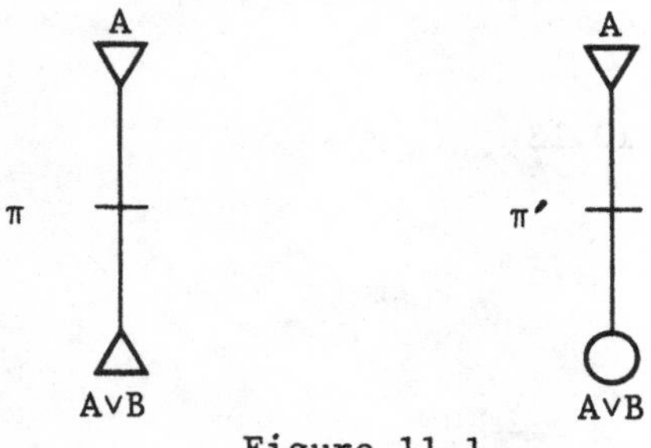


Figure 11.1

Similarly a *single-conclusion abstract subargument* of $\pi$ is an abstract subargument of $\pi$ with a conclusion. Thus $\pi'$ of Figure 11.1 is a (multiple-conclusion) abstract subargument of $\pi$, but not a single-conclusion subargument.

The remaining definitions of Chapter 7 are reworked in a similar way. Thus a single-conclusion argument $\pi$ from X to B is *valid* (or is a *proof of* B *from* X) if every partition that satisfies $R_\pi$ satisfies $\langle X_\pi, B_\pi \rangle$, i.e. if $X_\pi \vdash_{R_\pi} B_\pi$, and is *abstractly valid* (or is an *abstract proof of* B *from* X) if every single-conclusion argument abstractly isomorphic to it is valid. A *subproof* of a single-conclusion proof $\pi$ is a valid single-conclusion subargument of $\pi$, and an *abstract subproof* is an abstractly valid single-conclusion abstract subargument. A single-conclusion proof is *minimal* if it has no proper subproofs, and *concise* if it conforms to a minimal proof. There is no need to add a distinguishing adjective 'single-conclusion' to these definitions, since it is readily verified that for single-conclusion arguments each of the terms defined is extensionally equivalent to the corresponding multiple-conclusion one. In particular, every subproof (valid subargument) $\pi'$ of a single-conclusion proof $\pi$ is necessarily a single-conclusion argument and consequently is a subproof in the present sense too. For since $\pi'$ is valid $X_{\pi'} \vdash_{R_{\pi'}} Y_{\pi'}$ (where $R_{\pi'}$ is defined as a multiple-conclusion rule, as in Section 7.3), but since $\pi'$ is a subargument of $\pi$ no instance of $R_{\pi'}$ has zero conclusions and so $X_{\pi'} \nvdash_{R_{\pi'}} \Lambda$ by Theorem 5.26. Hence $Y_{\pi'} \neq \Lambda$, i.e. $\pi'$ has a conclusion.

The treatment of graph proofs in Sections 7.3-5 can accordingly be carried over wholesale to the single-conclusion case. For example, in place of Theorem 7.3 ($\pi$ is valid iff $X \vdash_R Y$ whenever $\pi$ is an argument from X to Y by R) we have that a single-conclusion argument $\pi$ is valid iff $X \vdash_R B$ whenever $\pi$ is an argument from X to B by R; and in place of Theorem 7.8 (if $\pi$ is a proof or abstract proof, so is any argument obtained by omitting edges from $\pi$) we have that if $\pi$ is a single-conclusion proof or abstract proof, so is any single-conclusion argument obtained by omitting edges from $\pi$; and so on. The only result that calls for any comment here is Theorem 7.4 (every graph proof is nonempty) whose analogue is trivial since single-conclusion arguments, unlike multiple-conclusion ones, are nonempty by definition.

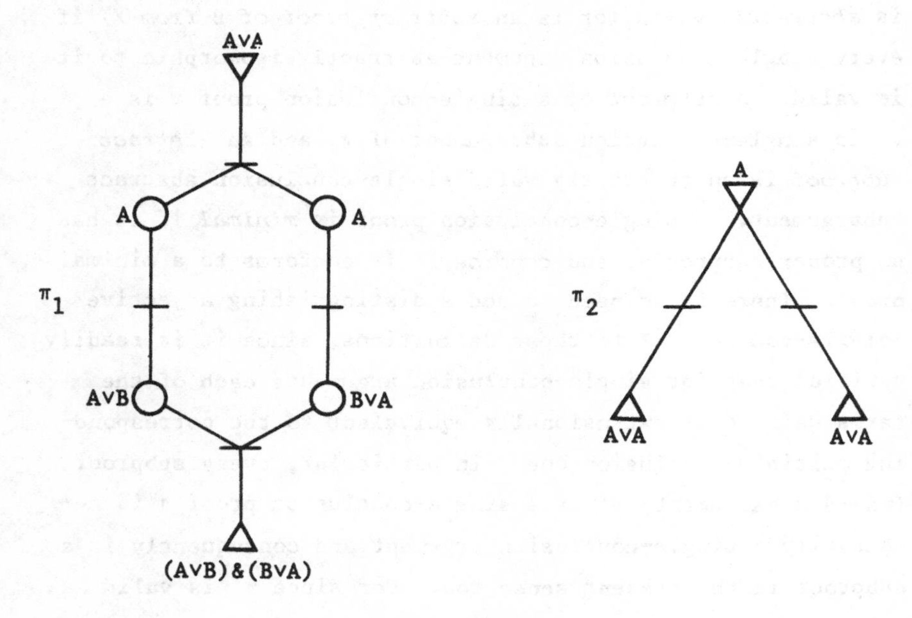


Figure 11.2

We pursue the separate treatment of single-conclusion graph proofs in the following sections, but we digress here to consider single-conclusion arguments as a special case in the context of multiple-conclusion ones. Even when all the rules in a multiple-conclusion calculus have singleton conclusions there will still be proofs by those rules which are not single-conclusion arguments. Figure 11.2 illustrates the point, $\pi_1$ having more than one edge going from a stroke, and $\pi_2$ having more than one vertex as a conclusion. We have already come across a similar situation in connection with tree proofs and sequence proofs in Section 3.1, and the following result is the analogue for graph proofs of Theorem 3.1.

*Theorem 11.1* Every proof (or abstract proof) of B from X by a single-conclusion rule R is a proof (an abstract proof) from X to {B} by the corresponding singleton-conclusion rule $R_1$; and every proof (or abstract proof) from X to Y by $R_1$ has a subproof which is a proof (an abstract proof) of B from X by R for some B in Y.

*Proof* It is immediate from the definitions that a proof (or abstract proof) of B from X by R is a proof (abstract proof) from X to {B} by $R_1$. Conversely, by 8.7 every proof from X to Y by $R_1$ has a simple subproof $\pi$. Every subproof of $\pi$ shares with $\pi$ the property that just one edge goes from each stroke, and it therefore only remains to show that $\pi$ has a subproof with just one conclusion. Let $a_1,\ldots,a_n$ be the vertices specified as conclusions of $\pi$. By 7.5, $\pi$ has a final standard vertex, and as this cannot be a stroke it must be a conclusion, so that $n\geq1$. For each i ($1\leq i\leq n$) let $\pi_i$ be obtained from $\pi$ by omitting all of $a_1,\ldots,a_n$ except $a_i$, together with the edges to them, the strokes from which these edges go, and all edges to these strokes. If any $\pi_i$ is valid it is the required subproof. Otherwise there is for each i a partition $\langle T_i,U_i\rangle$ which invalidates $\pi_i$, and we obtain a contradiction by showing that $\langle \bigcap T_i, \bigcup U_i\rangle$ invalidates $\pi$. Evidently each conclusion $a_j$ is an occurrence of a formula in $U_j$, and hence in $\bigcup U_i$. Each premiss of $\pi$ is distinct from each $a_i$ (since otherwise $\pi_i$ would contain a vertex specified simultaneously as premiss and conclusion, and hence be valid) and so occurs in each $\pi_i$ and must be an occurrence of a formula in $\bigcap T_i$. It remains to show that for each stroke c, if $X_c \subset \bigcap T_i$ then $B_c \in \bigcap T_i$. Let b be the relevant occurrence of $B_c$. If $b=a_j$ for some j then c is in $\pi_j$, and since $\langle T_j,U_j\rangle$ invalidates $\pi_j$ and $B_c \in U_j$, some member of $X_c$ is in $U_j$ and so not in $\bigcap T_i$. On the other hand if b is distinct from $a_1,\ldots,a_n$ then c is in each $\pi_i$. Since $\langle T_i,U_i\rangle$ invalidates $\pi_i$ it follows that if $X_c \subset T_i$ then $B_c \in T_i$ and hence if $X_c \subset \bigcap T_i$ then $B_c \in \bigcap T_i$ as required.

Similarly an abstract proof from X to Y by $R_1$ has a simple abstractly valid subproof $\pi$ by 8.7, and the abstract result follows by applying the formal one to an argument abstractly isomorphic to $\pi$ in which no formula occurs more than once, and using 7.7.

In the statement of the theorem we use the phrase 'subproof which is ... an abstract proof', and the theorem would fail if we had simply said 'abstract subproof', as the concise proof $\pi_1$ of Figure 11.2 shows. Although the property of being a single-conclusion argument is an abstract one, the property of being an argument by singleton-conclusion rules is not.

## 11.2 *Proofs with circuits or corners*

In Chapters 9 and 10 we introduced proofs containing circuits or corners or both, and it can be verified that all the theorems concerning them carry over to the single-conclusion case with the obvious adaptations. The scope of the theory is however drastically simplified, largely because

*Theorem 11.2* Every circuit in a single-conclusion argument is either directed or cornered (with a corner at a vertex initial in it).

*Proof* By 7.2, some vertex b is initial in any non-directed circuit $\gamma$ of a single-conclusion argument $\pi$, so there are two edges from b in $\gamma$. But $\pi$ contains only one edge from each stroke, so b is a formula vertex and $\gamma$ is cornered as required.

We can see the effect of this at several places in the exposition. Thus we demonstrated in Section 9.3 the existence of a whole class of corner-free proofs containing circuits, defining a cross-referenced-circuit argument to be a connected nonempty corner-free argument in which every circuit contains a cross-reference path. But a directed circuit can only contain directed paths, and in particular cannot contain a cross-reference path. Hence by Theorem 11.2 the only cross-referenced-circuit single-conclusion arguments are of the trivial variety in which there are no circuits at all.

Again, having introduced cornered-circuit proofs in Chapter 10, we demonstrated the existence of abstract proofs in which corner-free and even directed circuits play an essential part (Figure 10.11). No such proofs, abstract or otherwise, exist in the single-conclusion case:

*Theorem 11.3* Every single-conclusion proof has a cornered-circuit subproof.

*Theorem 11.4* Every concise single-conclusion proof is a cornered-circuit argument.

*Proofs* For 11.3, let $\pi$ be a minimal subproof of the given one. If $\pi$ is not a cornered-circuit argument then by 11.2 it has a directed circuit $\gamma$. For each stroke b in $\gamma$ let $\pi$-b be the subargument obtained by omitting b and the edges adjoining it. Since $\pi$ is minimal $\pi$-b is invalidated by a partition $\langle T_b, U_b \rangle$, and since $\pi$ is valid $B_b \in U_b$. But every formula which occurs in $\gamma$ is $B_b$ for some b in $\gamma$ and so belongs to $\bigcup U_b$. For each b in $\gamma$ some member of $X_b$ occurs in $\gamma$ and so is in $\bigcup U_b$ as we have seen. Hence $\langle \bigcap T_b, \bigcup U_b \rangle$ satisfies $\langle X_b, B_b \rangle$ for each b in $\gamma$, though it invalidates $\langle X_\pi, B_\pi \rangle$ since each $\langle T_b, U_b \rangle$ does so. Since $\pi$ is valid it follows that $\langle \bigcap T_b, \bigcup U_b \rangle$ invalidates $\langle X_c, B_c \rangle$ for some stroke c not in $\gamma$. So $B_c \in U_b$ for some b in $\gamma$ while $X_c \subset T_b$. Hence $\langle T_b, U_b \rangle$ satisfies $\pi$-b, contrary to hypothesis.

For 11.4 we note that by 11.3 every minimal single-conclusion proof, and hence every proof that conforms to one, is a cornered-circuit argument.

Finally, when discussing non-abstract proofs with circuits in Section 9.4, we made the obvious point that a non-standard proof can be concise but disconnected; but we also gave an example to show that the same is true even for standard proofs (Figure 9.18). No such examples are to be found in the single-conclusion case, for by Theorem 11.4 every concise single-conclusion proof is a cornered-circuit argument, and

*Theorem 11.5*† Every standard cornered-circuit single-conclusion argument is connected.

*Proof* Each of the components of a cornered-circuit single-conclusion argument $\pi$ has a final vertex by 7.2, and this must be a formula vertex since $\pi$ is a single-conclusion argument. But $\pi$ has only one conclusion, so that at most one of its components is standard; i.e. $\pi$ itself is either connected or non-standard.

Although every standard proof has a cornered-circuit subproof by Theorem 11.3, it cannot be inferred from Theorem 11.5 that this subproof is connected or even located exclusively in one component. This is shown in Figure 11.3, which depicts a standard proof that has no connected subproofs at all.

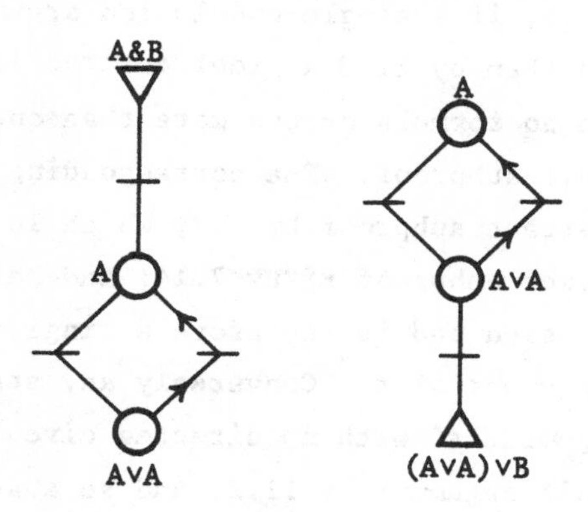


Figure 11.3

Simple syntactical criteria for a single-conclusion argument to be an abstract proof, or a concise abstract proof, emerge as corollaries of these results:

*Theorem 11.6*† A single-conclusion argument is abstractly valid iff it has a standard single-conclusion abstract subargument containing no directed circuits.

*Theorem 11.7*† A single-conclusion argument is a concise abstract proof iff it is standard, it contains no directed circuits, and any residue is connected.

*Theorem 11.8*† A single-conclusion argument is a concise abstract proof iff it is standard, it contains no directed circuits, and every corner is the unique corner in some circuit.

*Theorem 11.9*† A single-conclusion argument is a concise abstract proof iff it is standard, it contains no directed circuits, and every pair of distinct vertices is connected by a corner-free path.

*Proofs* For 11.6, if a single-conclusion argument $\pi$ is abstractly valid then by 11.3 a proof abstractly isomorphic to it in which no formula occurs more than once has a cornered-circuit subproof. The corresponding subargument of $\pi$ is an abstract subproof by 7.7, which in turn has a standard abstract subproof $\pi'$ by 7.14; and being valid, $\pi'$ has a conclusion and is therefore a single-conclusion abstract subargument of $\pi$. Conversely any standard single-conclusion argument $\pi'$ with no directed circuits is a cornered-circuit argument by 11.2, and so abstractly valid by 10.3. So any single-conclusion argument having $\pi'$ as an abstract subargument is abstractly valid.

11.7-9 are corollaries of 10.9-11, since by 11.2 and 11.4 both concise proofs and arguments containing no directed circuits are necessarily cornered-circuit ones, and, for 11.8, a standard argument without directed circuits is connected by 11.5.

## 11.3 *Articulated sequence proofs*

To provide the analogue for sequence proofs of the treatment of articulated tree proofs in Section 10.4 we need to ensure that an articulation of a sequence is a single-conclusion argument. Let $\sigma$ be a sequence $C_1,\ldots,C_k$; then an *articulation* of $\sigma$ qua sequence proof is obtained by specifying $C_k$ as conclusion and performing one of the following operations for each j: (i) specifying $C_j$ as a premiss or (ii) adding an inference stroke and an edge going from it to $C_j$ and edges going to it from the members of a subset of $C_1,\ldots,C_{j-1}$. Any such articulation is an *articulated sequence proof*. It is easily verified that the results of Section 10.4 carry over to the present case, and in particular only an abstract proof can be an articulated sequence proof. On the other hand, in contrast to Figure 10.11, the articulated sequence proofs do exhaust the concise abstract ones. We are moreover able to give a simple syntactical criterion for an argument to be an articulated sequence proof.

*Theorem 11.10* Every articulated sequence proof is a cornered-circuit abstract proof, and a sequence of formulae is a sequence proof of B from X by R iff it has an articulation which is a graph proof of B from X by R.

*Theorem 11.11*† A single-conclusion argument is an articulated sequence proof iff it has no directed circuits, every initial formula vertex is a premiss, and at most one edge goes to each formula vertex.

*Theorem 11.12* Every concise single-conclusion abstract proof is an articulated sequence proof.

*Proofs* The proof of 11.10 follows those of 10.17-18. For 11.11, it is evident that every articulated sequence proof has the properties cited in the theorem. (In particular, it is an articulated tree proof and so has no directed circuits by 10.17.) Conversely, if a single-conclusion argument $\pi$ has no directed circuits then by 7.1 the ancestral of the edge relation is a partial ordering, and by a classical result this may be embedded in a total ordering by means of which the formula vertices in particular may be arranged in a sequence $\sigma$. If also every formula vertex is either a premiss or (not being initial) has exactly one edge to it, then $\pi$ is an articulation of $\sigma$ as required.

The final clause in the statement of 11.11 would need

to be omitted if the definition of articulation were widened to allow multiple justification by allowing operation (ii) to be performed more than once on the same $C_j$. Justification by (i) and (ii) simultaneously is excluded by the requirement that premisses must be initial.

For 11.12, a concise abstract proof is standard and has no directed circuits, by 11.8. If more than one edge goes to some formula vertex, then by 11.8 the resulting corner is the unique corner in some circuit, contrary to 11.2. The result follows by 11.11.

When discussing symmetry in Section 7.6 we remarked that it was hopeless to try to construct a converse proof from a tree of formulae by simply reversing its ordering, since in general the result is not even a tree. By reversing the order of a finite sequence we do obtain another sequence, but this will still not be a converse proof since the converses of single-conclusion rules of inference are not single-conclusion rules, and for the same reason the converse of a single-conclusion graph argument is generally not a single-conclusion argument. If all the instances of a rule R have a single premiss, however, it is possible to define a single-conclusion rule which is in effect a converse of it, namely 'from {B} infer A if {A} R B'. This suggests an alternative way of making single-conclusion logic homogeneous: instead of broadening it to include multiple and zero conclusions, we might narrow it by excluding multiple and zero premisses.

The resulting *single-premiss single-conclusion calculi* will be ones in which consequence is a binary relation between formulae. Dilution has no part to play here, and overlap and cut become straightforward reflexivity ($A \vdash A$) and transitivity (if $A \vdash B$ and $B \vdash C$ then $A \vdash C$). A binary relation will thus be a consequence relation iff it is reflexive and transitive. Rules of inference will be of the form 'from A infer B if A R B', and $\vdash_R$ becomes the ancestral of R (so that Theorem 1.9 suggests an alternative definition of the ancestral of a relation, namely as its closure under reflexivity and transitivity).

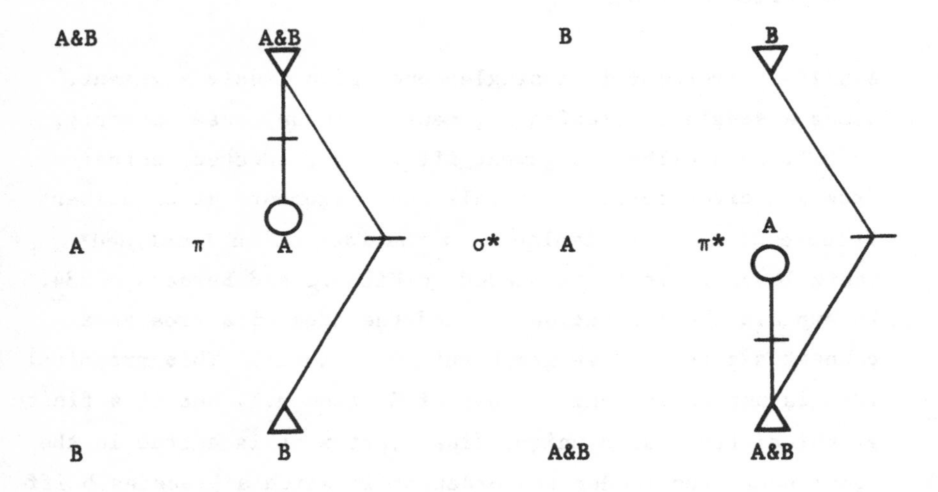


Figure 11.4

But even in the symmetrical context of a single-premiss single-conclusion calculus, the definition of sequence proof still does not permit the construction of a converse proof by reversing the ordering of a sequence of formulae. The reason is that the definition allows blind alleys to be explored on the way to the conclusion, in an unsymmetrical manner. For example, the single-premiss part of the classical calculus of conjunction is characterised by the rule 'from $A_1$&...&$A_m$ infer $B_1$&...&$B_n$ if each $B_j$ is an $A_i$ (irrespective of how the brackets are restored to form the conjunctions)'. Consequently the sequence σ shown in Figure 11.4 is a proof of B from A&B, and π is an appropriate articulation of it (cf. Figure 7.6). But the converse sequence σ* is not a proof at all in the converse calculus unless A happens to follow from B, and the converse graph proof $\pi^*$ is not an articulated sequence proof under any circumstances, by Theorem 11.11.

## 11.4 *Hilbert proofs*

A *Hilbert argument* is a single-conclusion Kneale argument. Since a single-conclusion argument is in any case nonempty, it will be a Hilbert argument iff it is connected, corner-free and circuit-free. We call such arguments after Hilbert because of their resemblance to the idea of an 'analysed' (aufgelösten) proof, expounded in Hilbert and Bernays, 1934. To explain the connection we need the idea of a *tree* as a connected circuit-free graph (cf. Ore, 1962). This graphical idea is not equivalent to that of Section 3.1, but if a finite graphical tree has a unique final vertex it is a tree in the other sense too (under the ordering in which a precedes b iff there is a directed path from b to a). This remark applies to Hilbert arguments, by the next theorem, but note that whereas branching in the tree proofs of Section 3.1 indicates the application of a rule with multiple conclusions, branching in a Hilbert argument indicates the application of a rule with multiple premisses. We also need the following results about Hilbert arguments.

*Theorem 11.13* Every Hilbert argument is a tree whose conclusion is its unique final vertex.

*Theorem 11.14*† A Hilbert argument is an articulated sequence proof iff it is standard.

*Theorem 11.15*† A Hilbert argument is an abstract proof iff it is standard.

*Proofs* For 11.13, let a be the conclusion of a Hilbert argument and let b be any other vertex. By 7.2, some vertex c is initial in the path connecting a and b. But c must be an endpoint, for otherwise it can neither be a stroke (since only one edge goes from each stroke in a single-conclusion argument) nor a formula vertex (since there are no corners in a Hilbert argument). It is not a, since there are no edges from a, and so it must be b, which is consequently not final in the argument. 11.14 is immediate from 11.11 and 11.13, and 11.15 is immediate from 8.3.

Hilbert and Bernays analyse only proofs of theorems, i.e. proofs without premisses. Each formula in such a sequence proof may be inferred by a rule (possibly in more than one way) from a set of preceding ones. We begin by transcribing the conclusion of the proof together with any one of the sets from which it may be inferred, and insert edges going to the conclusion from each member of the set in question. The process is repeated for each of these members in turn and so on, working backwards until it can be carried no further, the result being what Hilbert and Bernays call an *Auflösungsfigur*. Thus Figure 11.5 shows an Auflösungsfigur derived from the sequence A,B,C,D,E,F considered as a proof of F in a calculus with axioms A, B, C and rules ‘from B infer D’, ‘from C infer D’, ‘from D infer E’ and ‘from D,E infer F’.

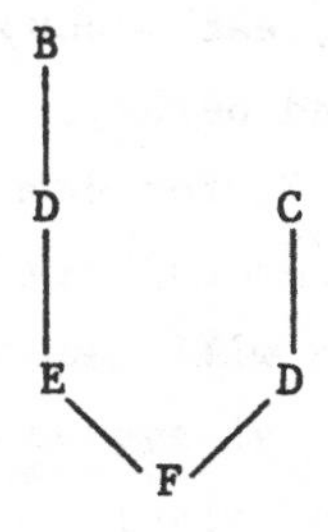


Figure 11.5

It follows from this account that an Auflösungsfigur is tantamount to a graph of formulae in the shape of a tree with the conclusion of the proof as its unique final vertex. Indeed this description can be used to define Auflösungsfiguren as an independent variety of proofs, though to avoid the ambiguity noted in connection with Figure 7.4 it is necessary to make the arbitrary stipulation that the formulae from which edges go to any one vertex shall invariably be construed as joint premisses of a single step. (The definition can also be extended to allow Auflösungsfiguren with premisses, though they will have an ambiguous structure unless a notation is introduced to distinguish premisses from axioms.)

It will now be obvious from Theorems 11.13 and 11.15 that every Auflösungsfigur can be transformed into an equivalent standard Hilbert proof from $\Lambda$ merely by the interpolation of inference strokes, the transformation being reversed by assimilating each stroke to the succeeding formula vertex. Despite this equivalence, however, and despite Theorem 11.14, a Hilbertian 'analysis' of a sequence proof is not to be equated with an articulation in the sense of Section 11.3. For if the sequence contains digressions - inferential trails that do not reach the conclusion - they will never be taken up in the process of constructing an Auflösungsfigur; while if one and the same occurrence of a formula is a premiss in a number of steps it will be replaced in the Auflösungsfigur by a number of occurrences, each occupying a different vertex, each used once only and perhaps inferred from different antecedents. In Figure 11.5, for example, A has been discarded and D duplicated. Accordingly an Auflösungsfigur extracted from a sequence $\sigma$ will generally not be an articulation of $\sigma$, though it will always be an articulation of some sequence obtained from $\sigma$ by omission of formulae and repetition of subsequences.

The theory of Kneale proofs in Sections 8.1-2 carries over wholesale to Hilbert proofs. The picture is further simplified by the fact that in the single-conclusion case (by contrast with Figure 9.17) every connected corner-free argument is necessarily circuit-free too. Thus although the Hilbert arguments do not exhaust the connected circuit-free arguments (Figure 9.19), they do exhaust the connected corner-free ones:

*Theorem 11.16* Every connected corner-free single-conclusion argument is a Hilbert argument.

*Proof* Let $\pi$ be a connected corner-free single-conclusion argument, so that (1) at most one edge goes from each vertex of $\pi$. Any circuit $\gamma$ in $\pi$ is a directed circuit by 11.2, and the conclusion b of $\pi$ is not in $\gamma$ since it adjoins only one edge. Since $\pi$ is connected, there is a path connecting b and each vertex in $\gamma$, and we choose $\alpha$ to be one of the shortest of these paths. By 7.2 some vertex c is initial in $\alpha$, and by (1), c must be the endpoint of $\alpha$ in $\gamma$. But $\{c\} = \alpha \cap \gamma$ by our choice of $\alpha$, and both $\alpha$ and $\gamma$ include an edge from c, contradicting (1).

On the other hand the negative results of Chapter 8 - the inadequacy of Kneale proofs and hence of circuit-free and of corner-free abstract proofs in general - are not reproduced in the single-conclusion case. For the restriction to one conclusion-vertex removes the threat of an infinite regress from the attempt to establish closure under cut by means of junction (Section 9.1): the distinction between junction and multiple junction vanishes and by analogy with Theorem 9.9 we have

*Theorem 11.17* Standard Hilbert proofs are adequate for single-conclusion rules.

*Proof* Let X ⊢ B mean that there is a standard Hilbert argument from X to B by R. As in 9.1 ⊢ contains R and is closed under overlap and dilution, and by joining replicas of an argument from X to A onto one from X,A to B at the premisses at which A occurs one sees that it is closed under cut for formulae too. Hence $\vdash_R \subset \vdash$ by 1.11, and conversely $\vdash \subset \vdash_R$ by 11.15.

Although the abstract Hilbert proofs are all standard, there are other Hilbert proofs which are not, notably the cross-referenced ones. Their definition is a special case of the definition of a cross-referenced Kneale proof given in Section 9.2. The vertices to be justified by cross-reference will all be initial, since by Theorem 11.13 there can be no non-standard final vertices in a Hilbert argument; but Figure 9.7 shows that there is the same need as in the multiple-conclusion case to arrange the cross-references in a suitable order if fallacies are to be avoided. The results of Sections 9.1-2 carry over to cross-referenced Hilbert proofs, but there is a further result which reflects the adequacy of the standard Hilbert proofs, and whose multiple-conclusion analogue therefore fails:

*Theorem 11.18* Every cross-referenced Hilbert proof of B from X by R is an abstract subargument of a standard Hilbert proof of B from X by R.

*Proof* Arguing as for 9.5, the class of cross-referenced Hilbert proofs is the closure of the standard Hilbert arguments under junction and election. It is sufficient to show that this is the class S of cross-referenced Hilbert proofs π such that whenever π is a proof of B from X by R, it is an abstract subargument of a standard Hilbert proof of B from X by R; i.e. to show that (i) S contains the standard Hilbert arguments and that S is closed under (ii) junction and (iii) election. But (i) is immediate, and for (iii) we need only observe that if $\pi'$ is obtained by election from $\pi$, then it is an abstract subargument of $\pi$. For (ii), suppose that $\pi$ is obtained by joining proofs $\pi_1$ and $\pi_2$ in S whose common vertex a is an occurrence of A which is the conclusion of $\pi_1$ and a premiss of $\pi_2$. Suppose that $\pi$ is a proof of B from X by R. Then $\pi_1$ is a proof of A from X by R, and $\pi_2$ of B from X,A by R. Moreover $\pi_1$ and $\pi_2$ are abstract subarguments of standard Hilbert proofs $\pi_1'$ and $\pi_2'$ of A from X and of B from X,A respectively. Let $\pi'$ be obtained by joining a replica of $\pi_1'$ at each occurrence of A as a premiss of $\pi_2'$, reserving $\pi_1'$ itself to join at a. Then $\pi'$ is the required standard Hilbert proof of B from X by R, containing $\pi$ as an abstract subargument.

Cross-referenced Hilbert proofs do not exhaust the Hilbert proofs, as the example $\pi$ in Figure 9.15 shows; but there remains the interesting possibility that, provided multiplicity of occurrence is taken into account in rules of inference (cf. Section 9.4), the cross-referenced Hilbert proofs may be exactly the concise ones.

# 12 · Infinite proofs

## 12.1 *Infinite arguments*

The previous exposition of graph proofs has assumed that all rules R are finite, i.e. that X R Y only if X and Y are finite. Following the pattern of Chapter 6, we consider now the effects of allowing rules of inference with infinitely many premisses and conclusions, and consequently abandoning the requirement that graph arguments must be finite. Throughout this chapter the words 'rule' and 'argument' are to be interpreted in this wider sense, but the other definitions in the preceding chapters are taken over unchanged except insofar as this interpretation requires. Proofs may now be infinite even when the rules in question are finite, but they are never essentially infinite in such a case, for we show that every proof by finite rules has a finite subproof. A similar result holds for abstract proofs, qualified by the fact that the property of being a proof by finite rules is not an abstract one (cf. Theorem 11.1).

*Theorem 12.1* Every proof (or abstract proof) by finite rules has a finite subproof (finite abstractly valid subproof).

*Proof* By the infinite analogue of 8.7 any proof by finite rules has a simple subproof $\pi$, each of whose strokes adjoins only finitely many edges. Let L be the calculus whose universe of formulae V is the union of the set $V_\pi$ of formulae occurring in $\pi$ and the set S of strokes in $\pi$, and which is characterised by the rule R 'from $X_b$,b infer $Y_b$ if $b \in S$'. Every partition of V that invalidates $\langle X_\pi \cup S, Y_\pi \rangle$ is of the form $\langle T \cup S, U \rangle$, where $\langle T,U \rangle$ is a partition of $V_\pi$ that invalidates $\langle X_\pi, Y_\pi \rangle$ and hence invalidates $R_\pi$, so that $\langle T \cup S, U \rangle$ invalidates R. Hence $X_\pi$,S $\vdash Y_\pi$; and since R, like $R_\pi$, is finite, it follows by 2.18 that $X',S' \vdash Y'$ for some finite subsets X', Y' and S' of $X_\pi$, $Y_\pi$ and S. Let $\pi'$ be a finite subargument of $\pi$ consisting of the strokes of S', the edges adjoining them and the formula vertices adjoining these edges, together with one occurrence of each member of X' and Y', these being its premisses and conclusions respectively. For any partition $\langle T,U \rangle$ of $V_\pi$ the partition $\langle T \cup S', U \cup (S-S') \rangle$ of V satisfies all the instances of R corresponding to strokes not in $\pi'$. So if $\langle T,U \rangle$ satisfies $R_{\pi'}$, then $\langle T \cup S', U \cup (S-S') \rangle$ satisfies R and hence satisfies $\langle X' \cup S', Y' \rangle$, so that $\langle T,U \rangle$ satisfies $\langle X',Y' \rangle$. It follows that $X_{\pi'} \vdash_{R_{\pi'}} Y_{\pi'}$, and hence that $\pi'$ is valid. Similarly any abstract proof by finite rules has a simple abstractly valid subproof $\pi$, and the abstract result follows by applying the formal one to an argument abstractly isomorphic to $\pi$ in which no formula occurs more than once, using the infinite analogue of 7.7.

Alternatively we may follow the lines of a well-known proof of the compactness of the classical propositional calculus (see e.g. Bell and Slomson, 1969). To bring out the parallel we shall suppose that there are only countably many formulae $A_1, A_2, \ldots$ (the argument in the general case is similar but proceeds by transfinite induction). Suppose then that every finite subargument of $\pi$ is invalid. We define inductively for each $n \geq 1$ a partition $\langle T_n, U_n \rangle$ of $\{A_m : m<n\}$ such that $T_1 \subset T_2 \subset \ldots \subset T_n \subset \ldots$ and $U_1 \subset U_2 \subset \ldots \subset U_n \subset \ldots$ and such that every finite subargument of $\pi$ is invalidated by a partition $\langle T,U \rangle$ for

which $T_n \subset T$ and $U_n \subset U$. (We may take $\langle T_1, U_1\rangle$ to be $\langle \Lambda, \Lambda\rangle$ and we must be able to take $\langle T_{n+1}, U_{n+1}\rangle$ either to be $\langle T_n \cup \{A_n\}, U_n\rangle$ or $\langle T_n, \{A_n\} \cup U_n\rangle$, for otherwise there would be finite subarguments $\pi_1$ and $\pi_2$ such that every partition $\langle T,U\rangle$ for which $T_n \subset T$ and $U_n \subset U$ satisfies $\pi_1$ if $A_n \in T$ and satisfies $\pi_2$ if $A_n \in U$, and hence satisfies $\pi_1 \cup \pi_2$, contrary to the definition of $\langle T_n, U_n\rangle$.) It follows that any finite subargument of $\pi$ is invalidated by $\langle T_m, U_m\rangle$ for sufficiently large m. Hence $\langle \bigcup T_n, \bigcup U_n\rangle$ invalidates all the finite subarguments of $\pi$ and so invalidates their union $\pi$.

The general theory of proofs developed in Chapter 7 carries over to the infinite case with two exceptions: the property that every proof has a minimal subproof, which is trivial in the finite case, and the abstract analogue of this result, expressed by Theorem 7.12. Although (as we show in Section 12.4 when discussing Theorem 11.3) both these results do carry over for infinite single-conclusion arguments, they do not do so in general: there is an infinite proof with no concise subproof and a fortiori no minimal subproof. It follows that the idea of relevance as developed in Section 10.3 has no serious application to the infinite case, for we should be forced to say that every element in this proof was both irrelevant and redundant.

*Theorem 12.2* There exists an abstract proof with no concise subproof.

*Proof* Let $A_1, A_2, \ldots$ enumerate without repetition the formulae of a countably infinite universe V. Let $\pi$ be an argument without premisses or conclusions, consisting of a single occurrence of each formula and strokes $a_n$ and $b_T$ for each $n \geq 1$ and each infinite subset T of V, with edges from $a_n$ to $A_m$ for all $m \geq n$, and from $A_m$ to $b_T$ for all $A_m$ in T. A subargument $\pi'$ of $\pi$ is valid iff (i) for infinitely many n, $a_n$ is in $\pi'$ and (ii) for every infinite T, $b_{T'}$ is in $\pi'$ for some $T' \subset T$. For if (i) fails then $a_n$ is not in $\pi'$ for any $n>k$, say, and $\langle\{A_1,\ldots,A_k\}, \{A_{k+1},\ldots\}\rangle$ invalidates $\pi'$; and if (ii) fails then for some infinite T we have that T′ overlaps V−T for all $b_{T'}$ in $\pi'$, and that $\langle T, V-T\rangle$ invalidates $\pi'$. Conversely, for any partition $\langle T,U\rangle$ we have $T \vdash_{R_{\pi'}} U$ if T is finite and (i) holds, or if T is infinite and (ii) holds; so if both hold then $\Lambda \vdash_{R_{\pi'}} \Lambda$ by cut, and $\pi'$ is valid. It follows at once that $\pi$ itself is a proof, and that it has no minimal subproof; and as no formula occurs in $\pi$ more than once, it is an abstract proof by the infinite analogue of 7.7, and has no concise subproof.

Theorem 7.12 is used to establish Theorem 7.14, that every abstract proof has a standard connected abstract subproof. The latter result carries over to the infinite case, but the failure of the former means that a new method of proof is necessary.

*Theorem 12.3* Every abstract proof has a standard connected abstract subproof.

*Proof* It is sufficient by the infinite analogue of 7.7 to show that a proof $\pi$ in which no formula occurs more than once has a standard connected subproof. We construct a (possibly transfinite) sequence of subarguments $\pi_o, \pi_1, \ldots, \pi_i, \ldots$ by taking $\pi_o$ to be $\pi$, and if i is a successor ordinal j+1, obtaining $\pi_i$ from $\pi_j$ by omitting all non-standard vertices, the edges adjoining these vertices, the strokes adjoining these edges, and the edges adjoining these strokes. If i is a limit ordinal, we take $\pi_i$

to be $\bigcap \pi_j$ over $j<i$, and specify a vertex as a premiss (or conclusion) iff it is a premiss (conclusion) of $\pi_j$ for every $j<i$. Considering the $\pi_i$ as sets of vertices $\ldots \subset \pi_{i+1} \subset \pi_i \subset \ldots \subset \pi_o$, so that for some $i$, $\pi_i = \pi_{i+1} = \pi'$, say; i.e. $\pi'$ is a standard subargument. Every member of the set $V_\pi$ of formulae occurring in $\pi$ is either a member of the set $V_{\pi'}$ of formulae occurring in $\pi'$ or is non-standard in $\pi_j$ for some $j<i$. For any partition $\langle T',U'\rangle$ of $V_{\pi'}$, let $\langle T,U\rangle$ be the partition of $V_\pi$ such that $T' \subset T$ and $U' \subset U$ and a formula occurring as a non-standard vertex a of $\pi_j$ is in T iff a is final in $\pi_j$ but not a conclusion. Then $\langle T,U\rangle$ satisfies $\pi$ since $\pi$ is valid, i.e. it invalidates $R_\pi$ or satisfies $\langle X_\pi,Y_\pi\rangle$. If it invalidates $R_\pi$ then it must invalidate $R_{\pi'}$, too, for each stroke b not in $\pi'$ is omitted in the construction of, say, $\pi_{j+1}$, and an edge E in $\pi_j$ joins it with a non-standard occurrence of a formula A, and A is in T or U according as E is from or to b. Similarly if $\langle T,U\rangle$ satisfies $\langle X_\pi,Y_\pi\rangle$ then it satisfies $\langle X_{\pi'},Y_{\pi'}\rangle$, since otherwise some non-standard premiss of some $\pi_j$ would be in U, or some non-standard conclusion in T, which is absurd. Thus $\langle T,U\rangle$, and hence $\langle T',U'\rangle$, satisfies $\pi'$, which is therefore valid. It is a standard subproof of $\pi$, and by the argument of 7.13, one of its components is a standard connected subproof.

Although we may use the infinite analogues of Theorems 7.5 and 7.6 to establish the validity or abstract validity of various kinds of argument by demonstrating the existence of a final vertex in every appropriate redirection, Lemma 7.2 no longer provides a sufficient condition for the existence of such a final vertex, and we must look for another. Let $a_1,E_1,a_2,\ldots,a_{n-1},E_{n-1},a_n,\ldots$ be an infinite sequence of alternate vertices and edges of a graph, in which each $E_n$ joins $a_n$ and $a_{n+1}$. If no edge or vertex occurs more than once, the corresponding subgraph $\{a_1,E_1,a_2,\ldots\}$ is an *infinite path*, with $a_1$ as its sole *endpoint*. If in addition each $E_n$ goes from $a_n$ to $a_{n+1}$, or each $E_n$ goes from $a_{n+1}$ to $a_n$, the infinite path is *directed*. It is not necessary for our purpose to discuss doubly infinite paths of the form $\{\ldots a_{-n},E_{-n},\ldots, a_o,E_o,\ldots,a_n,E_n,\ldots\}$.

*Lemma 12.4* A nonempty bipartite graph with no directed circuits and no directed infinite paths has at least one initial and at least one final vertex.

*Proof* If a graph $\pi$ has no directed circuits, then by 7.1 the ancestral $\geq$ of its edge relation is a partial ordering. If in addition $\pi$ has no directed infinite paths then every totally ordered subset S of the vertices has an upper bound in S, since otherwise there exists an infinite sequence $a_1,a_2,\ldots$ of vertices of S such that for each n there is a directed path $\alpha_n$ from $a_{n+1}$ to $a_n$, and (using the facts about $\geq$) $\bigcup\alpha_n$ is a directed infinite path. So by Zorn's lemma, $\pi$ has a maximal vertex (and by a similar argument, a minimal one) if it is nonempty. The result follows as for 7.2.

We may use Lemma 12.4 to establish the validity of a large class of arguments, but it is not an adequate class. To show this we consider the *partition rule* 'from T infer U if $\langle T,U\rangle$ is a partition'. By cut for V, it is evident that $\Lambda$ is a consequence of $\Lambda$ by the partition rule.

*Theorem 12.5* Any standard nonempty argument in which every circuit and every infinite path contains a corner or a cross-reference path is valid.

*Theorem 12.6* Any standard nonempty argument in which every circuit and every infinite path contains a corner is abstractly valid.

*Theorem 12.7* Standard proofs in which every circuit and every infinite path contains a corner or a cross-reference path are inadequate.

*Proofs* For 12.5, if a circuit or infinite path contains a corner or cross-reference path, then so does any formal redirection of it, which is consequently not directed; the result follows by 12.4 and the infinite analogue of 7.5. Similarly any abstract redirection of a circuit or infinite path containing a corner is not directed, and 12.6 follows from 12.4 and the infinite analogue of 7.6. For 12.7 we show that for countably infinite V, there is no proof of the required kind from $\Lambda$ to $\Lambda$ by the partition rule. Suppose that $\pi$ is a standard proof from $\Lambda$ to $\Lambda$ by the rule. Starting from an arbitrary stroke $a_1$, we construct a sequence $a_1, E_1, a_2, \ldots, a_n, E_n, \ldots$ taking $E_n$ to be an edge to or from an occurrence of a formula with no previous occurrence in the sequence if $a_n$ is a stroke, and an edge to or from $a_n$ according as $E_{n-1}$ is from or to $a_n$ if $a_n$ is a formula vertex. Such an edge may always be found by the nature of the rule if $a_n$ is a stroke, and by the standardness of the proof otherwise. The subgraph $\{a_1, E_1, a_2, \ldots\}$ contains a circuit (if some stroke appears twice in the sequence) or is an infinite path (otherwise) which contains neither a corner nor a cross-reference path.

Despite this negative result, the class of infinite proofs as a whole, and the class of infinite abstract proofs, are adequate for infinite rules, just as the class of finite proofs is adequate for finite rules. On the other hand countable proofs are not adequate, even assuming a countable universe, as Theorem 12.9 shows.

*Theorem 12.8* Abstract proofs are adequate.

*Theorem 12.9* Every proof from $\Lambda$ to $\Lambda$ by the partition rule has more vertices than V has formulae.

*Proofs* Suppose that $X \vdash_R Y$, and let $\pi$ be the graph whose vertices consist of an occurrence of every formula and a stroke for every instance of R, with edges going to each stroke from the premisses of the relevant instance and from it to the conclusions. Taking the premisses and conclusions of $\pi$ to be the occurrences of members of X and Y respectively, we have $X_\pi = X$ and $Y_\pi = Y$ and $R_\pi = R$, so that $X_\pi \vdash_{R_\pi} Y_\pi$. Since no formula occurs more than once in $\pi$, it satisfies all the conditions to be an abstract proof except that its premisses are not necessarily initial nor its conclusions final; and by the infinite analogue of 7.9 we obtain the required abstract proof from X to Y by R by omitting edges and strokes as appropriate, and so establish 12.8.

For 12.9, every partition $\langle T,U \rangle$ satisfies such a proof $\pi$, so that $\langle T,U \rangle$ must be an instance of $R_\pi$. Hence $\pi$ must include at least one stroke corresponding to each partition, and if there are n formulae there are $2^n$ partitions.

## 12.2 *Kneale proofs*

We begin by establishing necessary and sufficient conditions for a Kneale argument to be abstractly valid.

*Theorem 12.10* A Kneale argument is an abstract proof iff it is standard and has no infinite paths.

*Proof* A standard Kneale argument with no infinite paths is abstractly valid by 12.6. Conversely, an abstract Kneale proof $\pi$ is standard by the argument of 8.3, and we note that every Kneale argument has an abstract redirection in which any given stroke a is the unique final vertex. (It is obtained by changing the direction of each edge $\langle b,c \rangle$ iff b lies on the path joining a and c.) If $\alpha$ is an infinite path $\{a_1, E_1, \ldots\}$ of $\pi$ (where we may assume that $a_1$ is a stroke), let $\pi'$ be the argument obtained from $\pi$ by deleting the edges $E_n$ and the formula vertices $a_{2n}$. Each component of $\pi'$ is a Kneale argument containing just one of the strokes $a_{2n+1}$ of $\alpha$, and so has an abstract redirection in which this stroke is the unique final vertex. But $\alpha$ itself has an abstract redirection with no final vertex, and these various redirections can be combined to yield an abstract redirection of $\pi$ with no final vertex, so that by the infinite analogue of 7.6 $\pi$ is not abstractly valid.

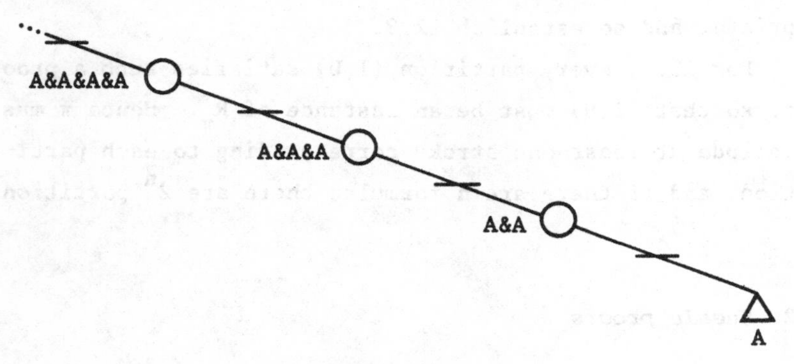


Figure 12.1

In the finite case we saw that standard Kneale arguments are abstractly valid but inadequate (Theorems 8.3 and 8.11). In

the infinite case, by contrast, they need not even be valid (Figure 12.1), but those that are valid constitute an adequate class of proofs (Theorem 12.12). There is a similar situation when the other half of the condition for abstract validity is taken on its own (Figure 9.7 and Theorem 12.13).

*Theorem 12.11* Abstract Kneale proofs are inadequate.

*Theorem 12.12* Standard Kneale proofs are adequate.

*Theorem 12.13* Kneale proofs with no infinite paths are adequate.

*Proofs* 12.11 follows at once from 12.10 and 12.7. For 12.12, suppose that $X \vdash_R Y$, so that by 12.8 there exists an abstract proof $\pi_1$ from X to Y by R. Let $\pi_2$ be the argument obtained by replacing each premiss (and conclusion) that adjoins more than one edge by separate premisses (conclusions), one adjoining each of the edges in question. (Figure 12.2 illustrates the process.) Then $\pi_2$ is an abstract proof also, since a partition $\langle T,U \rangle$ invalidating an argument abstractly isomorphic to $\pi_2$ must include the premisses in T and the conclusions in U, and hence must invalidate an argument abstractly isomorphic to $\pi_1$. By 12.3, $\pi_2$ has a standard connected abstract subproof $\pi_3$ in which corners occur only at intermediate vertices. Let $\pi_4$ be obtained by replacing each such vertex c and the edges adjoining it by an occurrence of the same formula for each choice of a stroke b from which an edge goes to c and a stroke b′ to which an edge goes from c, with an edge to the new occurrence from b, and an edge from it to b′. Then $\pi_4$ is corner-free, and like $\pi_3$ it is connected and standard. Although $\pi_4$ may not be abstractly valid, $R_{\pi_4}$ =

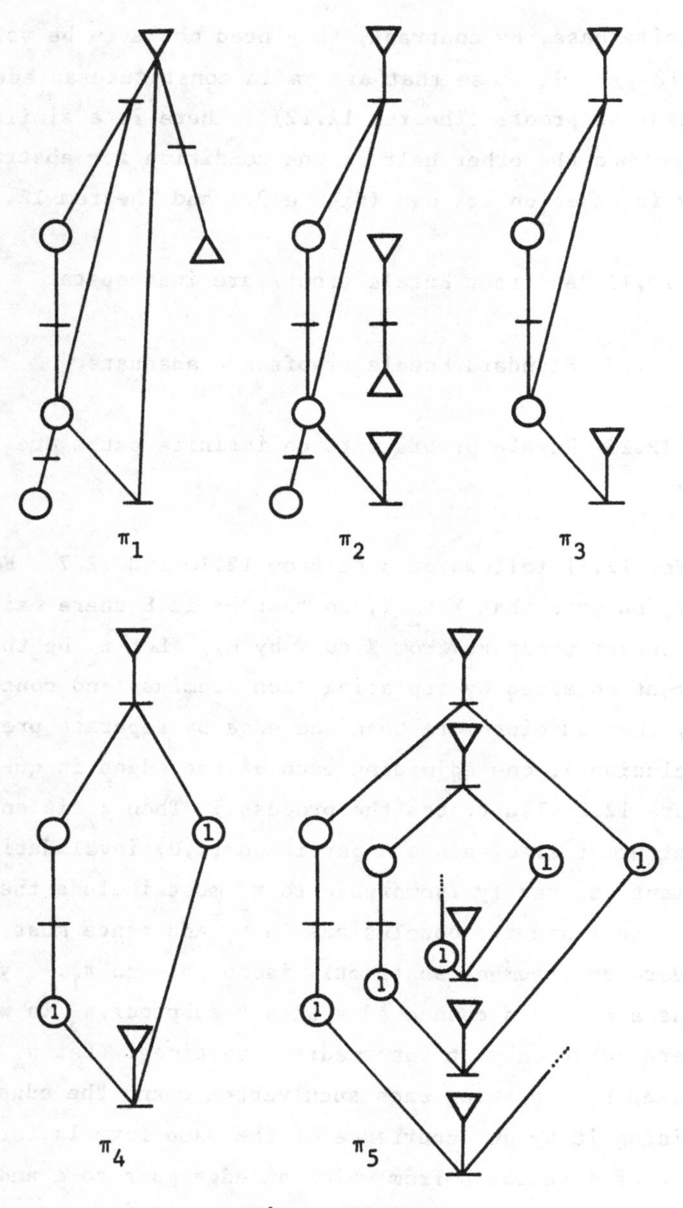


Figure 12.2

$R_{\pi_3}$, so that it is a proof from X to Y by R. Let $a_1$ be any vertex of $\pi_4$, and consider the set S of sequences $a_1, E_1, \ldots, a_n$ of (not necessarily distinct) vertices and edges in which $n \geq 1$ and each $E_i$ joins $a_i$ and $a_{i+1}$ and is distinct from $E_{i-1}$. We use S to index the vertices of an argument $\pi_5$, (the vertex indexed by) $a_1, E_1, \ldots, a_n$ taking the same role in $\pi_5$ - be it a stroke, an occurrence of a formula A, a premiss or a conclusion - as $a_n$ does in $\pi_4$. An edge joins $a_1, E_1, \ldots, a_n, E_n, a_{n+1}$ and its initial segment $a_1, E_1, \ldots, a_n$, directed to or from $a_1, E_1, \ldots, a_n$ according as $E_n$ is to or from $a_n$ in $\pi_4$; and these are the only edges in $\pi_5$. Since $\pi_4$ is connected, each of its vertices appears as the final term in at least one vertex $a_1, E_1, \ldots, a_n$ of $\pi_5$, and $a_n$ and $a_1, E_1, \ldots, a_n$ have isomorphic contexts in the sense that the edges to and from them may be placed in a one-to-one correspondence, the other vertices of corresponding edges being occurrences of the same formula or stroke. So $\pi_5$, like $\pi_4$, is standard and corner-free, $X_{\pi_5} = X_{\pi_4}$, $Y_{\pi_5} = Y_{\pi_4}$ and $R_{\pi_5} = R_{\pi_4}$. Moreover $\pi_5$ is connected, since all its vertices are connected by a path to the sequence $a_1$, and circuit-free since in any finite subset of the vertices, a vertex with a longest index can be joined by an edge with at most one member of the subset. So $\pi_5$ is a standard Kneale proof from X to Y by R as required.

For 12.13, suppose that $X \vdash_R Y$, so that by 12.12 there is a standard Kneale proof $\pi$ from X to Y by R. We assume that $\pi$ has at least one edge, since the result is otherwise immediate, and let $a_1$ be any stroke of $\pi$. Let $\pi_n$ be for each n a replica of the union of the paths $\{a_1, E_1, \ldots, a_{2n}\}$ of $\pi$ with $a_1$ as endpoint and $2n-1$ edges. We take these replicas to be disjoint apart from a common vertex as $a_1$, and specify each vertex as a premiss or a conclusion if the corresponding vertex of $\pi$ is so specified. Each $\pi_n$ is a Kneale argument from X to Y by R and it contains no path with more than $4n-2$ edges. So their union $\pi'$ is a Kneale argument from X to Y by R with no infinite paths, and since $X_{\pi'} = X_\pi$ and $Y_{\pi'} = Y_\pi$ and $R_{\pi'} = R_\pi$, $\pi'$ is valid as required.

The construction of $\pi_4$ in 12.12 depends on our decision to ignore multiplicity of occurrence in connection with rules of inference. If we require in the application of a rule that the premisses and conclusions shall occur with a fixed multiplicity, 12.12 and 12.13 both fail and indeed the entire class of Kneale proofs becomes inadequate by 12.9; for if V is countable so is any connected corner-free proof by the partition rule, when the rule is construed so that no formula occurs more than once in any instance.

We still lack a syntactical description of an adequate class of Kneale proofs. We have seen that for standard arguments the unrestricted presence of infinite paths is too weak a condition, but a ban on them is too strong; and similarly for arguments with no infinite paths the unrestricted presence of non-standard vertices is obviously too weak and standardness too strong. Theorem 12.5 suggests that an appropriate intermediate condition might be found by considering arguments which are *recurring* in the sense that every infinite path contains a cross-reference path; and another plausible condition is obtained by extending the idea of a *cross-referenced* argument to suit infinite arguments by allowing transfinite sequences of non-standard vertices in the definition of Section 9.2. The various combinations of these conditions are summarised in the table below, in which ‘sound’ means that every argument of the relevant kind is valid, and ‘complete’ means that whenever $X \vdash_R Y$ there exists a valid argument of the relevant kind from X to Y by R. It will be seen that none of the resulting kinds of argument determines an adequate class of proofs, but the dotted line shows where, if anywhere, such a class might be found. The entries are justified by the foregoing discussion, by Theorems 12.16-18 below, and by the example in Figure 12.3 of a pattern of recurring cross-referenced Kneale argument every instance of which is invalidated by ⟨V,Λ⟩.

CONDITION ON PATHS

CONDITION ON VERTICES

| Kneale arguments | no infinite paths | recurring | unconditional |
|---|---|---|---|
| standard | sound* incomplete | sound incomplete** | unsound complete |
| cross--referenced | sound incomplete** | unsound *** | unsound complete |
| unconditional | unsound complete | unsound complete | unsound complete |

*and all abstractly valid ** but complete for finite rules
*** not known whether complete or not

*Theorem 12.14* Every cross-referenced Kneale argument from X to Y by R with no infinite paths is a subargument of a standard recurring Kneale argument from X to Y by R.

*Theorem 12.15* Every standard recurring Kneale argument has a cross-referenced Kneale subargument with no infinite paths.

*Theorem 12.16* Every standard recurring Kneale argument is valid.

*Theorem 12.17* Every cross-referenced Kneale argument with no infinite paths is valid.

*Theorem 12.18* Standard recurring Kneale proofs and cross-referenced Kneale proofs with no infinite paths determine the same relation of deducibility, and are adequate for finite rules but not in general.

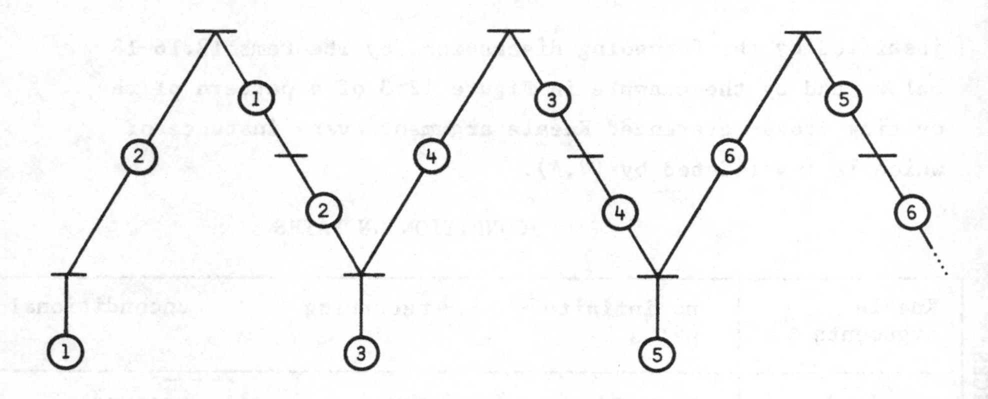

Figure 12.3

*Proofs* For 12.14, if $\pi$ is a cross-referenced Kneale argument with no infinite paths from X to Y by R, its non-standard vertices can be arranged in a sequence $a_1, a_2, \ldots$ such that there are cross-reference paths $(a_1, b_1), (a_2, b_2), \ldots$ where (1) no $b_i$ appears in $(a_j, b_j)$ for $j>i$ unless $b_i = b_j$, and (2) each $b_i$ is standard. We lose no generality by assuming (2), since for each non-standard $b_i$, $(a_i, b_j)$ is, for some standard $b_j$ such that $j<i$, a cross-reference path containing no standard $b_k$ for $k<j$ unless $b_k = b_j$. For if not, let i be the least index for which no such path exists. Then $b_i = a_{i'}$, for some $i'<i$ by (1), and by our choice of i there is a standard $b_j$ such that $j \leq i'$ and $(a_{i'}, b_j)$ is a cross-reference path containing no standard $b_k$ for $k<j$ unless $b_k = b_j$. But $(a_i, b_j)$ is then after all a cross-reference path, since otherwise $b_j$ is in $(a_i, b_j) \cup (b_j, a_{i'}) = (a_i, b_i)$, contradicting (1); and unless $b_k = b_j$ it contains no standard $b_k$ for $k<j$, since $b_k$ is neither in $(a_{i'}, b_j)$ nor by (1) in $(a_i, a_{i'})$. We now obtain the required sequence of cross-reference paths satisfying (2) by taking the subsequence of the $(a_j, b_j)$ for which $b_j$ is standard, and inserting in order after each $(a_j, b_j)$ the new cross-reference paths of the form $(a_i, b_j)$.

For each i, $\pi$ is the junction of two Kneale arguments containing $b_i$, and (3) we let $\pi(i)$ be that one which does not contain $a_i$ (see Figure 12.4). We define a sequence $\pi_1,\pi_2,\ldots$ of arguments such that $\pi_1 = \pi$ and each $\pi_{n+1}$ is obtained from $\pi_n$ as illustrated in Figure 12.4 by simultaneously joining a replica of $\pi(i)$ at each non-standard vertex $a_i$ or its replica (first promoting the vertex to be a premiss or conclusion to satisfy the definition of junction). Let $\pi'$ be the argument whose graph is the union of those of the $\pi_n$, a vertex being a premiss or conclusion in $\pi'$ iff it is so in one of the $\pi_n$. Then by (2) $\pi'$ is a standard Kneale argument from X to Y by R which has $\pi$ as a subargument. It remains only to show that $\pi'$ is recurring. Suppose that $\pi'$ contains an infinite path $\alpha$ containing no cross-reference path. Since by a simple inductive argument none of the $\pi_n$ contains an infinite path, the intersection of $\alpha$ and each of the $\pi_n$ with which it has an edge in common must be a path with a non-standard endpoint, and hence $\alpha$ contains an infinite path which consists (for some $i_1,i_2,\ldots$) of replicas of $(b_{i_1},a_{i_2})$, $(b_{i_2},a_{i_3}),\ldots$ taken in order, where (4) each $(b_{i_m},a_{i_{m+1}})$ is in $\pi(i_m)$. If $i_m \geq i_k$ for all $m \geq k$, then for all $m \geq k$ (5) $(b_{i_m},a_{i_{m+1}})$ is in $\pi(i_k)$. (For if not, let m be the least integer $\geq k$ for which (5) fails. Then $m>k$ by (4), so that $a_{i_m}$ is in $\pi(i_k)$, and as by (1) $b_{i_k}$ does not lie on $(b_{i_m},a_{i_m})$ unless $b_{i_m}=b_{i_k}$, it follows that $b_{i_m}$ also is in $\pi(i_k)$. If $b_{i_m} \neq b_{i_k}$ then $(b_{i_m},a_{i_{m+1}})$ has a subpath $(b_{i_m},b_{i_k})$ in $\pi(i_k)$, and $\alpha$ has a subpath replicating $(b_{i_k},a_{i_{k+1}}),\ldots,(b_{i_m},b_{i_k})$ which is a cross-reference path since both $(b_{i_k},a_{i_{k+1}})$ and $(b_{i_m},b_{i_k})$ are in $\pi(i_k)$; and if $b_{i_m}=b_{i_k}$ then since $(b_{i_k},a_{i_m})$ is a cross-reference path in $\pi(i_k)$, the subpath of $\alpha$ replicating $(b_{i_k},a_{i_{k+1}}),\ldots,(b_{i_{m-1}},a_{i_m})$ is a cross-reference path, the edges adjoining its endpoints replicating those adjoining the endpoints of $(b_{i_k}, a_{i_m})$.) If $i_m > i_n$ for all m such that $k<m<n$, then for all m such that $k \leq m<n$ (6) $(b_{i_m}, a_{i_{m+1}})$ and $\pi(i_n)$ are disjoint. (For if not, let m be the greatest integer such that $k \leq m<n$ for which (6) fails. Then $a_{i_{m+1}}$ is not in $\pi(i_n)$, by (3) if $m+1=n$ and otherwise because $(b_{i_{m+1}}, a_{i_{m+2}})$ and $\pi(i_n)$ are disjoint and $b_{i_n}$ is not in $(a_{i_{m+1}}, b_{i_{m+1}})$ by (1). But $(b_{i_m}, a_{i_{m+1}})$ overlaps $\pi(i_n)$, and so has a subpath $(b_{i_n}, a_{i_{m+1}})$ not in $\pi(i_n)$; i.e. $b_{i_n}$ is initial in $(b_{i_n}, a_{i_{m+1}})$ iff $a_{i_n}$ is initial. Hence the subpath of $\alpha$ replicating $(b_{i_n}, a_{i_{m+1}})$, $(b_{i_{m+1}}, a_{i_{m+2}}), \ldots, (b_{i_{n-1}}, a_{i_n})$ is a cross-reference path.) Now let $j_1, j_2, \ldots$ be the subsequence of $i_1, i_2, \ldots$ such that $j_1$ is the least index, and $j_{k+1}$ the least to occur subsequent to $j_k$. Then by (5), $b_{j_m}$ is in $\pi(j_k)$ whenever $m \geq k$, and by (6), $b_{j_k}$ is not in $\pi(j_{k+1})$. It follows that $b_{j_{k+1}} \neq b_{j_k}$, and that $\{(b_{j_k}, b_{j_{k+1}}) - b_{j_{k+1}}\}$ is a family of disjoint sets, since $(b_{j_k}, b_{j_{k+1}}) - b_{j_{k+1}}$ and $\pi(j_{k+1})$ are disjoint, but whenever $m>k$, $b_{j_m}$ and $b_{j_{m+1}}$ - and hence $(b_{j_m}, b_{j_{m+1}}) - b_{j_{m+1}}$ - are in $\pi(j_{k+1})$. The union of this family is an infinite path in $\pi$ replicating $(b_{j_1}, b_{j_2})$, $(b_{j_2}, b_{j_3}), \ldots$, contrary to hypothesis.

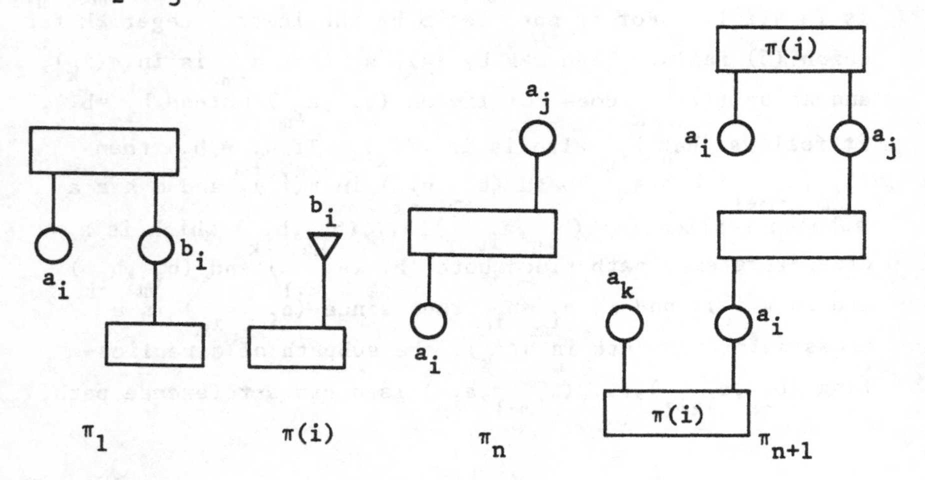


Figure 12.4

For 12.15, let $c_1$ be any stroke of a standard recurring Kneale argument $\pi$ (the result being immediate for an argument without strokes), and let $\pi'$ be the union of $c_1$ and the paths $(c_1,c)$ that contain no cross-reference path which does not have c as an endpoint. Evidently $\pi'$ is connected and nonempty and, specifying each vertex as a premiss or conclusion iff it is so specified in $\pi$, $\pi'$ is a subargument of $\pi$. If $\pi'$ contains an infinite path, it contains an infinite path $c_1,\ldots,c_n,\ldots$ with $c_1$ as endpoint; but this contains a cross-reference path $c_i,\ldots,c_j$ since $\pi$ is recurring, so that $c_{j+1}$ is not in $\pi'$, contrary to hypothesis. Each non-standard vertex $a_i$ of $\pi'$ is an endpoint of a path $(c_1,a_i)$ which contains a cross-reference path $(b_i,a_i)$. These vertices may be arranged in a sequence so that whenever $a_i$ precedes $a_j$, $(c_1,b_j)$ has at least as many edges as $(c_1,b_i)$. So if $a_i$ precedes $a_j$ then $b_i$ cannot appear in $(b_j,a_j)$ unless $b_i$=$b_j$; and hence $\pi'$ is a cross-referenced Kneale proof as required.

12.16 is a corollary of 12.5. For 12.17, every cross-referenced Kneale argument $\pi$ with no infinite paths is by 12.14 and 12.16 a subargument of a proof from $X_\pi$ to $Y_\pi$ by $R_\pi$, whence $X_\pi \vdash_{R_\pi} Y_\pi$ as required. For 12.18 the identity of the deducibility relations follows from 12.14-17. Finite cross-referenced Kneale proofs are adequate for finite rules by 9.7, and a fortiori so are cross-referenced Kneale proofs with no infinite paths. The inadequacy of standard recurring Kneale proofs in general follows from 12.7.

The term ‘recurring proof’ is suggested by analogy with the way in which a recurring decimal like 2.1473473473... can be written as $2.1\dot{4}7\dot{3}$. The analogy is closest when the rules are finite, for any finite cross-referenced Kneale argument can be extended to a standard recurring one as in Theorem 12.14, the justifying sequence of cross-reference paths functioning analogously to the dots above the decimal digits by indicating which parts of the finite proof or decimal must be replicated and joined in order to generate the intended infinite one. Conversely any standard recurring Kneale argument can be curtailed to form a finite cross-referenced one, as we now show.

*Lemma 12.19* Every standard Kneale argument has a simple nonempty standard subargument.

*Theorem 12.20* Every standard recurring Kneale proof by finite rules has a finite cross-referenced Kneale subproof.

*Proofs* For 12.19, let c be a formula vertex of a standard Kneale argument $\pi$ (the lemma being immediate for an argument without formula vertices). Let $\pi'$ be a simple subargument obtained by deleting edges as in the proof of 8.7, such that for each stroke b the edge in (c,b) adjacent to b is retained. The component of $\pi'$ which contains c is also a nonempty simple subargument of $\pi$, and it is standard since by construction $\pi'$ contains every edge which adjoins c or a formula vertex connected with c by a path.

For 12.20 we apply first 12.19 and then 12.15 to show that every standard recurring Kneale proof by finite rules has a cross-referenced Kneale subproof which has no infinite paths and in which only finitely many edges adjoin each vertex. By König's lemma, any graph with these two properties is finite. Although in general there are infinite proofs by finite rules, this argument shows that if we were to take multiplicity of occurrence into account in connection with rules of inference, the only abstract Kneale proofs by finite rules would be the finite ones.

Recurring proofs provide yet another solution to the problem of junction and cut posed in Section 9.1. We saw there that the attempt to neutralise unwanted premisses and conclusions led to an unending series of accretions, violating the requirement that a proof must be finite. Where infinite proofs are admissible, infinitely many replicas of finite standard Kneale proofs may be joined simultaneously as in the proof of Theorem 12.14, and the resulting argument is an infinite standard recurring Kneale proof. More generally, it is easy to show that deducibility by standard recurring Kneale proofs (or, equivalently, by cross-referenced Kneale proofs with no infinite paths) contains the rules, finite or otherwise, and is closed under overlap, dilution and cut for formulae, and under $\text{cut}_1$, $\text{cut}_2$ and $\text{cut}_3$ as defined in Section 2.1; we conjecture that this deducibility relation is the closure of the rules under overlap, dilution and $\text{cut}_3$. By contrast deducibility by abstract Kneale proofs is not even closed under cut for formulae, since

*Theorem 12.21* Abstract Kneale proofs are inadequate even for finite rules.

*Proof* By 12.3 and the infinite analogue of 8.7, every abstract Kneale proof $\pi$ has a subproof which is a simple abstract Kneale proof, and by König's lemma this must be finite if $R_\pi$ is finite. The result follows by 8.11.

For single-conclusion rules, on the other hand, abstract Hilbert proofs are adequate as in the finite case. By Theorem 12.10, a Hilbert argument is abstractly valid iff it is standard and has no infinite paths, and

*Theorem 12.22* Abstract Hilbert proofs are adequate for single-conclusion rules.

*Proof* The argument of 11.17 carries over to the infinite case. For cut for sets we simultaneously join onto an abstract Hilbert proof of B from X,Z a replica of an abstract proof of A from X at each premiss which is an occurrence of a formula A in Z. The resulting argument is a Hilbert one, and may be shown to be an abstract proof by adapting the proof of 9.2 to allow for simultaneous junction of the kind invoked above.

## 12.3 *Cornered-circuit proofs*

Not every standard cornered-circuit argument is abstractly valid (or even valid, as Figure 12.1 shows), but those that are constitute an adequate class of proofs.

*Theorem 12.23* Standard cornered-circuit abstract proofs are adequate.

*Proof* Since the proofs of 10.17 and 10.18 carry over to the infinite case, every tree proof from X to Y by R has an articulation which is a cornered-circuit abstract proof from X to Y by R, and this has a standard abstract subproof by 12.3. The result follows from the adequacy of tree proofs, established in 6.1.

All standard cornered-circuit arguments without infinite paths are abstract proofs (Theorem 12.6), and it seems likely that many analogies between them and the finite standard cornered-circuit proofs can be established by using Theorem 12.6 as a substitute for 10.3. For example, we would expect them to be the closure of the abstract Kneale proofs under identification of premisses and conclusions and junction (where the latter is generalised to allow simultaneous junction as in Theorem 12.22). Although they are not an adequate class of proofs (Theorem 12.7) they are adequate for finite rules (Theorem 10.4). Even for infinite rules they determine a deducibility relation which is closed under cut for formulae (and is, we conjecture, the relation alluded to in Theorem 12.18). For example, let L be a conventional system of first-order arithmetic using an induction axiom scheme, and let L′ be a multiple-conclusion one obtained by adding (i) the rule of complete induction 'from A(0),A(1),... infer (x)A(x)' and (ii) all true statements of equality or inequality not containing variables to (iii) a predicate calculus based on PC. It is not hard to show that the axioms and rules of L can be reproduced in L′ using only standard cornered-circuit proofs with no infinite paths, and hence that there is a proof of this kind for every theorem of L.

We shall not undertake the search for a purely syntactical description of an adequate class of cornered-circuit proofs. We close by suggesting analogues of the universal tree $\tau_V$ of Section 6.1. Because graph arguments possess an unambiguous structure, none of them (save those without strokes) can be a proof by every R, but any that share with a universal tree the property of being a proof from Λ to Λ by the partition rule are also the basis of proofs for every occasion, in the following sense.

*Theorem 12.24* If π is a proof (or abstract proof) from Λ to Λ by the partition rule, then whenever $X \vdash_R Y$ there is a proof (an abstract proof) from X to Y by R whose graph is a subgraph of that of π.

*Proof* By the infinite analogue of 7.9, we obtain a proof (abstract proof) from X to Y by specifying every occurrence of a member of X in $\pi$ as a premiss and every occurrence of a member of Y as a conclusion, and deleting every stroke (with its adjoining edges) from which there is an edge to a premiss or to which there is an edge from a conclusion. For each remaining stroke b, we have $X \subset X_b$ and $Y \subset Y_b$, so that by dilution $X_b \vdash_R Y_b$ and by the infinite analogue of 2.17 $X'_b$ R $Y'_b$ for some subsets $X'_b$ and $Y'_b$ of $X_b$ and $Y_b$. We may therefore obtain an argument by R by omitting edges, its validity (abstract validity) being ensured by the infinite analogue of 7.8.

Theorem 12.23 ensures that there exist cornered-circuit proofs to which this result applies, and two in particular (one abstract, the other not) are worth mentioning for their remarkable symmetry. Let $\pi_1(V)$ have as vertices a stroke $b_{\langle T,U\rangle}$ for each partition $\langle T,U\rangle$, and two occurrences $a_A$ and $c_A$ of each formula A, with an edge from $a_A$ to $b_{\langle T,U\rangle}$ if $A \in T$, and from $b_{\langle T,U\rangle}$ to $c_A$ if $A \in U$. We specify no premisses or conclusions. For $\pi_2(V)$ we consider any well-ordering of V, and write X/A for the set of formulae in X which precede A in it. The vertices of $\pi_2(V)$ are to comprise an occurrence $b_{\langle X,Y\rangle}$ of A for each A and each partition $\langle X,Y\rangle$ of V/A; and an occurrence $b_{\langle T,U\rangle}$ of the inference stroke for each partition $\langle T,U\rangle$, with an edge to it from $b_{\langle T/A,U/A\rangle}$ for each A in T, and from it to $b_{\langle T/A,U/A\rangle}$ for each A in U. As before, we specify no premisses or conclusions; indeed every formula vertex is intermediate. It is easy to show that the partial ordering determined by the ancestral of the edge relation of $\pi_2(V)$ (Lemma 7.1 and Theorem 12.26) is a simple ordering, and for a finite universe it corresponds to the order in which the discs are moved in the classic puzzle of the Tower of Hanoi.

Figure 12.5 illustrates $\pi_1(V)$ and $\pi_2(V)$ when V has just two members, of which A precedes B. The properties of $\pi_1(V)$ and

$\pi_2(V)$ are summed up in the following results, the details of whose proofs are left to the reader.

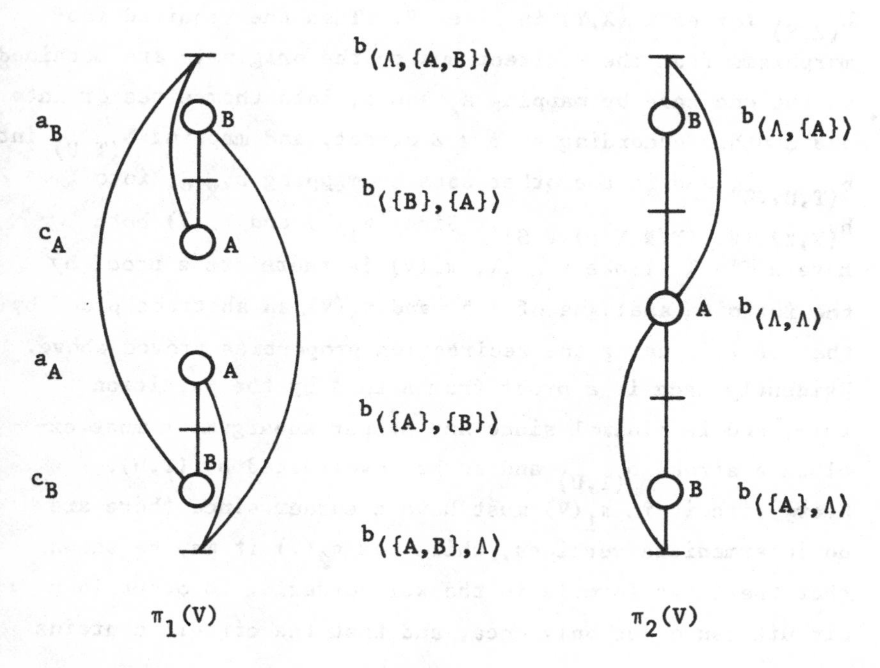


Figure 12.5

*Theorem 12.25* $\pi_1(V)$ is a minimal cornered-circuit proof from $\Lambda$ to $\Lambda$ by the partition rule, and every formal redirection of $\pi_1(V)$ is a replica of it.

*Theorem 12.26* $\pi_2(V)$ is a minimal cornered-circuit abstract proof from $\Lambda$ to $\Lambda$ by the partition rule, and every abstract redirection of $\pi_2(V)$ is a replica of it.

*Outline of proofs* If $\langle X,Y\rangle$ is a partition of some set, we write $\langle X,Y\rangle.Z$ for $\langle (X-Z)\cup(Y\cap Z),\ (Y-Z)\cup(X\cap Z)\rangle$, obtained by transposing the members of Z in $\langle X,Y\rangle$. For the redirection properties, consider the formal redirection of $\pi_1(V)$ obtained by redirecting the edges adjoining occurrences of formulae in a set Z, and the abstract redirection of $\pi_2(V)$ obtained by redirecting the edges adjoining formula vertices $b_{\langle X,Y\rangle}$ for each $\langle X,Y\rangle$ in a set S. Then the required isomorphisms from the redirections to the originals are obtained in the one case by mapping $a_A$ and $c_A$ into themselves or into one another according as $A \notin Z$ or not, and mapping $b_{\langle T,U\rangle}$ into $b_{\langle T,U\rangle.Z}$; and in the other case by mapping $b_{\langle X,Y\rangle}$ into $b_{\langle X,Y\rangle.\{B:\ \langle X/B,Y/B\rangle \in S\}}$. Since $\pi_1(V)$ and $\pi_2(V)$ both have a final stroke $b_{\langle V,\Lambda\rangle}$, $\pi_1(V)$ is therefore a proof by the infinite analogue of 7.5, and $\pi_2(V)$ an abstract proof by that of 7.6, using the redirection properties proved above. Evidently each is a proof from $\Lambda$ to $\Lambda$ by the partition rule, and is minimal since any proper subargument must exclude a stroke $b_{\langle T,U\rangle}$ and so be invalidated by $\langle T,U\rangle$. Every circuit in $\pi_1(V)$ must have a corner since there are no intermediate vertices, while for $\pi_2(V)$ it may be shown that the first formula in the well-ordering to occur in a circuit can occur only once, and that the circuit contains a corner at that occurrence.

## 12.4 *Marking of theorems*

Following the pattern of Section 6.2, we sketch here proofs of the results of Chapters 7-11 which carry over to the case of infinite rules and graphs; and we provide disproofs of those, already marked with a †, which do not.

| | |
|---|---|
| 7.1-2 | These results do not involve rules or arguments. |
| 7.3-11 | Proofs are as in the finite case. |
| 7.12† | See Theorem 12.2. |
| 7.13 | Proof is as in the finite case. |
| 7.14 | See Theorem 12.3. |
| 7.15 | Proof is as in the finite case. |

8.1† Figure 12.1 provides a counterexample.

8.2 Proof is as in the finite case.

8.3† See Theorem 12.10.

8.4†-6† Figure 12.6 shows a standard Kneale proof which is not an abstract proof.

8.7 Proof is as in the finite case.

8.8† Figures 12.1 and 12.6 provide counterexamples.

8.9 By 12.12 there is a standard Kneale proof from X to Y by R, where R are the rules for PC, iff $X \vdash_R Y$. By compactness, therefore, it is sufficient to consider finite X and Y. We proceed by induction on the complexity of $X \cup Y$, measured as the total number of connectives in its members. If X and Y overlap or the complexity is zero, the result is immediate. Otherwise we select any formula in X or in Y which is not a variable, and proceed by cases of which we consider only a typical one. Suppose we select $A \supset B$ in X, and let $X' = X-(A \supset B)$. By the rules, $B \vdash_R A \supset B$ and $\Lambda \vdash_R A, A \supset B$, so that if $A \supset B, X' \vdash_R Y$ then by cut and the induction hypothesis, there are standard simple normal Kneale proofs $\pi_1$ and $\pi_2$ from $B, X'$ to $Y$ and from $X'$ to $A, Y$. By joining (as in 12.14) a possibly infinite set of replicas of $\pi_1$, $\pi_2$ and the proof from $A, A \supset B$ to $B$ of Figure 7.10, we obtain the required standard simple normal Kneale proof from $A \supset B, X'$ to $Y$.

8.10† We shall show in Section 18.1 that the rules for PC characterise a counterpart of the classical calculus, and it then follows from 12.12 that there is a standard Kneale proof from $\Lambda$ to B by the rules for PC whenever B is a tautology.

8.11† See Theorem 12.12.

8.12-13 Suppose that for a universe of two formulae, there is a corner-free or circuit-free abstract proof $\pi$

from $\Lambda$ to $\Lambda$ by the partition rule. By 12.3 and the infinite analogue of 8.7 we may suppose that $\pi$ is simple, standard and connected, so that exactly two edges adjoin each stroke. If $\pi$ is corner-free then exactly two edges adjoin each vertex, and $\pi$ is either a circuit or the union of two infinite paths, and being corner-free cannot be abstractly valid. If $\pi$ is circuit-free then, taking b to be any stroke, it has an abstract redirection in which there is no final vertex, obtained by making b initial, and ensuring that each stroke c is not final by appropriately directing the edges through the formula vertex just beyond c from b; and so by the infinite analogue of 7.6, $\pi$ is not abstractly valid.

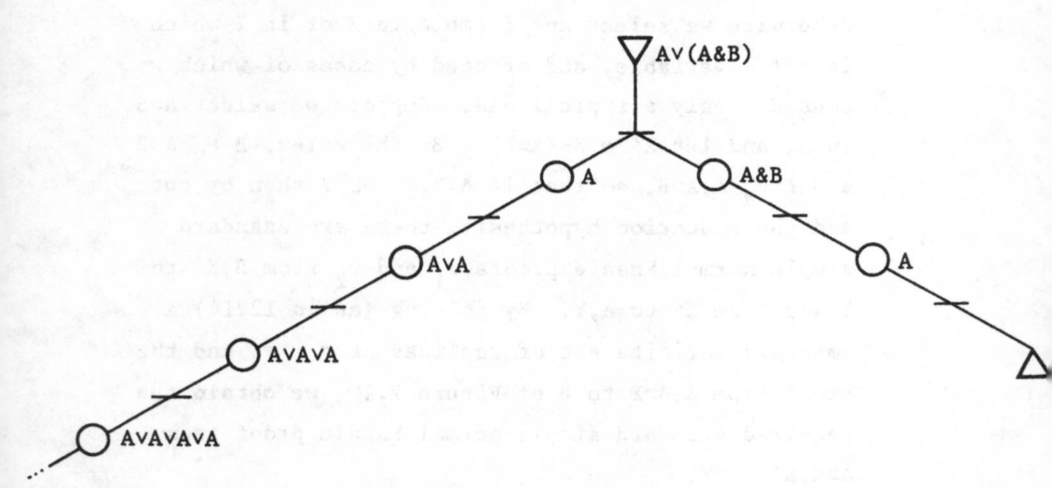


Figure 12.6

9.1-3 Proofs are as in the finite case.

9.4 The result is a corollary of 12.12 and the infinite analogues of 9.2 and 9.3.

9.5[†] Figure 12.7 provides an example of a cross-referenced Kneale proof not in the closure (though it is in the closure under simultaneous junction of the kind envisaged in Theorem 12.22).

9.6$^{\dagger}$ Figure 12.1 provides a counterexample.

9.7 The result is a corollary of 12.12.

9.8 Proof is as in the finite case.

9.9 The result is a corollary of 12.12 and the infinite analogue of 9.8.

9.10$^{\dagger}$ Figure 12.1 provides a counterexample.

9.11 The result is a corollary of 12.12. (Standard cross-referenced-circuit arguments without infinite paths are valid by 12.5, but inadequate by 12.7.)

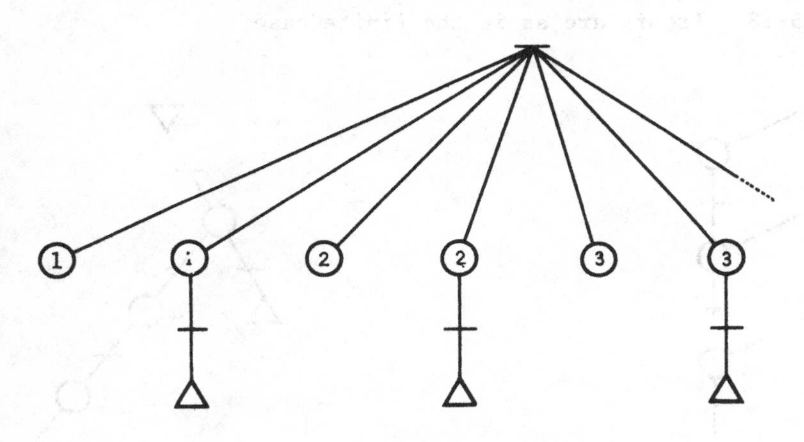


Figure 12.7

10.1 Proof is as in the finite case.

10.2 The result is a corollary of 12.12 and the infinite analogues of 10.1 and 9.2.

10.3$^{\dagger}$ Figure 12.1 provides a counterexample.

10.4 See Theorem 12.23.

10.5$^{\dagger}$ Figure 12.8 provides an example of a standard cornered-circuit argument not in the closure.

10.6$^{\dagger}$ Figure 12.8 provides a counterexample.

10.7 Proof is as in the finite case.

10.8† Figure 12.9 provides a counterexample, $\mu$ and $\mu'$ both being residues of $\pi$.

10.9†-11† Figure 12.1 provides an example of a cornered-circuit argument which satisfies the condition of each of the theorems, but is not a proof.

10.12†-13† Even a standard Kneale proof need not be concise, as Figure 12.6 shows.

10.14† Figure 12.10 provides an example of a standard cornered-circuit abstract proof in which all but five of the vertices are irrelevant and abstractly irrelevant.

10.15-18 Proofs are as in the finite case.

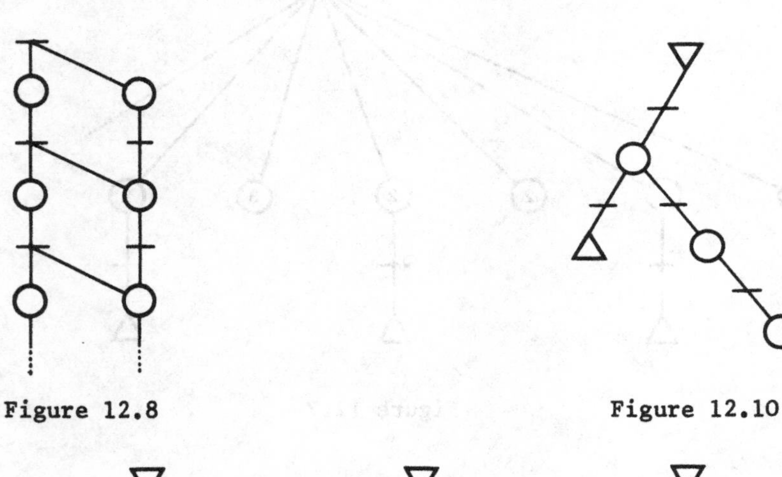

Figure 12.8

Figure 12.10

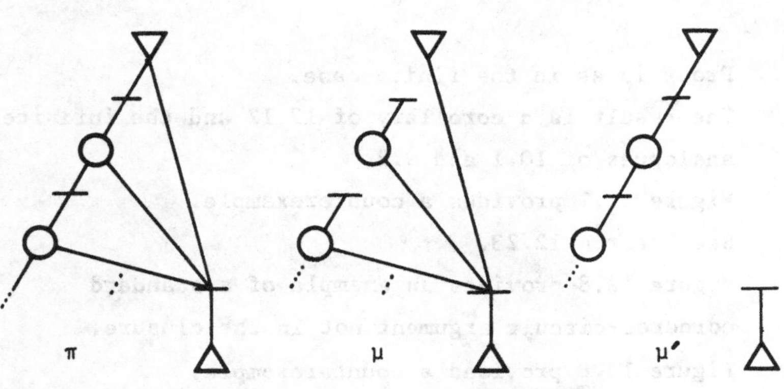


Figure 12.9

11.1-4 Proofs are as in the finite case. For 11.3 we need to show that (in contrast to 12.2) every single-conclusion proof $\pi$ has a minimal subproof $\pi'$. By the infinite analogue of 1.14 there is a sequence proof $\sigma$ of $B_\pi$ from $X_\pi$ by $R_\pi$ in which no formula occurs more than once (for otherwise we may obtain such a proof by omitting everything subsequent to the first occurrence of $B_\pi$ and every occurrence of each other formula subsequent to its first). By the infinite analogue of 11.10 there is an articulation $\pi_1$ of $\sigma$ which is a proof of $B_\pi$ from $X_\pi$ by $R_\pi$, and we let $\pi_1'$ be its standard subargument based on the union of its conclusion b and the directed paths to b. Evidently $\pi_1'$ is an articulation of the subsequence of $\sigma$ formed by its formula vertices, and so is a subproof of $\pi_1$. Moreover $\pi_1'$ is minimal, since any proper subargument has a non-standard formula vertex c, and is invalidated by $\langle T,U \rangle$ where U is the set of formulae which occur in the directed paths of $\pi_1$ from c to b. We therefore take $\pi'$ to be a subargument of $\pi$ such that $X_{\pi'} = X_{\pi_1'}$ and $R_{\pi'} = R_{\pi_1'}$, but no proper subargument of $\pi'$ has this property. (It follows by the argument of 7.12 that every single-conclusion abstract proof has a concise abstract subproof.)

11.5† Figure 12.11 provides a counterexample.

11.6†-9† Figure 12.1 provides an example of a single-conclusion argument which satisfies the conditions of each of the theorems, but is not a proof.

11.10 Proof is as in the finite case.

11.11† Figure 12.1 provides a counterexample.

11.12 A concise single-conclusion abstract proof $\pi_1$ conforms to a minimal one $\pi$ in which no formula occurs more than once. Proceeding as for 11.3, there exists a sequence proof $\sigma$ of $B_\pi$ from $X_\pi$ by $R_\pi$ consisting exclusively of formulae in $\pi$, with no formula occurring more than once. It is easy to see that an articulation of $\sigma$ qua proof of $B_\pi$ from $X_\pi$ by $R_\pi$ is (a replica of) a subproof of $\pi$, and so a replica of $\pi$ itself, since $\pi$ is minimal. So $\pi$, and hence $\pi_1$, is an articulated sequence proof.

11.13 Proof is as in the finite case.

11.14†-15† Figure 12.1 provides an example of a standard Hilbert argument which is not an articulated sequence proof or even a proof.

11.16 Proof is as in the finite case.

11.17 See Theorem 12.22.

11.18 Whether or not the cross-referenced Hilbert proof $\pi$ has infinite paths, the construction of Theorem 12.14 determines a standard Hilbert argument of which $\pi$ is an abstract subargument, and this must consequently be a proof also.

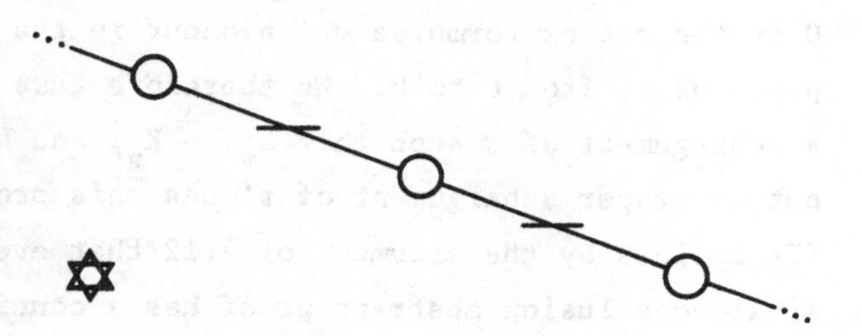

Figure 12.11

# Part III · Many-valued logic

# 13 · Many-valued calculi

## 13.1 *Definitions*

A *propositional calculus* satisfies two conditions, one on its formulae and the other on its consequence relation. The first condition postulates the existence of a countably infinite set of atomic formulae, traditionally called *propositional variables*, together with a finite number of *connectives* each of some specified valency (singulary, binary, k-ary, etc.). Each propositional variable is to be a formula, and if F is a k-ary connective and $A_1,\ldots,A_k$ are formulae then $FA_1\ldots A_k$ is to be a formula. Conversely every formula must admit a unique finite decomposition in terms of the variables and connectives. As long as this condition is met it does not matter what the propositional variables are, and we use p, q, etc. to stand for unspecified propositional variables. The notation $FA_1\ldots A_k$ is likewise a metalinguistic one, intended to cover a variety of notations (Polish, Russellian, etc.) by which the application of a connective can be expressed in particular calculi. In giving actual examples of calculi we assume a Russellian notation for them, and sometimes use dots to replace parentheses, e.g. $A\supset.A\supset.A\supset B$ for $A\supset(A\supset(A\supset B))$. We write this formula also as $A\supset^3 B$ and define $A\supset^n B$ similarly for other n.

The second condition on a propositional calculus is that consequence must be closed under substitution. A *substitution* s is any mapping of formulae into formulae such that $s(FA_1\ldots A_k) = Fs(A_1)\ldots s(A_k)$. Every mapping of the propositional variables into the formulae thus determines a substitution and vice versa. We say that ⊢ is *closed under substitution* if, for every substitution s, s(X) ⊢ s(Y) whenever X ⊢ Y. The corresponding condition for a single-conclusion calculus is that s(X) ⊢ s(B) whenever X ⊢ B.

A *matrix* is constituted by an arbitrary set of elements called *truth-values* or simply *values*, some or none or all of which are *designated*, and a number of *basic functions* (singulary, binary, k-ary, etc.) defined over the truth-values and taking truth-values as their values. The definitions that follow presuppose a one-one correspondence between the connectives of a propositional calculus and the basic functions of a matrix, a k-ary connective being paired with a k-ary function. We therefore use the same symbol to denote a connective and the associated function. A *valuation* v is a mapping of the formulae into the truth-values such that $v(FA_1 \ldots A_k) = F(v(A_1), \ldots, v(A_k))$ for each F. Every assignment of truth-values to the propositional variables thus determines a valuation and vice versa, and moreover v(A) depends only on the values assigned to those variables that actually figure in A. Each valuation partitions the formulae into those that receive designated and undesignated values, and we say that the valuation *satisfies* or *invalidates* ⟨X,Y⟩ or ⟨X,B⟩ according as the resulting partition satisfies or invalidates it. (Valuations must be distinguished from partitions, since different valuations may produce the same partition.) Every set of valuations thus characterises a consequence relation: in particular the set of all valuations in a matrix M characterises a consequence relation which we call *consequence in* M and write $\vdash_M$. We say that M *characterises* the calculus whose consequence relation is $\vdash_M$. In other words, the condition for M to characterise L is that X ⊢ Y iff, for every v, either v(A) is undesignated for some A in X or v(A) is designated for some A in Y. Similarly the condition for M to characterise a single-conclusion calculus L is that X ⊢ B iff, for every v, either v(A) is undesignated for some A in X or v(B) is designated.

A (multiple- or single-conclusion) calculus is *many-valued* if it is characterised by some matrix. We do not exclude matrices with two (or even fewer) values from this definition: even if we did, the classical calculus would still turn out to be many-valued, as we show in Theorem 18.5.

## 13.2 *Examples*

We begin with the weakest possible calculi, in which consequence reduces to mere overlap, and with calculi at the other extreme, in which all or nearly all inferences are valid.

*Minimum calculi.* A *minimum* calculus is one with the weakest possible consequence relation for its universe of formulae. Since the overlap relation is closed under overlap, dilution and cut and is obviously contained in every such relation, it follows that there is a minimum calculus for each universe, and that in it we have X ⊢ Y iff X overlaps Y. Similarly in a minimum single-conclusion calculus we have X ⊢ B iff B ∈ X. A minimum calculus is evidently characterised by the universal set of partitions, and by vacuous rules of inference.

Every minimum propositional calculus is many-valued, by Theorem 15.2 in the multiple-conclusion case and Theorems 15.6 and 15.7 in the single-conclusion one. This is noteworthy as the first of our results to depend on the exclusion of 'propositional constants', for in order to describe a matrix as characterising a calculus containing such a constant F, we should presumably have to associate F with a constant truth-value b, so that v(F) is b for every valuation v, and according as b is designated or undesignated this implies that $\Lambda \vdash F$ or $F \vdash \Lambda$, neither of which can obtain in a minimum calculus. To the best of our knowledge no philosopher of logic has succeeded in justifying the inclusion of constants in a propositional calculus, but generally speaking it makes no mathematical difference, since for most purposes a calculus L containing a constant F is equivalent to a calculus L′ in which F is a singulary connective and in which all formulae of the form FA are synonymous in the sense to be defined in Section 16.1. For example, L′ is many-valued if L is, for to each matrix M which characterises L and in which F is associated with the value b, there corresponds a matrix M′ which characterises L′ and in which F is associated with the constant function $\lambda x.b$. The property of being a minimum calculus, however, is not preserved in the passage from L to L′, for even when L is a minimum calculus L′ contains distinct synonymous formulae and this is not possible in a minimum calculus.

*Singular calculi*. A calculus is *singular* if $A \vdash B$ for every A and B. The only partitions that can satisfy such a consequence relation are $\langle \Lambda, V \rangle$ and $\langle V, \Lambda \rangle$, and we can therefore distinguish four types of singular calculus, characterised by (i) the empty set of partitions, (ii) $\{\langle V, \Lambda \rangle\}$, (iii) $\{\langle \Lambda, V \rangle\}$ and (iv) $\{\langle V, \Lambda \rangle, \langle \Lambda, V \rangle\}$. They are respectively (i) *inconsistent* calculi, in which $X \vdash Y$ for every X and Y, and calculi in which (ii) $X \vdash Y$ iff Y is nonempty, (iii) $X \vdash Y$ iff X is nonempty, and (iv) $X \vdash Y$ iff X and Y are both nonempty.

If we now define a *singular matrix* as one whose values are all designated or all undesignated, we can distinguish three types of singular matrix: (i) empty matrices, (ii) nonempty matrices with exclusively designated values, and (iii) nonempty matrices with exclusively undesignated values. Singular propositional calculi of types (i)-(iii) are characterised by singular matrices of the corresponding type and so are many-valued. This is another result that depends on the exclusion of propositional constants. If constants are present, the requirement that each constant be paired with a truth-value rules out the empty matrix and prevents an inconsistent calculus from being many-valued. The point does not arise in the single-conclusion case, where inconsistent calculi are characterised by nonempty as well as by empty matrices and consequently are many-valued whether they contain constants or not. Since a singular calculus can only be characterised by a singular matrix (for if M has both a designated and an undesignated value then $p \nvdash_M q$ for distinct p and q), and since these are pre-empted by the calculi of types (i)-(iii), it follows that calculi of type (iv) are not many-valued.

Every single-conclusion consequence relation is satisfied by $\langle V,\Lambda\rangle$, so there are only two kinds of singular calculus here, namely *inconsistent* calculi in which $X \vdash B$ for every X and B, and calculi in which $X \vdash B$ iff X is nonempty. An example of the second kind is the calculus of contonktion (Prior, 1960), characterised by the rules 'from A infer A-tonk-B' ~~and~~ 'from A-tonk-B infer B'. Both kinds are many-valued, inconsistent calculi being characterised by singular matrices of types (i) or (ii), and the others by those of type (iii).

Our other examples of many-valued calculi are characterised by matrices taken from the literature, though for expository reasons we sometimes depart from the original notation. We shall also discuss *fragments* of these calculi, i.e. calculi based on a subset of their connectives and characterised by the corresponding fragment of the matrix. For this reason we treat $\supset$, &, $\vee$ and ~ as separate connectives even where some of them are definable in terms of others.

*The classical matrix* has two truth-values, t and f. The only designated value is t and the basic functions $\supset$, &, $\vee$, ~ are given by the familiar truth-tables.

*Kleene's matrix* (Kleene, 1952, Section 64) has three values, $0, \frac{1}{2}, 1$. The only designated value is 0 and the basic functions are $x \supset y = \min(1-x, y)$, $x \& y = \max(x, y)$, $x \vee y = \min(x, y)$, and $\sim x = 1-x$.

*Łukasiewicz matrices* (Łukasiewicz, 1920, 1930). The truth-values of an m-valued Łukasiewicz matrix ($m \geq 3$) are $0, \frac{1}{m-1}, \frac{2}{m-1}, \ldots, 1$, while those of the two infinite matrices are the rational or real numbers in the interval [0,1]. In each case the only designated value is 0 and the basic functions are $x \supset y = \max(0, y-x)$, $x \& y = \max(x, y)$, $x \vee y = \min(x, y)$, and $\sim x = 1-x$.

*Gödel matrices* may have as truth-values any simply ordered set with a least member 0, greatest member 1, and at least one other member. The only designated value is 0 and the basic functions are given by: $x \supset y = 0$ if $x \geq y$ and $x \supset y = y$ otherwise, $x \& y = \max(x, y)$, $x \vee y = \min(x, y)$, and $\sim x = 0$ if $x = 1$ and $\sim x = 1$ otherwise. The finite matrices of this family were devised by Gödel (1932) but the extension to the infinite case is due to Dummett (1959).

The following truth-tables make it easy to compare Kleene's matrix with the 3-valued Łukasiewicz and Gödel ones, the tables for & and $\vee$ being the same in every case. As in later examples, the designated values are indicated by an asterisk.

| & | 0 | ½ | 1 |
|---|---|---|---|
| *0 | 0 | ½ | 1 |
| ½ | ½ | ½ | 1 |
| 1 | 1 | 1 | 1 |

| v | 0 | ½ | 1 |
|---|---|---|---|
| *0 | 0 | 0 | 0 |
| ½ | 0 | ½ | ½ |
| 1 | 0 | ½ | 1 |

| ⊃ | 0 | ½ | 1 |
|---|---|---|---|
| *0 | 0 | ½ | 1 |
| ½ | 0 | ½ | ½ |
| 1 | 0 | 0 | 0 |

| ~ | |
|---|---|
| *0 | 1 |
| ½ | ½ |
| 1 | 0 |

Kleene

| ⊃ | 0 | ½ | 1 |
|---|---|---|---|
| *0 | 0 | ½ | 1 |
| ½ | 0 | 0 | ½ |
| 1 | 0 | 0 | 0 |

| ~ | |
|---|---|
| *0 | 1 |
| ½ | ½ |
| 1 | 0 |

Łukasiewicz

| ⊃ | 0 | ½ | 1 |
|---|---|---|---|
| *0 | 0 | ½ | 1 |
| ½ | 0 | 0 | 1 |
| 1 | 0 | 0 | 0 |

| ~ | |
|---|---|
| *0 | 1 |
| ½ | 1 |
| 1 | 0 |

Gödel

## 13.3 *Compactness*

Minimum and singular calculi are obviously compact. The next theorem establishes the compactness of all our examples taken from the literature, except for the infinite Łukasiewicz and Gödel calculi which are dealt with in the remaining theorems. All the results apply to multiple- and single-conclusion calculi alike.

*Theorem 13.1* Every calculus characterised by a finite matrix is compact.

*Proof* Given a calculus L characterised by a finite matrix with values $c_1,\dots,c_m$ we construct a rudimentary meta-calculus $L_1$ for it as follows. The formulae of $L_1$ are of the form a(A) and des(A) - read 'A takes value a' and 'A takes a designated value' - for each formula A of L and each value a. It has rules R1 'from $\Lambda$ infer $c_1(A),\ldots,c_m(A)$', R2 'from a(A),b(A) infer $\Lambda$ if a$\neq$b', R3 'from $a_1(A),\ldots,a_k(A)$ infer $F(a_1,\ldots,a_k)(FA_1\ldots A_k)$' for each F, R4 'from a(A) infer des(A) if a is designated' and R5 'from a(A),des(A) infer $\Lambda$ if a is undesignated'. For each valuation v of the formulae of L in the matrix let $\langle T_v,U_v\rangle$ be the partition of the formulae of $L_1$ such that a(A) $\in T_v$ iff v(A)=a, and des(A) $\in T_v$ iff v(A) is designated. It is easy to verify that $\langle T_v,U_v\rangle$ satisfies R1-5. Conversely, for each partition $\langle T,U\rangle$ which satisfies R1-5 there is by R1-2 a mapping v of the formulae of L into the truth-values such that a(A) $\in$ T iff v(A)=a, and by R3 v is a valuation. Moreover by R4-5 des(A) $\in$ T iff v(A) is designated, and hence $\langle T,U\rangle = \langle T_v,U_v\rangle$. Now X $\vdash$ Y iff every valuation v satisfies $\langle X,Y\rangle$, i.e. (using Frobenius notation) iff $\langle T_v,U_v\rangle$ satisfies $\langle$des(X),des(Y)$\rangle$ for each v, i.e. iff every partition satisfying R1-5 satisfies $\langle$des(X),des(Y)$\rangle$, i.e. iff des(X) $\vdash_1$ des(Y). But $\vdash_1$ is compact by 2.18 and hence $\vdash$ too is compact.

For single-conclusion L the analogous result can be proved similarly, since by the same argument X $\vdash$ B iff des(X) $\vdash_1$ {des(B)}. For other proofs of it see Łoś and Suszko, 1958, and Shoesmith and Smiley, 1971.

*Theorem 13.2* The rational- and real-valued Łukasiewicz matrices characterise distinct non-compact calculi.

*Theorem 13.3* The rational- and real-valued Gödel matrices characterise the same compact calculus.

*Theorem 13.4* No infinite well-ordered Gödel matrix characterises a compact calculus.

*Proofs* For 13.2 we need a result of McNaughton, 1951, which we state without proof:

*Lemma 13.5* For each closed interval S with rational endpoints in [0,1] there exists a formula S(p) in which p is the only variable and ⊃ and ~ the only connectives, such that for every valuation v in the rational- or real-valued Łukasiewicz matrix, v(S(p))=0 iff v(p) ∊ S.

Applying the lemma to $S_n = [0,\frac{1}{n}]$ it follows that $S_1(p),S_2(p),\ldots \vdash p$; but for any finite subset of these premisses the corresponding inference can be invalidated by taking v(p) sufficiently small (cf. Hay, 1963, p.84). To show that the two matrices characterise different calculi, let $S_1,S_2,\ldots$ be a series of nested closed intervals with rational endpoints that converges on an irrational number. Then $S_1(p),S_2(p),\ldots \vdash q$ in the rational-valued but not the real-valued case.

For 13.3, let $p_1,p_2,\ldots$ be an enumeration of the variables, and let I be the set of all valuations in the rational-valued Gödel matrix and I′ the set of valuations v′ such that, for every n, $v'(p_n)$ can be expressed as a fraction with denominator $2^n$. Let $\vdash_I$ and $\vdash_{I'}$ be the consequence relations characterised by I and I′ respectively. Then for each v in I there exists a valuation v′ in I′ such that, for every A and B, (i) v′(A)=0 iff v(A)=0, (ii) v′(A)=1 iff v(A)=1, and (iii) v′(A) ≥ v′(B) iff v(A) ≥ v(B). To show this we note that by the definition of I′ we may define $v'(p_n)$ inductively on n so that (i) $v'(p_n)=0$ iff $v(p_n)=0$, (ii) $v'(p_n)=1$ iff $v(p_n)=1$, and (iii) for each i<n, $v'(p_n) \geq v'(p_i)$ iff $v(p_n) \geq v(p_i)$ and $v'(p_n) \leq v'(p_i)$ iff $v(p_n) \leq v(p_i)$. This supplies the basis for a proof of the result by induction on the complexity of A and B, while the induction step exploits the fact that each of v(A⊃B), v(A&B), v(A∨B), v(~A) must be one of 0, 1, v(A), v(B). As a corollary of this result v(A) is designated iff v′(A) is designated, so that $\vdash_I = \vdash_{I'}$. A

similar argument shows that the real-valued matrix characterises the same consequence relation. But $\vdash_{I'}$ can be shown to be compact by the method of 13.1, taking R1 to be the rule 'from $\Lambda$ infer $\{m.2^{-n}(A): 0 \leq m \leq 2^n\}$ if A has no variables other than $p_1,\ldots,p_n$'.

For 13.4 let $p, p_1, p_2, \ldots$ be distinct variables and let $X = \bigcup(p_{n+1} \supset p_n) \supset p$. To invalidate $\langle X, p \rangle$ requires a valuation such that $v(p_{n+1}) < v(p_n)$ for each n, contradicting the assumption of well-ordering; hence $X \vdash p$. On the other hand if $X'$ is a finite subset of X let j exceed each index n for which $p_n$ appears in a member of $X'$. Then a valuation v such that $v(p_1) > v(p_2) > \ldots > v(p_j) > v(p) > 0$ invalidates $\langle X', p \rangle$, and hence $X' \nvdash p$.

Dummett, 1959, gave axioms for a single-conclusion calculus LC which he showed to be characterised by the Gödel matrix of order type $\omega+1$ in the restricted sense that all the theorems of LC are valid in the matrix and vice versa. Theorem 13.4 shows that this matrix cannot characterise consequence in LC, but on the other hand both the rational- and real-valued Gödel matrices do so, by Theorem 13.3 and

*Theorem 13.6* Dummett's LC is the only compact single-conclusion calculus characterised by an infinite Gödel matrix.

*Proof* Let M be an infinite Gödel matrix such that the single-conclusion relation $\vdash_M$ is compact, and let $\vdash_{LC}$ stand for consequence in LC. Dummett (1959, Theorem 4) shows that for every infinite Gödel matrix the valid formulae are just the theorems of LC; so that $\Lambda \vdash_M A$ iff $\Lambda \vdash_{LC} A$. The deduction theorem holds for $\supset$ with respect to $\vdash_M$ by 14.1 below, and holds in LC by the standard proof; hence $A_1,\ldots,A_m \vdash_M B$ iff $\Lambda \vdash_M A_1 \supset \ldots \supset . A_m \supset B$, i.e.

iff $\Lambda \vdash_{LC} A_1 \supset \ldots \supset . A_m \supset B$, i.e. iff $A_1, \ldots, A_m \vdash_{LC} B$. Since $\vdash_M$ and $\vdash_{LC}$ are both compact, it follows for all X that $X \vdash_M B$ iff $X \vdash_{LC} B$, as required.

# 14 · Matrices

## 14.1 *Matrix functions*

In this chapter we develop the elements of an algebraic theory of matrices, the results of which will be needed later.

A *truth-function* is any function (singulary, binary, k-ary) defined over the truth-values of a matrix and taking truth-values for its values. A truth-function is a *matrix function* if it can be obtained from the basic functions and the projective functions $\lambda x_1 \ldots x_n . x_i$ by a finite number of operations of composition. Thus if f is a k-ary matrix function and $g_1, \ldots, g_k$ are n-ary matrix functions, $\lambda x_1 \ldots x_n . f(g_1(x_1, \ldots, x_n), \ldots, g_k(x_1, \ldots, x_n))$ is also a matrix function. In general the matrix functions will not exhaust the truth-functions: if they do the matrix is said to be *functionally complete*. Let $p_1, \ldots, p_n$ be distinct propositional variables that include all those in the formula A, and let f be the n-ary truth-function such that $f(x_1, \ldots, x_n)$ is the value taken by A when each $p_i$ is assigned the value $x_i$. It is easy to see that the matrix functions are precisely the functions associated with formulae in this way, and their logical significance derives from this. For example, where two matrices differ in their basic functions but have the same matrix functions, the corresponding calculi will merely be versions of each other, based on a different choice of connectives.

A *submatrix* M′ of M is determined by any subset of the values of M that is closed under the basic functions (or, equivalently, under the matrix functions). A value of M′ is to be designated in M′ iff it is designated in M, and the basic functions are to be those of M restricted to the subset in question. In particular a *proper* submatrix of M is a (possibly empty) submatrix other than M itself, and the submatrix *generated* by a subset of values S is the submatrix whose set of values is the closure of S under the basic (or matrix) functions of M.

Just as any relation between a matrix and a calculus presupposes a pairing of the basic functions with the connectives, so in relating two matrices we presuppose that they are of the same similarity class, i.e. that each k-ary basic function f of either matrix has been paired off with a k-ary basic function f′ of the other. This pairing induces a pairing of the matrix functions such that paired functions are associated with the same formulae in the manner described above. Both pairings are one-one if the functions are counted intensionally, though generally not if they are counted extensionally, since there will generally be several ways of building the same matrix function out of the basic ones, and indeed the same function may occur several times in the list of basic functions.

A (many-many) *correspondence* between two matrices is a relation between their respective truth-values such that each value of either matrix is related to at least one value of the other. A correspondence $x \leftrightarrow x'$ *preserves structure* if, for every $a, a', a_1, a_1', \ldots$ such that $a \leftrightarrow a'$, $a_1 \leftrightarrow a_1'$, etc., (i) a is designated iff a′ is designated and (ii) $f(a_1, \ldots, a_k) \leftrightarrow f'(a_1', \ldots, a_k')$ for each pair of basic functions or, equivalently, of matrix functions. A structure-preserving correspondence between M and M′ is a *homomorphism* of M onto M′ if it is many-one, and an *isomorphism* if it is one-one. Using the notation of functions for a many-one correspondence or mapping, the conditions for a mapping h of the values of M onto those of M′ to be a homomorphism are thus: (i) a is designated iff h(a) is designated and (ii) $h(f(a_1,\dots,a_k)) = f'(h(a_1),\dots,h(a_k))$ for each pair of basic functions.

To illustrate these ideas, let M be any Gödel matrix and let a be any truth-value other than 1; let $M_a$ differ from M only in that all values $\le a$ are designated, and let $h_a$ be the mapping of truth-values into truth-values such that $h_a(x) = 0$ if $x \le a$ and otherwise $h_a(x) = x$. Then it can easily be verified that 0 and the values greater than a make up the values of a submatrix $M'_a$ of M, and $h_a$ is a homomorphism of $M_a$ onto $M'_a$. We may use this to show that

*Lemma 14.1* The deduction theorem holds in the single-conclusion calculus characterised by a Gödel matrix M.

*Proof* If $X \nvdash_M A \supset B$ then some valuation v in M invalidates $\langle X, A \supset B \rangle$, so that $v(A) < v(B) \le 1$. But v invalidates $\langle X \cup \{A\}, B \rangle$ in $M_{v(A)}$, so that $h_{v(A)}v$ invalidates $\langle X \cup \{A\}, B \rangle$ in $M'_{v(A)}$ and hence in M, showing that $X, A \nvdash_M B$ as required.

## 14.2 *Separability*

Let $a_1,\dots,a_n$ and $b_1,\dots,b_n$ be n-tuples of truth-values; then an n-ary truth-function f *separates* them if $f(a_1,\dots,a_n)$ is designated and $f(b_1,\dots,b_n)$ is undesignated or vice versa. Two sequences are *separable* if some matrix function separates them; otherwise they are *inseparable*. In the latter case we write $a_1,\dots,a_n \approx b_1,\dots,b_n$.

*Theorem 14.2* Inseparability is an equivalence relation.

*Theorem 14.3* If $a_1,\ldots,a_n \approx b_1,\ldots,b_n$ then, for each i, $a_i$ is designated iff $b_i$ is designated.

*Theorem 14.4* If $a_1,\ldots,a_n \approx b_1,\ldots,b_n$ and $i_1,\ldots,i_r$ are drawn (possibly with repetitions) from $1,\ldots,n$ then $a_{i_1},\ldots,a_{i_r} \approx b_{i_1},\ldots,b_{i_r}$.

*Theorem 14.5* If $a_1,\ldots,a_n,b_1,\ldots,b_n$ belong to a submatrix M′ of M then $a_1,\ldots,a_n \approx b_1,\ldots,b_n$ in M′ iff $a_1,\ldots,a_n \approx b_1,\ldots,b_n$ in M.

*Theorem 14.6* If $x\leftrightarrow x'$ is a structure-preserving correspondence between M and M′ then $a_1,\ldots,a_n \approx b_1,\ldots,b_n$ in M iff $a'_1,\ldots,a'_n \approx b'_1,\ldots,b'_n$ in M′. In particular, if h is a homomorphism of M onto M′ then $a_1,\ldots,a_n \approx b_1,\ldots,b_n$ iff $h(a_1),\ldots,h(a_n) \approx h(b_1),\ldots,h(b_n)$.

*Proofs* Evidently $\approx$ is reflexive, symmetric and transitive, establishing 14.2. For 14.3, if the two sequences are inseparable then in particular the projective functions fail to separate them. For 14.4 we note that if f separates $a_{i_1},\ldots,a_{i_r}$ and $b_{i_1},\ldots,b_{i_r}$ then $\lambda x_1\ldots x_n.f(x_{i_1},\ldots,x_{i_r})$ separates $a_1,\ldots,a_n$ and $b_1,\ldots,b_n$ and is easily seen to be a matrix function if f is. For 14.5, the matrix functions f′ of M′ are precisely the restrictions to M′ of the matrix functions f of M, and f and f′ separate the same sequences in M′. Finally, for 14.6, let $f\leftrightarrow f'$ be the correspondence between the matrix functions of M and M′ induced by the pairing of their respective basic functions; then f separates $a_1,\ldots,a_n$ and $b_1,\ldots,b_n$ iff f′ separates $a'_1,\ldots,a'_n$ and $b'_1,\ldots,b'_n$.

The *dimension* $\dim(a_1,\ldots,a_n)$ of a sequence $a_1,\ldots,a_n$ of truth-values is the least number of distinct values from which a sequence $c_1,\ldots,c_n$ can be formed (possibly with repetitions) such that $c_1,\ldots,c_n \simeq a_1,\ldots,a_n$.

*Theorem 14.7* If each of $a_1,\ldots,a_n$ occurs in $b_1,\ldots,b_m$ then $\dim(a_1,\ldots,a_n) \leq \dim(b_1,\ldots,b_m)$.

*Theorem 14.8* If $a_1,\ldots,a_n \simeq b_1,\ldots,b_n$ and $\dim(a_1,\ldots,a_n) \geq \dim(b_1,\ldots,b_{n+k})$ then there exist $i_1,\ldots,i_k$ ($1 \leq i_j \leq n$ for each $j$) such that $a_1,\ldots,a_n,a_{i_1},\ldots,a_{i_k} \simeq b_1,\ldots,b_{n+k}$.

*Proofs* For 14.7, suppose $\dim(b_1,\ldots,b_m) = r$; then $b_1,\ldots,b_m \simeq c_1,\ldots,c_m$, say, where at most $r$ of the $c_i$ are distinct. By 14.4 $a_1,\ldots,a_n$ is therefore inseparable from a sequence formed from values in $c_1,\ldots,c_m$, and hence $\dim(a_1,\ldots,a_n) \leq r$.

For 14.8, suppose that $b_1,\ldots,b_{n+k} \simeq c_1,\ldots,c_{n+k}$, where just $r$ of the $c_i$ are distinct and $r \leq \dim(a_1,\ldots,a_n)$. Then by 14.4 $a_1,\ldots,a_n \simeq b_1,\ldots,b_n \simeq c_1,\ldots,c_n$, so that all $r$ of the distinct $c_i$ occur among $c_1,\ldots,c_n$; i.e. for each $j$ ($1 \leq j \leq k$) there exists $i_j$ ($1 \leq i_j \leq n$) such that $c_{n+j} = c_{i_j}$. Hence by 14.4 $b_1,\ldots,b_{n+k} \simeq c_1,\ldots,c_n,c_{i_1},\ldots,c_{i_k} \simeq a_1,\ldots,a_n,a_{i_1},\ldots,a_{i_k}$.

Two values $a$ and $b$ are *congruent* in a matrix if $a,x_1,\ldots,x_n \simeq b,x_1,\ldots,x_n$ for every $n$ and every $n$-tuple of values $x_1,\ldots,x_n$. The use of the term 'congruent' is justified by Theorem 14.10 below, but we need to show first that congruence implies inseparability, in the sense that two sequences which are congruent term-by-term are inseparable.

*Theorem 14.9* If $a_i$ is congruent to $b_i$ for each $i$, then $a_1,\ldots,a_n \simeq b_1,\ldots,b_n$.

*Theorem 14.10* Congruence is a structure-preserving equivalence relation.

*Theorem 14.11* If $x \leftrightarrow x'$ is a structure-preserving correspondence between M and M′ then a and b are congruent in M iff a′ and b′ are congruent in M′. In particular if h is a homomorphism of M onto M′, a and b are congruent iff h(a) and h(b) are.

*Theorem 14.12* If M is generated by $c_1,\ldots,c_m$ then a and b are congruent in M iff $a,c_1,\ldots,c_m \approx b,c_1,\ldots,c_m$.

*Theorem 14.13* Two values of a submatrix M′ of M are congruent in M′ if congruent in M.

*Proofs* For 14.9, if $a_i$ and $b_i$ are congruent then in particular $a_i,b_1,\ldots,b_{i-1},a_{i+1},\ldots,a_n \approx b_i,b_1,\ldots,b_{i-1},a_{i+1},\ldots,a_n$. Hence by 14.4 we have successively $a_1,a_2,\ldots,a_n \approx b_1,a_2,\ldots,a_n \approx \ldots \approx b_1,\ldots,b_i,a_{i+1},\ldots,a_n \approx \ldots \approx b_1,\ldots,b_n$.

For 14.10 we need to show that (i) congruence is an equivalence relation, (ii) two congruent values are both designated or both undesignated, and (iii) if each pair $a_i$ and $b_i$ are congruent, so are $f(a_1,\ldots,a_k)$ and $f(b_1,\ldots,b_k)$ for each basic function f. Of these (i) is immediate from 14.2 and (ii) from 14.3, while for (iii) we note that if a mat_ix function g were to separate $f(a_1,\ldots,a_k),c_1,\ldots,c_n$ and $f(b_1,\ldots,b_k),c_1,\ldots,c_n$, the matrix function $\lambda x_1\ldots x_k y_1\ldots y_n.g(f(x_1,\ldots,x_k),y_1,\ldots,y_n)$ would separate $a_1,\ldots,a_k,c_1,\ldots,c_n$ and $b_1,\ldots,b_k,c_1,\ldots,c_n$, and so by 14.9 $a_i$ and $b_i$ would be incongruent for some i. We note that in general there are a number of congruence relations - structure-preserving equivalence relations - on a matrix (cf. Wójcicki, 1974), lying between identity at one extreme and what we have called congruence at the other.

For 14.11, if $a,x_1,\ldots,x_n \neq b,x_1,\ldots,x_n$ then by 14.6 $a',x_1',\ldots,x_n' \neq b',x_1',\ldots,x_n'$ and vice versa. For 14.12, if a matrix function g separates $a,x_1,\ldots,x_n$ and $b,x_1,\ldots,x_n$, where each $x_i$ is of the form $f_i(c_1,\ldots,c_m)$ for a matrix function $f_i$, then $\lambda yy_1\ldots y_m.g(y,f_1(y_1,\ldots,y_m),\ldots, f_n(y_1,\ldots,y_m))$ is a matrix function separating $a,c_1,\ldots,c_m$ and $b,c_1,\ldots,c_m$; and the converse is immediate. Finally, 14.13 follows immediately from 14.5.

The converse of Theorem 14.13 fails: that is to say, although by Theorem 14.5 two sequences of values of a submatrix M′ of M are inseparable in M′ iff inseparable in M, two values of M′ may be congruent in M′ but incongruent in M. Thus in Figure 14.1, the values 1 and 3 are congruent in $M_3$ by Theorem 14.11, for the mapping h such that h(0)=0 and h(1)=h(3)=1 is a homomorphism of $M_3$ onto $M_4$. Also $M_3$ is a submatrix of $M_2$; but 1 and 3 are not congruent in $M_2$, for 3⊃2 is designated while 1⊃2 is not, so that 1,2 ≠ 3,2.

| ⊃ | 0 | 1 | 2 |
|---|---|---|---|
| *0 | 0 | 1 | 2 |
| 1 | 0 | 0 | 1 |
| 2 | 0 | 0 | 0 |

$M_1$

| ⊃ | 0 | 1 | 2 | 3 |
|---|---|---|---|---|
| *0 | 0 | 1 | 2 | 3 |
| 1 | 0 | 0 | 1 | 0 |
| 2 | 0 | 0 | 0 | 0 |
| 3 | 0 | 0 | 0 | 0 |

$M_2$

| ⊃ | 0 | 1 | 3 |
|---|---|---|---|
| *0 | 0 | 1 | 3 |
| 1 | 0 | 0 | 0 |
| 3 | 0 | 0 | 0 |

$M_3$

| ⊃ | 0 | 1 |
|---|---|---|
| *0 | 0 | 1 |
| 1 | 0 | 0 |

$M_4$

Figure 14.1

### 14.3 *Equivalence*

A *simple* matrix is one in which congruence is the identity relation, so that no two distinct values are congruent. In

Figure 14.1 for example, $M_1$ is simple, for each of 1 and 2 is incongruent to 0 by Theorem 14.3 while 1 and 2 are incongruent because ⊃ separates 1,2 and 2,2. Similarly it can be verified that $M_2$ and $M_4$ are simple, but $M_3$ is not simple since, as we have seen, 1 and 3 are distinct but congruent values. Any congruence relation defined over an algebra induces a quotient algebra on the resulting set of equivalence classes, the two being related by a natural homomorphism. In particular the relation of congruence as we have defined it for a matrix M determines the *quotient matrix* $|M|$ whose values are the various equivalence classes $|a|$ = {x in M: x is congruent to a}, where $|a|$ is designated in $|M|$ iff a is designated in M, and the basic functions of $|M|$ are of the form $\lambda |x_1| \ldots |x_k| . |f(x_1,\ldots,x_k)|$.

*Theorem 14.14* The only homomorphisms of a simple matrix are isomorphisms, and the only structure-preserving correspondences between M and a simple matrix M′ are homomorphisms of M onto M′.

*Theorem 14.15* There is a homomorphism of M onto a simple matrix M′ iff M′ is isomorphic to $|M|$.

*Theorem 14.16* Every quotient matrix is simple.

*Theorem 14.17* A simple matrix is isomorphic to its own quotient.

*Proofs* 14.14 follows immediately from 14.11. For 14.15 we note that the natural mapping $x \to |x|$ combines with any isomorphism between $|M|$ and M′ to produce a homomorphism of M onto M′. Conversely, if h is a homomorphism of M onto M′, the relation $h(x) \leftrightarrow |x|$ is a one-one correspondence between M′ and $|M|$ if M′ is simple, by 14.11; and since it is easily verified that it is in any case a homomorphism,

M′ is isomorphic to |M| as required. The two remaining theorems follow at once from 14.15 and 14.11.

Two matrices are *equivalent* if there is a structure-preserving correspondence between them. In Figure 14.1 for example, $M_3$ and $M_4$ are equivalent, since, as we have seen, there is a homomorphism of $M_3$ onto $M_4$.

*Theorem 14.18* Two matrices are equivalent iff their quotients are isomorphic. In particular, two simple matrices are equivalent iff they are isomorphic.

*Theorem 14.19* If M and M′ are equivalent, every submatrix of M is equivalent to a submatrix of M′.

*Proofs* For 14.18, a structure-preserving correspondence a↔a′ between two matrices induces a correspondence |a| ↔ |a′| between their quotients, and this is an isomorphism by 14.14 and 14.16. Conversely, given an isomorphism between |M| and |M′|, the relation which holds between a and a′ if |a| corresponds to |a′| is the required structure-preserving correspondence between M and M′. The corollary for the simple case follows by 14.17.

For 14.19, let a↔a′ be a structure-preserving correspondence between M and M′. For each submatrix $M_1$ of M the set of values a′ of M′ such that a↔a′ for some a in $M_1$ is seen to be closed under the basic functions and so defines a submatrix $M_1'$, and the restriction of ↔ to $M_1$ and $M_1'$ is the required structure-preserving correspondence between them.

The classical matrix and all its fragments are obviously simple, as is any non-singular two-valued matrix. (Singular matrices with two or more values are never simple.) All

Łukasiewicz and Gödel matrices are simple, for in each case ⊃ separates a,b from b,b whenever $a<b$, so that any two distinct values are incongruent. Similarly any fragment of any of these matrices is simple as long as ⊃ is present; but when ⊃ is absent all the values other than 0 and 1 become congruent to each other, so that every fragment without ⊃ is equivalent to a three-valued one. This explains why in the literature there is no family of Kleene matrices, i.e. matrices differing from the Łukasiewicz ones only in that $a \supset b =_{df} {\sim}a \vee b$; for every such matrix is merely equivalent to the three-valued one.

## 14.4 *Monadicity*

We say that two values a and b are inseparable if the one-member sequences a and b are inseparable. Inseparability is generally a broader relation than congruence, for by Theorem 14.9 congruent values are always inseparable but not vice versa. In Figure 14.1 for example, $1 \approx 3$ in $M_2$ since the only singulary matrix functions available are the identity $\lambda x.x$ and the function $\lambda x.x \supset x$ with constant value 0; but as we saw 1 and 3 are not congruent since $1,2 \not\approx 3,2$. A matrix is *monadic* if congruence and inseparability coincide for it, i.e. if inseparable values are always congruent.

*Theorem 14.20* In a monadic matrix, $a_1,\ldots,a_n \approx b_1,\ldots,b_n$ iff $a_i \approx b_i$ for each i.

*Theorem 14.21* Any submatrix M′ of a monadic matrix M is monadic, and values of M′ are congruent in M′ iff they are congruent in M.

*Theorem 14.22* Any matrix equivalent to a monadic matrix is monadic.

*Theorem 14.23* Any submatrix of a simple monadic matrix is simple.

*Proofs* 14.20 is immediate from 14.4 and 14.9. 14.21 is immediate from 14.5 and 14.13: it shows that the converse of 14.13 holds for monadic matrices although not in general, as we saw in Section 14.2. 14.22 is immediate from 14.6 and 14.11. 14.23 is immediate from 14.21: the discussion of $M_2$ and $M_3$ in Section 14.3 shows that the result fails for non-monadic matrices.

*Theorem 14.24* Every functionally complete matrix is monadic.

*Theorem 14.25* Every two-valued matrix is monadic.

*Theorem 14.26* Every matrix with no non-singulary basic functions is monadic.

*Proofs* For 14.24, either the matrix is singular and every value congruent to every other, or else any two values can be separated by a suitable matrix function. Likewise for 14.25, either the matrix is singular or the values are separated by the identity function. For 14.26 we note that every matrix function here is of the form $\lambda x_1 \ldots x_k . g(x_i)$ for a singulary matrix function g, and if such a function separates $a, c_2, \ldots, c_k$ and $b, c_2, \ldots, c_k$ then i=1 and g separates a and b.

The classical matrix and its fragments are monadic by Theorem 14.25, and the table below shows the situation with regard to the other examples from the literature.

| | Kleene | Łukasiewicz | Gödel |
|---|---|---|---|
| Complete matrix or fragments with ⊃ and ~ | Monadic | Monadic | 3-valued monadic<br>Rest not monadic |
| Fragments with ⊃ but not ~ | Monadic | Not monadic | Not monadic |
| Fragments without ⊃ | Monadic | Monadic | Monadic |

To justify the 'monadic' entries for the top row we note that the undesignated values in Kleene's or the three-valued Gödel matrix are separated by ~, while by Lemma 13.5 any two values in a Łukasiewicz matrix are separated by a McNaughton function constructed out of ⊃ and ~. Similarly for the middle row the undesignated values in the Kleene fragments are separated by $\lambda x.x \supset x$, while in each case in the bottom row all the undesignated values other than 1 are congruent and are either separated from 1 by ~ or, if ~ is absent, are congruent to it too. For the 'not monadic' entries we take a and b to be distinct undesignated values (for the Gödel cases in the top row we also need to assume that neither a nor b is 1); then a and b are incongruent, but the submatrices generated by a and b respectively are isomorphic, and it is left to the reader to deduce that a and b are inseparable.

Monadicity is a special case of the concept of the degree of a matrix. By Theorem 14.20 two sequences of values in a monadic matrix are inseparable iff every subsequence of length at most 1 of either is inseparable from the corresponding subsequence of the other. Similarly for each of the non-monadic matrices listed in the table it can be shown that two sequences are inseparable iff corresponding subsequences of length at most 2 are always inseparable. Generally, then, we define the *degree* of a matrix to be the least number r such that two sequences of values are inseparable iff corresponding subsequences of length at most r are always inseparable. Using Theorem 14.4, the degree is the least number r such that, for all n and n-tuples $a_1,\ldots,a_n$ and $b_1,\ldots,b_n$ if $a_{i_1},\ldots,a_{i_r} \approx b_{i_1},\ldots,b_{i_r}$ whenever $i_1,\ldots,i_r$ are drawn (possibly with repetitions) from $1,\ldots,n$, then $a_1,\ldots,a_n \approx b_1,\ldots,b_n$. If no such r exists we say that the degree is infinite. It can be shown, for example, that matrices characterising minimum calculi have in general an infinite degree. The definition also allows for a matrix to have degree 0, namely when there exist no separable sequences. Thus a matrix has degree 0 iff it is singular. Our definition of monadicity included such matrices as monadic, but for all non-singular matrices monadicity and being of degree 1 coincide.

It can be shown that the degree of a finite matrix is finite and that there exist matrices of infinite degree and of every finite degree (indeed, matrices characteristic for the classical calculus, for every infinite Boolean matrix (Section 18.3) has infinite degree, while a Boolean matrix of $2^n$ values has degree n if $n<3$ and degree $n-1$ if $n\geq3$). It can also be shown that the degree of a submatrix of a matrix M cannot exceed that of M, that equivalent matrices have the same degree and so do any matrices that characterise the same calculus. We prove only the first of these results, since it is the only one we actually require. We therefore show that the degree of an m-valued matrix cannot exceed m, this being the best possible result in the sense that for each m there can be shown to exist m-valued matrices of degree exactly m.

*Theorem 14.27* In an m-valued matrix, if $a_{i_1},\ldots,a_{i_m} \approx b_{i_1},\ldots,b_{i_m}$ whenever $i_1,\ldots,i_m$ are drawn (possibly with repetitions) from $1,\ldots,n$, then $a_1,\ldots,a_n \approx b_1,\ldots,b_n$.

*Proof* We argue by induction on n. If $n \le m$ the result is immediate from 14.4. If $n > m$ then by the induction hypothesis $a_1, \ldots a_{i-1}, a_{i+1}, \ldots, a_n \approx b_1, \ldots, b_{i-1}, b_{i+1}, \ldots, b_n$ for each i; in particular it follows from 14.4 that

(1) $a_2, a_2, a_3, a_4, a_5, \ldots, a_n \approx b_2, b_2, b_3, b_4, b_5, \ldots, b_n$ (assuming $n \ge 2$)
(2) $b_2, b_2, b_2, b_4, b_5, \ldots, b_n \approx a_2, a_2, a_2, a_4, a_5, \ldots, a_n$ (assuming $n \ge 3$)
(3) $a_1, a_2, a_2, a_4, a_5, \ldots, a_n \approx b_1, b_2, b_2, b_4, b_5, \ldots, b_n$ (assuming $n \ge 3$)
(4) $b_2, b_2, b_3, b_3, b_5, \ldots, b_n \approx a_2, a_2, a_3, a_3, a_5, \ldots, a_n$ (assuming $n \ge 4$)
(5) $a_1, a_2, a_3, a_3, a_5, \ldots, a_n \approx b_1, b_2, b_3, b_3, b_5, \ldots, b_n$ (assuming $n \ge 4$).

We seek to show that $a_1, \ldots, a_n \approx b_1, \ldots, b_n$. Since $n > m$ two of the $a_i$ are identical, say $a_1 = a_2$, and two of the $b_i$ are identical. We need only consider three representative cases. In the first, $b_1 = b_2$ and the result is implicit in (1) above. In the second $b_2 = b_3$, say, and the result is obtained by 14.2 from (1), (2) and (3). In the third case, $b_3 = b_4$, say, and the result is obtained by 14.2 from (1), (4) and (5).

# 15 · Many-valuedness

## 15.1 *Cancellation*

It is sometimes held that one proposition cannot entail another unless there is some connection of meaning between them. We are interested in a related principle, namely that unless there is a connection of meaning between two propositions there cannot be a connection between their truth or falsity. This represents a generalisation of the entailment principle as far as contingent propositions are concerned, but leaves open such questions as whether p&~p entails q or whether p entails q∨~q. For suppose that there is no connection of meaning between two propositions. By our principle it follows that the truth or falsity of either one is not constrained or conditioned by that of the other; i.e. if they can be true separately they can be true simultaneously, if they can be false separately they can be false simultaneously, and if one can be true and the other false separately they can be so simultaneously. When the propositions concerned are contingent each of these conditionals has a true antecedent, and so we may conclude from the last of them that it is possible for either proposition to be true while the other is false. It therefore follows - on any account of entailment - that neither entails the other. When either of the propositions is necessary or impossible, however, there is no way of drawing such a conclusion from our principle.

Our principle refers to a pair of propositions, but it can be made to cover an arbitrary family instead of a pair and arbitrary sets of propositions instead of individual propositions. We wish to formulate this generalised version in the context of a propositional calculus. Starting with formulae, therefore, we express the idea of connection of meaning by saying that two formulae are *unconnected* if no variable appears in both of them. We then say that two sets are *unconnected* if no formula in one is connected with any formula in the other, and a family of sets is *disconnected* if it is nonempty and no two members of it are connected. (Any generalisation can at best have a conventional significance in the case of an empty family. Our decision not to count an empty family as being disconnected has the effect of making the 'cancellation' principle, as we shall call it, trivially true in the empty case. The contrary decision would make no difference where consistent calculi are concerned, but would mean that inconsistent calculi ceased to satisfy the principle. The analogous condition for single-conclusion calculi is the same whether we count the empty family as disconnected or not.)

In a disconnected family $\{Z_i\}$ there are no variables in common between formulae belonging to different $Z_i$. The generalised principle is that where a family $\{Z_i\}$ is disconnected there can be no connection between the truth or falsity of formulae belonging to different $Z_i$. Following our previous explanation, this means that if for each i it is possible for all the formulae in a subset $X_i$ of $Z_i$ to be true while those of another subset $Y_i$ are false, it must be possible for all the formulae in $\bigcup X_i$ to be true while all those in $\bigcup Y_i$ are false. It is obviouslv sufficient to limit the assertion of this condition to the case in which $X_i \cup Y_i = Z_i$. In the context of a calculus the possibilities for truth and falsehood are dictated by the possible states of affairs, i.e. the partitions which characterise the calculus, and the condition can therefore be reformulated in terms of the consequence relation of the calculus by writing '$X \nvdash Y$' instead of 'it is possible for all the formulae in X to be true while all those in Y are false'. Doing this, and contraposing the result, we say finally that a calculus permits *cancellation* if, whenever $\{X_i \cup Y_i\}$ is a disconnected family such that $\bigcup X_i \vdash \bigcup Y_i$, then $X_i \vdash Y_i$ for some i.

We can best derive the analogous condition for single-conclusion calculi by considering what must obtain in the single-conclusion part of a calculus L that permits cancellation. We first state the special case of cancellation in L in which one of the $Y_i$ consists of a single formula while the others are empty, namely: if the family consisting of $X \cup \{B\}$ and various $X_i$ is disconnected and $X, \bigcup X_i \vdash \{B\}$, then either $X \vdash \{B\}$ or, for some i, $X_i \vdash \Lambda$ (i.e. some $X_i$ is inconsistent). By Theorems 5.31 and 5.32 inconsistency in L corresponds to formal inconsistency rather than mere inconsistency in its single-conclusion part, but in the present context the latter two are equivalent, for if $X_i$ is unconnected with B then at least one variable does not occur in any member of $X_i$, and we may go on to show exactly as in the proof of Theorem 5.33 that $X_i$ is inconsistent iff it is formally inconsistent. It therefore makes no difference which we choose, and we say that a single-conclusion calculus permits *cancellation* if, whenever $\{X_i\}$ is a disconnected family of sets which are neither (formally) inconsistent nor connected with X or B, and $X, \bigcup X_i \vdash B$, then $X \vdash B$.

Not every propositional calculus permits cancellation; examples in the single-conclusion case are Johansson's Minimalkalkül and Lewis's S1-S3 (see Shoesmith and Smiley, 1971, Theorem 6). We do however have

*Theorem 15.1* Every many-valued (multiple- or single-conclusion) calculus permits cancellation.

*Proof* Let $\{X_i \cup Y_i\}$ be a disconnected family in a calculus characterised by a matrix M. If $X_i \nvdash_M Y_i$ for each i, some valuation $v_i$ invalidates $\langle X_i, Y_i \rangle$. Since the value of a formula depends only on the values assigned to the variables that actually occur in it, the various $v_i$ can be combined to produce a single valuation v such that $v(A) = v_i(A)$ for each i and each A in $X_i \cup Y_i$. Hence v invalidates $\langle \bigcup X_i, \bigcup Y_i \rangle$ as required. The single-conclusion result is proved similarly.

Our immediate interest in cancellation is as a possible criterion for many-valuedness. Theorem 15.1 shows it to be a necessary condition; is it also sufficient?

## 15.2 *Compact calculi*

In a compact calculus (though not in general) the cancellation property is equivalent to the special case in which there are only two sets in the family, namely: if $X_1 \cup Y_1$ and $X_2 \cup Y_2$ are unconnected and $X_1, X_2 \vdash Y_1, Y_2$, then $X_1 \vdash Y_1$ or $X_2 \vdash Y_2$. For if $\bigcup X_i \vdash \bigcup Y_i$ then by compactness and dilution the same is true when the unions are taken over some finite subfamily, whence by repeated application of the special case $X_i \vdash Y_i$ for some i. Cancellation for a compact single-conclusion calculus is similarly shown to be equivalent to this special case: if $X, Y \vdash B$ and Y is neither (formally) inconsistent nor connected with X or B, then $X \vdash B$.

*Theorem 15.2* A compact propositional calculus is many-valued iff it permits cancellation.

*Proof* A many-valued calculus permits cancellation by 15.1. For the converse, if L is a compact propositional calculus permitting cancellation, let $L_1$ be obtained by adding to it an arbitrary number of propositional variables, consequence in $L_1$ being defined as the closure of consequence in L under substitution and dilution. That is to say, $X \vdash_1 Y$ iff there exist sets $X'$ and $Y'$ of formulae of L and a substitution s in $L_1$ such that $s(X') \subset X$ and $s(Y') \subset Y$ and $X' \vdash Y'$. Evidently $\vdash_1$ is closed under overlap, dilution and substitution, and where X and Y are sets of formulae in L, $X \vdash_1 Y$ iff $X \vdash Y$.

*Lemma 15.3* $\vdash_1$ is a consequence relation.

To prove the lemma it is sufficient by 2.10 to show that $\vdash_1$ is compact and closed under cut for formulae. For compactness, suppose that $X \vdash_1 Y$, so that $X' \vdash Y'$ and $s(X') \subset X$ and $s(Y') \subset Y$ for appropriate $X'$,$Y'$ and s. By hypothesis L is compact, so there exist finite subsets $X''$ and $Y''$ of $X'$ and $Y'$ such that $X'' \vdash Y''$; and $s(X'')$ and $s(Y'')$ are thus finite subsets of X and Y such that $s(X'') \vdash_1 s(Y'')$ as required. For cut, suppose that $X,A \vdash_1 Y$ and $X \vdash_1 A,Y$. By compactness there exist finite subsets $X_1$,$X_2$ of X and $Y_1$,$Y_2$ of Y such that $X_1,A \vdash_1 Y_1$ and $X_2 \vdash_1 A,Y_2$. Since only finitely many variables occur in $X_1,X_2,Y_1,Y_2,A$, there exists a substitution s in $L_1$ which interchanges each of these variables with a distinct variable of L. Then $s(X_1),s(A) \vdash s(Y_1)$ and $s(X_2) \vdash s(A),s(Y_2)$, and by cut and dilution in L it follows that $s(X_1),s(X_2) \vdash s(Y_1),s(Y_2)$. But $s(s(X_1) \cup s(X_2)) = X_1 \cup X_2 \subset X$, and similarly $s(s(Y_1) \cup s(Y_2)) \subset Y$, so that $X \vdash_1 Y$ as required.

*Lemma 15.4* If $\{X_i \cup Y_i\}$ is a disconnected family of countable sets, and $X_i \nvdash_1 Y_i$ for each i, then some partition $\langle T,U \rangle$ of the formulae of $L_1$ is such that $T \nvdash_1 U$ and $X_i \subset T$ and $Y_i \subset U$ for each i.

To prove the lemma, it is sufficient by 15.3 to show that $L_1$ permits cancellation. If $\bigcup X_i \vdash_1 \bigcup Y_i$ there exist $X'$ and $Y'$ in L and s in $L_1$ such that $s(X') \subset \bigcup X_i$ and $s(Y') \subset \bigcup Y_i$ and $X' \vdash Y'$. Let $X'_i = \{A\colon A \in X'$ and $s(A) \in X_i\}$ and $Y'_i = \{A\colon A \in Y'$ and $s(A) \in Y_i\}$. Then $X' = \bigcup X'_i$ and $Y' = \bigcup Y'_i$ and $\{X'_i \cup Y'_i\}$ is a disconnected family if $\{X_i \cup Y_i\}$ is one. It follows by cancellation in L that $X'_i \vdash Y'_i$ for some i, and hence $X_i \vdash_1 Y_i$ as required.

In proving the theorem we can set aside the case of inconsistent calculi which, as we have seen in Section 13.2, are characterised by the empty matrix and so are many-valued. In the case of a consistent calculus L we construct a matrix whose truth-values are the formulae of $L_1$, where this is obtained in the way described above by adding to L a variable $p_{\langle X,Y\rangle}$ corresponding to each combination of a variable p of L and sets X and Y of formulae of L. The truth-function corresponding to the k-ary connective F is $\lambda A_1 \ldots A_k . FA_1 \ldots A_k$, and the designated values are defined as follows. For each X and Y of L there is a substitution $s_{\langle X,Y\rangle}$ in $L_1$ which interchanges each variable p of L with $p_{\langle X,Y\rangle}$. Since $s_{\langle X,Y\rangle}(s_{\langle X,Y\rangle}(A)) = A$ for each A in L, it follows that $s_{\langle X,Y\rangle}(X) \vdash_1 s_{\langle X,Y\rangle}(Y)$ only if $X \vdash Y$. Since L is consistent the family $\{s_{\langle X,Y\rangle}(X) \cup s_{\langle X,Y\rangle}(Y)\}$, indexed by the set of pairs $\langle X,Y\rangle$ for which $X \nvdash Y$, is nonempty and so disconnected. Hence by 15.4 there is a partition $\langle T,U\rangle$ of the formulae of $L_1$ such that $T \nvdash_1 U$ and $s_{\langle X,Y\rangle}(X) \subset T$ and $s_{\langle X,Y\rangle}(Y) \subset U$ for each pair $\langle X,Y\rangle$ for which $X \nvdash Y$. The members of T are to be the designated values.

The valuations in this matrix are precisely the substitutions s of formulae of $L_1$ for variables of L. If $X \vdash Y$ then for every such s we have $s(X) \vdash_1 s(Y)$, and since $T \nvdash_1 U$ it follows that either s(X) overlaps U or s(Y) overlaps T. Thus if $X \vdash Y$ every valuation satisfies

$\langle X,Y\rangle$. Conversely if $X \nvdash Y$ then $s_{\langle X,Y\rangle}(X) \subset T$ and $s_{\langle X,Y\rangle}(Y) \subset U$, and so $s_{\langle X,Y\rangle}$ invalidates $\langle X,Y\rangle$. Hence the matrix characterises L as required.

We explore in the next section the possibility of extending Theorem 15.2 to non-compact calculi, but the theorem as it stands has an application to calculi in general. We recall that every calculus has a maximum compact subcalculus (Theorem 2.11), and show that

*Theorem 15.5* The maximum compact subcalculus of a many-valued calculus is many-valued.

*Proof* Let L′ be the maximum compact subcalculus of a many-valued calculus L. By hypothesis $\vdash$ is closed under substitution, and since by 2.12 $X \vdash' Y$ iff there exist finite subsets $X'$ and $Y'$ of X and Y such that $X' \vdash Y'$, it follows that $\vdash'$ too is closed under substitution and so L′ too is a propositional calculus. To show that L′ permits the special case of cancellation, suppose that $X_1,X_2 \vdash' Y_1,Y_2$, where $X_1 \cup Y_1$ is unconnected with $X_2 \cup Y_2$. By 2.12 these sets have finite subsets such that $X_1',X_2' \vdash Y_1',Y_2'$, whence by 15.1 $X_1' \vdash Y_1'$ or $X_2' \vdash Y_2'$ and so $X_1 \vdash' Y_1$ or $X_2 \vdash' Y_2$, as required. Since L′ is compact it follows that it permits cancellation in general and so is many-valued by 15.2.

A similar result holds for single-conclusion calculi. For example, let L be the single-conclusion calculus characterised by Łukasiewicz's rational-valued matrix, and let $L_{\aleph_0}$ (Łukasiewicz and Tarski, 1930) consist of the theorems of L with modus ponens. It can be shown that $L_{\aleph_0}$ is the maximum compact subcalculus of L, and so it too is many-valued, though not characterised by the same matrix (cf. Theorem 13.2).

## 15.3 *Stability*

As a first step towards extending Theorem 15.2 to cover non-compact calculi, we note that the assumption of compactness is only required in the proof of Lemma 15.3. We call a calculus 'stable' if it satisfies this lemma, whether or not it is compact; thus L is *stable* iff the closure of consequence in L under substitution and dilution, following the addition of an arbitrary number of propositional variables, is always a consequence relation. The same definition is applicable to single-conclusion calculi without any verbal change, and in each case we have

*Theorem 15.6* A stable (multiple- or single-conclusion) propositional calculus is many-valued iff it permits cancellation.

> *Proof* In the multiple-conclusion case the result has already been established by that part of the proof of 15.2 that follows the proof of 15.3. In the single-conclusion case we adapt the argument of 15.2 by taking the values of the matrix to be the formulae of the calculus $L_1$ obtained by adding to L a variable $p_X$ corresponding to each combination of a variable p of L and a set X of formulae of L. A value A is designated if $\bigcup s_X(X) \vdash_1 A$, where the union is taken over all consistent X and where $s_X$ interchanges p and $p_X$ for every p in L. For each consistent X the value $s_X(B)$ is designated iff $X \vdash B$, and the result follows as before.

At this point, however, there appears a striking difference between multiple- and single-conclusion calculi with regard to stability:

*Theorem 15.7* Every single-conclusion propositional calculus is stable.

*Theorem 15.8* Not every propositional calculus is stable.

*Proofs* For 15.7, let $L_1$ be obtained in the prescribed way by adding variables to the single-conclusion calculus L. As in the proof of 15.2 $\vdash_1$ is closed under overlap and dilution. For cut, suppose that $X,Z \vdash_1 B$ and that $X \vdash_1 A$ for every A in Z. By the definition of $\vdash_1$ there exist countable subsets $X'$ and $Z'$ of X and Z such that $X',Z' \vdash_1 B$, and for each A in $Z'$ there exists a countable subset $X_A$ of X such that $X_A \vdash_1 A$. Only countably many variables appear in these subsets, and we lose no generality by assuming that these are all in L, since $\vdash_1$ is closed under substitution and hence under any permutation of the variables. It follows by cut for $Z'$ in L that $X \vdash_1 B$.

An immediate corollary of 15.6 and 15.7 is that a single-conclusion propositional calculus is many-valued iff it permits cancellation. This was originally proved without explicitly invoking stability in Shoesmith and Smiley, 1971. Ryszard Wójcicki subsequently pointed out to us that Łoś and Suszko, 1958, give a similar but fallacious proof using as the target condition their property of 'uniformity', this being tantamount to the special case of cancellation in which the family $\{X_i\}$ consists of a single set. Wójcicki (1969) had himself derived a necessary and sufficient condition by adding to uniformity the property that the members of any disconnected family $\{X_i\}$ of 'mutually uniform' sets are mutually uniform with $\bigcup X_i$; where by the mutual uniformity of X and Y he means that $Z,X' \vdash B$ iff $Z,Y' \vdash B$, for all $X'$ obtained from X by substitution of distinct variables for variables, all $Y'$ obtained similarly from Y, and all Z and B such that $Z \cup \{B\}$ is unconnected with $X' \cup Y'$.

For 15.8, let L be a calculus in which there are formula schemes n(A) and n(A,B) for each n, such that no formula is of both forms and no formula can be expressed in either form in more than one way. For example, L might have the vocabulary of the classical calculus, with n(A) defined as $\sim(A\supset^{n+1}A)$ and n(A,B) as n(A)⊃B. For each set S of natural numbers we write S(A) for {n(A): n ∈ S} and S(A,B) for {n(A,B): n ∈ S}. Consequence in L is characterised by the rules 'from $N_1(A),N_2(A,B)$ infer $N_1(A,B),N_2(A)$' for each partition $\langle N_1,N_2\rangle$ of the set N of natural numbers. $L_1$ is obtained in the prescribed way by adding to L a variable $p_S$ for each combination of a variable p of L and a subset S of N. We need

*Lemma 15.9* If $T \vdash_1 U$, where ⟨T,U⟩ is any partition of the set of formulae of $L_1$, there exist A and B and a partition $\langle N_1,N_2\rangle$ of N such that (i) $N_1(A) \subset T$, (ii) $N_2(A) \subset U$, (iii) $N_1(A,B) \subset U$, and (iv) $N_2(A,B) \subset T$.

To prove the lemma, we note that if $T \vdash_1 U$ there exist s and X and Y in L such that $s(X) \subset T$ and $s(Y) \subset U$ and $X \vdash Y$. Let $\langle T_o,U_o\rangle$ be the partition of the formulae of L such that $s(T_o) \subset T$ and $s(U_o) \subset U$. Then $X \subset T_o$ and $Y \subset U_o$, so $T_o \vdash U_o$ and $U_o$ is an immediate consequence of $T_o$ by virtue of 2.17. So there exist C and D in L and a partition $\langle N_1,N_2\rangle$ of N such that $N_1(C)\cup N_2(C,D) \subset T_o$ and $N_2(C)\cup N_1(C,D) \subset U_o$, and the lemma follows by taking A = s(C) and B = s(D).

To prove the theorem we note that if p is any variable of L then $N_1(p),N_2(p,p_{N_1}) \vdash_1 N_1(p,p_{N_1}),N_2(p)$ for each partition $\langle N_1,N_2\rangle$ of N. If L is stable $L_1$ is closed under cut for N(p), so $\bigcup N_2(p,p_{N_1}) \vdash_1 \bigcup N_1(p,p_{N_1})$ where the unions are taken over all partitions of N. By the definition of $\vdash_1$ the same must be true when the unions are taken over some countable set S of these partitions, and so let $\langle N_1',N_2'\rangle$ be a partition of N that is not in S, and such that $0 \in N_2'$. Let

$\langle T,U \rangle$ be the partition of the formulae of $L_1$ such that $T = N_1'(p) \cup \bigcup N_2(p,p_{N_1})$, where the union is taken over $\langle N_1,N_2 \rangle$ in S. Then $T \vdash_1 U$, so there exist A, B and $\langle N_1,N_2 \rangle$ satisfying (i)-(iv) of 15.9. By (i) and (iv) A = p, so by (i) and (ii) $\langle N_1,N_2 \rangle = \langle N_1',N_2' \rangle$. By (iv) $O(p,B) \in T$, so $B = p_{N_1''}$ for some $\langle N_1'',N_2'' \rangle$ in S. Hence by (iii) and (iv) $\langle N_1'',N_2'' \rangle = \langle N_1',N_2' \rangle \notin S$, providing the required contradiction and establishing the theorem.

We note finally that 15.4 holds for this choice of L. For suppose that $X_i \nvdash_1 Y_i$ for each i, where $\{X_i \cup Y_i\}$ is a disconnected family of countable sets. Using a substitution which maps each of the variables in $X_i,Y_i$ into a distinct variable of L we may show by cut in L that for each i there is a partition $\langle T_i,U_i \rangle$ of those formulae of $L_1$ that involve only variables occurring in members of $X_i$ or $Y_i$, which is such that $X_i \subset T_i$ and $Y_i \subset U_i$ and $T_i \nvdash_1 U_i$. Let $T = \bigcup T_i \cup \{n(A,B): N(A) \subset \bigcup T_i \text{ and } n(A,B) \notin \bigcup U_i\}$ and let $\langle T,U \rangle$ be the corresponding partition of the formulae of $L_1$. Then $\bigcup X_i \subset \bigcup T_i \subset T$ and $\bigcup Y_i \subset \bigcup U_i \subset U$. If $T \vdash_1 U$ there exist A and B and $\langle N_1,N_2 \rangle$ such that (i)-(iv) of 15.9 hold. Since $T_i \nvdash U_i$ for each i, the variables in A and B cannot all come from the same $T_i \cup U_i$, so no formula in N(A,B) is in $\bigcup T_i$ or in $\bigcup U_i$. It follows that $N(A,B) \subset T$ or $N(A,B) \subset U$ according as $N(A) \subset T$ or not; and this is the required contradiction with (i)-(iv).

It follows from Theorem 15.7 that to find examples of instability in single-conclusion calculi we must look among those whose vocabulary is too large or too small to fall under our definition of a propositional calculus. Why do unstable calculi with orthodox vocabularies occur only in the multiple-conclusion case? The strategy of Theorem 15.8 was to take a countably infinite set Z - in the example Z is N(p) - and for each *partition* of Z to find $X_i$ and $Y_i$ such that $X_i,Z_1 \vdash_1 Z_2,Y_i$, establishing this by substitutions from L which introduce distinct variables into the $X_i \cup Y_i$ corresponding to distinct partitions. If $\vdash_1$ were closed under cut we should be able to derive $\bigcup X_i \vdash_1 \bigcup Y_i$, but we ensured in our example that this was not obtainable by substitution from L, using the fact that Z has $2^{\aleph_0}$ partitions but L has only $\aleph_0$ variables, where of course $2^{\aleph_0} > \aleph_0$. The corresponding single-conclusion strategy would have to be to obtain $X_i \vdash_1 A$ for each *member* A of Z, from which, with $Z \vdash_1 B$, we should be able to derive $\bigcup_i X_i \vdash_1 B$ by cut. But although Z has $2^{\aleph_0}$ partitions it only has $\aleph_0$ members, so that even if each $X_i$ includes as many as $\aleph_0$ variables $\bigcup X_i$ cannot involve more than $\aleph_0^2$ variables, and $\aleph_0^2 \not> \aleph_0$. For cut to fail in $L_1$ in the single-conclusion case we must either postulate uncountably many connectives in L (so that Z can be chosen to have $2^{\aleph_0}$ members), or only finitely many variables. A simple example of the second sort is the fragment L of the classical calculus based on just two variables, p and q. We write AB for A&(B∨~B), so that $A \vdash AB$ and $AB \vdash A$. Let $L_1$ be obtained by adding a third variable r; then $pq \vdash_1 p$ and $p \vdash_1 pr$ by substitution from L, but $pq \nvdash_1 pr$ since there is no way of deriving this from L by substitution and dilution alone.

Theorems 15.6-7 establish cancellation as a necessary and sufficient condition for the many-valuedness of single-conclusion calculi. But the existence of multiple-conclusion calculi which are unstable and so fall outside the scope of Theorem 15.6 raises the possibility that there are such calculi which permit cancellation but are not many-valued. For example, the calculus we shall describe in Theorem 18.9 is not many-valued, but we do not know whether it permits cancellation. (The calculus of Theorem 15.8 does nothing to settle the matter, for it is many-valued. Indeed the proof of Theorem 15.2 was chosen with an eye to showing this; for although Lemma 15.3 fails, only the proof of Lemma 15.4 is affected, and an ad hoc substitute is noted in the proof of Theorem 15.8.) We are therefore left with an open question: is cancellation a necessary and sufficient condition for many-valuedness? Answering it is likely to involve a closer look at the effects of stipulating a countable vocabulary for propositional calculi, a stipulation which affects other properties such as compactness (cf. last paragraph of Section 16.1) and categoricity (cf. Section 17.1).

# 16 · Counterparts

## 16.1 *Many-valued counterparts*

Since matrix evaluation is only a special case of the general method of characterising a calculus by means of a set of partitions, it follows as in Section 5.4 that when a calculus L is characterised by a matrix M, a set of formulae will be inconsistent in L iff it is unsatisfiable in M, i.e. iff there is no valuation which gives a designated value to every member of the set. For single-conclusion calculi there is no such general equivalence between inconsistency (of any kind) and unsatisfiability, as Theorem 5.25 shows, but we can establish one with respect to matrix evaluation in all but one anomalous case:

*Theorem 16.1* If a consistent single-conclusion calculus L is characterised by a matrix M, a set of formulae is formally inconsistent in L iff it is unsatisfiable in M.

*Proof* If L is consistent, M has an undesignated value a. Let p be any variable, and let s be a substitution based on a one-one mapping of the set of all variables onto those other than p. If there is a valuation v which satisfies X, let v′ be such that v′(p)=a and v′(s(q)) = v(q) for every variable q. Then v′ invalidates ⟨s(X),p⟩, and hence s(X) ⊬ p and X is not formally inconsistent. Conversely if X is not formally inconsistent then s(X) ⊬ B for some s and

B, and so some valuation v invalidates ⟨s(X),B⟩, whence vs satisfies X.

We have already defined sign for multiple- and single-conclusion propositional calculi, and we now define it for matrices. A matrix is *positive* if it has a nonempty submatrix consisting exclusively of designated values, and *negative* otherwise.

*Theorem 16.2* A calculus (or a consistent single-conclusion calculus) characterised by a matrix M has the same sign as M.

*Theorem 16.3* Every many-valued counterpart of a consistent single-conclusion calculus L has the same sign as L.

*Proofs* For 16.2, if V is satisfied by an assignment v then v(V) generates a nonempty submatrix of designated values, while V is satisfiable in any such submatrix and so in M. Thus M is negative iff V is unsatisfiable. For a multiple-conclusion L, therefore, M is negative iff V is inconsistent, i.e. iff L is negative; and for a consistent single-conclusion L, by 16.1, M is negative iff V is formally inconsistent, i.e. iff L is negative. 16.3 is an immediate corollary of 16.2.

Inconsistent single-conclusion calculi are an exception to each of Theorems 16.1-3, but it is interesting that the anomaly has nothing to do with the admission of empty matrices. An empty matrix is negative and every set of formulae is unsatisfiable in it, matching the fact that an inconsistent calculus is negative and every set is formally inconsistent. The anomalous matrices are the nonempty ones with exclusively designated values, for the proof of Theorem 16.1 requires that if M has any values at all (a necessary and sufficient condition for some set - even if only the empty one - to be satisfiable) it must have an undesignated one, and it is by violating this requirement that we obtain positive characteristic matrices and positive many-valued counterparts for negative single-conclusion calculi.

The single-conclusion counterpart of a many-valued calculus is obviously many-valued, being characterised by any matrix that characterises the original. On the other hand a many-valued single-conclusion calculus may have counterparts which are not many-valued - indeed it may have counterparts which are not even propositional calculi, their vocabulary being suitable but their consequence relation not being closed under substitution. This can be shown from considerations of sign. For example, the minimum counterpart of the classical propositional calculus is also a propositional calculus (Theorem 16.4) but is positive (Theorem 5.25) and so not many-valued (Theorem 16.3). And, taking any positive fragment of the classical calculus, its maximum counterpart is negative (Theorems 5.11 and 5.25) and so not a propositional calculus (Theorem 5.31). It follows from the first example that the minimum counterpart, which has so far stood at one end of all the other ranges of counterparts, cannot in general be the minimum many-valued counterpart, as Theorem 16.5 shows.

*Theorem 16.4* Every single-conclusion propositional calculus has a minimum propositional counterpart, namely its minimum counterpart.

*Theorem 16.5* Every many-valued single conclusion calculus L has a minimum many-valued counterpart, namely its minimum counterpart (for positive or inconsistent L) or the negative variant of its minimum counterpart (for negative consistent L).

*Proofs* For 16.4, it is evident from 5.4 that $\vdash_{\cap}$ is closed under substitution if $\vdash$ is. For 16.5, the minimum counterpart of an inconsistent L is many-valued, being characterised by any nonempty matrix with exclusively designated values; and hence it is the required minimum many-valued counterpart. For consistent L, let M be the matrix constructed in Theorem 15.6, and let Y be nonempty. If $X \nvdash_{\cap} Y$ then $X \nvdash B$ for each B in Y, so that X is consistent in L and $s_X(A)$ is designated in M iff $X \vdash A$. In particular $s_X(A)$ is designated for every A in X but not for any A in Y, so that $s_X$ invalidates $\langle X,Y \rangle$ and $X \nvdash_M Y$. Conversely, if $X \vdash_{\cap} Y$ then $X \vdash B$ for some B in Y, so that $X \vdash_M B$ and hence $X \vdash_M Y$. Thus if Y is nonempty $X \vdash_{\cap} Y$ iff $X \vdash_M Y$, and so by 5.24 M characterises either the minimum counterpart of L or its negative variant. Hence by 16.3 if L is positive its minimum counterpart is many-valued and is therefore the minimum many-valued counterpart. Likewise if L is negative the negative variant of its minimum counterpart is many-valued, and since it is the minimum negative counterpart it is therefore the minimum many-valued one, by 16.3.

At the other end of the range of counterparts we lack any alternative description, comparable to the description of $\vdash_{\cup}$ in Theorem 5.5, of the unions of the propositional or many-valued counterparts of a single-conclusion calculus L. We can define $X \vdash_{(\cup)} Y$ as meaning that $s(X) \vdash_{\cup} s(Y)$ for every substitution s, but $\vdash_{(\cup)}$ is not in general the union of the consequence relations of the propositional counterparts of L (see proof of Theorem 16.8), and the closest we can get is to show that

*Lemma 16.6* If $\vdash_{(\cup)}$ is a consequence relation it defines the maximum propositional counterpart of L.

*Theorem 16.7* If a single-conclusion propositional calculus L has a maximum counterpart it has a maximum propositional counterpart $L_{(\cup)}$.

*Proofs* For 16.6, let L′ be any propositional counterpart of L and let $X \vdash' Y$. Then for every substitution s we have $s(X) \vdash' s(Y)$ and hence $s(X) \vdash_{\cup} s(Y)$. Thus $\vdash' \subset \vdash_{(\cup)}$, and the result follows since $\vdash_{(\cup)}$ is evidently closed under substitution and using 16.4 $\vdash_{\cap} \subset \vdash_{(\cup)} \subset \vdash_{\cup}$.

For 16.7, suppose that L has a maximum counterpart $L_{\cup}$. Obviously $\vdash_{(\cup)}$ is closed under overlap and dilution. For cut, if $X,Z_1 \vdash_{(\cup)} Z_2,Y$ for every partition $\langle Z_1,Z_2 \rangle$ of Z then $s(X),s(Z_1) \vdash_{\cup} s(Z_2),s(Y)$ for every s. But every partition of s(Z) has the form $\langle s(Z_1),s(Z_2) \rangle$, and since by hypothesis $\vdash_{\cup}$ is closed under cut it follows that $s(X) \vdash_{\cup} s(Y)$ and so $X \vdash_{(\cup)} Y$. Thus $\vdash_{(\cup)}$ is a consequence relation and the result follows by 16.6.

$L_{(\cup)}$ therefore exists if L is compact (by 5.11), but it does not follow that $L_{(\cup)}$ will be compact. For example, let L be the minimum calculus with the classical vocabulary; taking Y to be $\{p_o \supset^n p_n : n \geq 1\}$, where $p_o, p_1, \ldots$ enumerate the variables, it is easy to show that $\Lambda \vdash_{(\cup)} Y$ but $\Lambda \nvdash_{(\cup)} Y'$ for any finite subset Y′ of Y. On the other hand, as we claim in Section 20.3, $L_{(\cup)}$ is compact when L is Heyting's calculus, and indeed $L_{(\cup)}$ is the maximum compact counterpart L′ of L. For if $X \vdash_{(\cup)} Y$ then in particular $s(X) \vdash_{\cup} s(Y)$ where $s(p_i)=p_{i+1}$ for each i, and by 5.14 $s(X),s(Y)\supset p_o \vdash p_o$. Since L is compact $s(X'),s(Y')\supset p_o \vdash p_o$ for some finite subsets X′ and Y′ of X and Y, and substituting $p_i$ for $p_{i+1}$ and an arbitrary B for $p_o$ we have $X',Y'\supset B \vdash B$. By 5.14 $X' \vdash_{\cup} Y'$, and as in 2.11 and 2.12 $X \vdash' Y$. Conversely it is clear from 5.35 that the positive variant of L′ is propositional, so if $X \vdash' Y$ for nonempty Y then $X \vdash_{(\cup)} Y$; while if $X \vdash' \Lambda$ then X is formally inconsistent in L by 5.30 and 5.33, whence $s(X) \vdash_U \Lambda$ for all s and so $X \vdash_{(U)} \Lambda$.

We shall meet plenty of examples in Section 16.2 of single-conclusion calculi for which a maximum propositional counterpart and a maximum many-valued counterpart exist and coincide with one another. On the other hand

*Theorem 16.8* There exists a many-valued single-conclusion calculus which has no maximum propositional counterpart.

*Theorem 16.9* There exists a many-valued single-conclusion calculus which has a maximum propositional counterpart but not a maximum many-valued counterpart.

*Theorem 16.10* There exists a many-valued single-conclusion calculus whose maximum propositional counterpart and maximum many-valued counterpart exist but are distinct.

*Proofs of 16.8 and 16.9.* For 16.8, take the formulae of the classical propositional calculus and let $X \vdash B$ iff $B \in X$ or $A \supset^n A \in X$ for some A and infinitely many n. It is easily verified that $\vdash$ is closed under overlap, dilution, cut and substitution, and that it permits cancellation and so defines a many-valued calculus L. For each $m>1$ let $X \vdash_m Y$ iff X and Y overlap, or $\{A \supset A, A \supset^m A\} \subset Y$ for some A, or $A \supset^n A \in X$ for some A and infinitely many n. Evidently $X \vdash_m \{B\}$ iff $X \vdash B$, and $\vdash_m$ is obviously closed under overlap, dilution and substitution. For cut, suppose that $X \nvdash_m Y$ and let $T = X \cup \{A \supset A\colon A \supset A \notin Y\} \cup \{A \supset^m A\colon A \supset^m A \notin Y\}$; then $X,T \nvdash_m V-T,Y$ and hence the result by 2.2. Thus each $\vdash_m$ defines a propositional counterpart of L. Moreover, given any A, for each partition $\langle T,U \rangle$ either $A \supset^m A \in U$ for some $m>1$ or $A \supset^n A \in T$ for infinitely many n; and in either case $T \vdash_m U, A \supset A$

for some m. Hence if L had a maximum propositional counterpart L′ we would have T ⊢′ U,A⊃A for every partition ⟨T,U⟩. By cut it would follow that Λ ⊢′ A⊃A, contradicting the assumption that L′ is a counterpart of L.

The example used in proving 16.8 serves also to show that $\vdash_{(U)}$ is not in general the union of the consequence relations of the propositional counterparts of L. For it is easily verified from 5.5 that Λ $\vdash_U$ B,A⊃B for all A and B, and hence that Λ $\vdash_{(U)}$ q,p⊃q for distinct variables p and q. But if we were to have Λ ⊢′ q,p⊃q in any propositional counterpart L′ of L, then Λ ⊢′ $A\supset^n A, A\supset^{n+1} A$ for each n. Using the definition of ⊢ it would follow that $Z_1$ ⊢′ $Z_2$,B for every partition $\langle Z_1, Z_2\rangle$ of $\{A\supset^n A\colon n\geq 1\}$, and hence by cut that Λ ⊢′ B. But Λ ⊬ B, so L′ would not after all be a counterpart of L.

For 16.9, let L be the minimum single-conclusion calculus with the classical vocabulary. The rule 'from A∨B infer ~A,B⊃B' is non-iterable, so by 4.9 consequence and immediate consequence by this rule coincide. From this we can verify that the calculus characterised by the rule both permits cancellation and is a counterpart of L, and since it is compact it follows by 15.2 that it is a many-valued counterpart. A similar argument shows that the rule 'from Λ infer ~A,A∨B' also characterises a many-valued counterpart of L. Hence if L′ was a maximum many-valued counterpart of L (or even a maximum many-valued compact one), we would have A∨B ⊢′ ~A,B⊃B and Λ ⊢′ ~A,A∨B, whence by cut Λ ⊢′ ~A,B⊃B. But by 5.1 Λ ⊬′ ~A and Λ ⊬′ B⊃B, so that L′ does not permit cancellation, contradicting 15.1. On the other hand, since L is compact, it has a maximum propositional counterpart by 5.11 and 16.7.

To prove Theorem 16.10 we need a result about synonymity. Two formulae A and B are *synonymous* in a (multiple- or single-conclusion) propositional calculus if they are interreplaceable salva consequence, i.e. if $C \vdash C'$ and $C' \vdash C$ whenever $C'$ is obtained from C by replacing some or all of the occurrences of A by B. The same definition can be applied to any kind of calculus for which a suitable idea of replacement has been defined, and for propositional calculi there is an equivalent criterion in terms of substitutions: A and B are synonymous if $C(A) \vdash C(B)$ and $C(B) \vdash C(A)$ whenever $C(A)$ and $C(B)$ are obtained from the same formula C by substituting A and B respectively for the same variable.

*Lemma 16.11* If L is characterised by a matrix M, formulae A and B are synonymous in L iff $v(A)$ and $v(B)$ are congruent for every valuation v in M.

*Proof* If $v(A)$ and $v(B)$ are congruent then by 14.10 so are $v(C(A))$ and $v(C(B))$ for every C, so that v satisfies $\langle C(A),C(B)\rangle$ and $\langle C(B),C(A)\rangle$. Hence if $v(A)$ and $v(B)$ are congruent for every v then $C(A) \vdash_M C(B)$ and $C(B) \vdash_M C(A)$, so that A and B are synonymous in L. Conversely, if $v(A)$ and $v(B)$ are incongruent for some v then $v(A),x_1,\ldots,x_n \neq v(B),x_1,\ldots,x_n$ for some $x_1,\ldots,x_n$. Let $p_0,p_1,\ldots,p_n$ be distinct variables not appearing in A or B, and let $v'(p)=v(p)$ if p appears in A or B while $v'(p_i)=x_i$ for each $i\neq 0$. Then $v'(A),v'(p_1),\ldots,v'(p_n) \neq v'(B), v'(p_1),\ldots,v'(p_n)$, and so some matrix function separates these sequences. Hence there is an associated formula C with variables drawn from $p_0,p_1,\ldots,p_n$, as described in Section 14.1, such that if $C(A)$ and $C(B)$ are obtained from C by substituting A and B respectively for $p_0$, either $v'(C(A))$ is designated and $v'(C(B))$ is undesignated or vice versa. Hence either $C(A) \nvdash_M C(B)$ or $C(B) \nvdash_M C(A)$, so that A and B are not synonymous.

| ∨ | 0 | 1 | 2 | 3 |
|---|---|---|---|---|
| *0 | 0 | 0 | 0 | 0 |
| 1 | 0 | 2 | 0 | 0 |
| 2 | 0 | 0 | 1 | 0 |
| 3 | 0 | 0 | 0 | 3 |

*Proof of 16.10* Let L be the single-conclusion calculus characterised by the above matrix M. L has a maximum many-valued counterpart, characterised by M. To show this it is sufficient to show that any simple matrix M′ that characterises L has a submatrix isomorphic to M, for then $\vdash_{M'} \subset \vdash_M$. We know from 17.21 below that M′ will have just one designated value 0′, and since $\Lambda \nvdash (p\vee p)\vee p$ it must also have an undesignated value 3′ of the form $(a'\vee a')\vee a'$ for some a′. Since $(p\vee p)\vee p \nvdash p$ there must be an undesignated value 1′ such that $(1'\vee 1')\vee 1'=0'$. Let $2'=1'\vee 1'$, so that $2'\vee 1'=0'$. Then $2'\neq 1'$, but 2′ is undesignated since $p\vee p \vdash p$. Since $p \vdash p\vee q$, $0'\vee b'=0'$ for every value b′. Also $(p\vee p)\vee p \vdash ((q\vee q)\vee q)\vee p$ and $(p\vee p)\vee p \vdash ((q\vee q)\vee q)\vee(p\vee p)$, so that, assigning 1′ to p and a′ to q, $3'\vee 1'=3'\vee 2'=0'$. Since $((p\vee p)\vee p)\vee((p\vee p)\vee p)$ and $(p\vee p)\vee p$ are synonymous by 16.11 applied to M, it follows by 16.11 that $3'\vee 3'$ and 3′ are congruent in M′, so that $3'\vee 3'=3'$ and hence $3'\neq 1'$ and $3'\neq 2'$. Similarly, since p and $(p\vee p)\vee(p\vee p)$ are synonymous, $2'\vee 2'=1'$. Finally, since $p\vee q \vdash q\vee p$, each of $2'\vee 3'$, $1'\vee 3'$, $1'\vee 2'$ and $b'\vee 0'$ is 0′, so that 0′,1′,2′,3′ form a submatrix isomorphic to M.

L has a maximum propositional counterpart, by 5.11, 13.1 and 16.7. If this were many-valued, i.e. if it coincided with the maximum many-valued counterpart, it would be characterised by M. To show that this is not so, let $X \vdash_1 Y$ iff $X \vdash_{M'} Y$ for every proper submatrix M′ of M; then $\vdash_1$ is a consequence relation, and moreover $X \vdash_1 \{B\}$ iff $X \vdash B$. For any valuation v which invalidates

$\langle X,B \rangle$ in M assigns a value to each of the variables in B that lies in the proper submatrix M′ generated by 0 and v(B); and the valuation which assigns the same values to these variables and 0 to every other invalidates $\langle X,\{B\} \rangle$ in M′. Thus $\vdash_1$ defines a propositional counterpart of L, and so if M characterised the maximum propositional counterpart we should have $\vdash_1 \subset \vdash_M$. But it is easily verified that $p \vee q, (p \vee p) \vee q \vdash_1 p, q$, while a valuation v such that $v(p)=3$ and $v(q)=1$ shows that this inference is not valid in M.

A compact single-conclusion propositional calculus L will always possess a maximum propositional counterpart $L_{(\cup)}$ by Theorems 16.7 and 5.11, and a maximum compact propositional counterpart defined from $L_{(\cup)}$ as in Lemma 2.12. Similarly if a compact many-valued single-conclusion calculus possesses a maximum many-valued counterpart it will also have a maximum compact many-valued counterpart, by Theorem 15.5; but it need not possess either, as the proof of Theorem 16.9 shows.

We note finally that the same matrix may characterise a compact single-conclusion calculus and a non-compact multiple-conclusion one, so that a compact many-valued single-conclusion calculus can have non-compact many-valued counterparts. An example is this matrix M.

| | *0 | 1 | 2 | ... | n | ... |
|---|---|---|---|---|---|---|
| ◇ | 0 | 0 | 1 | ... | n-1 | ... |

The single-conclusion calculus characterised by M (which is in fact the ‘possibility’ fragment of any of the standard modal calculi weaker than S4, being characterised by the rule ‘from A infer ◇A’) is compact; but the multiple-conclusion one is not, since $\Lambda \vdash_M A, \Diamond A, \Diamond^2 A, \ldots$ although

$\Lambda \nvdash_M A, \ldots, \Diamond^n A$ for any n. Drawing the obvious distinctions, one would say that M was premiss-compact but not compact, where the latter implies both premiss- and conclusion-compactness. (We may note that the submatrix {0,1,2} of M, which characterises the possibility fragment of S6 or S7, is compact in the full sense. Also, by adding to M an undesignated value $\omega$ such that $\Diamond\omega=\omega$, we obtain a compact matrix M′ which matches the singleton-conclusion rule 'from A infer $\{\Diamond A\}$'. Thus a submatrix of a non-compact matrix may be compact, and a submatrix of a compact one may be non-compact.)

It is still necessary to be cautious, however, in treating these properties as being intrinsic to a matrix, for it can happen that a matrix characterises a compact calculus when a countable vocabulary of propositional variables is assumed, but not when uncountably many variables are allowed. For example, we showed in Theorem 13.3 that the rational-valued Gödel matrix characterises compact single- and multiple-conclusion calculi, but this ceases to be true if there are uncountably many variables $p_i$, for then we have $\bigcup\{((p_i \supset p_j)\&(p_j \supset p_i)) \supset p_1\} \vdash p_1$ when the union is taken over all i and j such that $i \neq j$, but not when it is taken over any finite subfamily.

## 16.2 *Principal matrices*

We say that a matrix M is *principal* if the single-conclusion calculus characterised by M possesses a maximum many-valued counterpart and M also characterises this counterpart. A principal matrix thus occupies a special place among those that characterise a given single-conclusion calculus, and it is interesting that all our examples from the literature turn out to be principal.

*Theorem 16.12* Non-singular two-valued matrices, Kleene's matrix, the finite, rational- and real-valued Łukasiewicz matrices, and the finite, rational-valued, real-valued and well-ordered Gödel matrices are all principal.

*Proof* If M is any one of these matrices and L the corresponding single-conclusion calculus, we prove the stronger result that M characterises the maximum propositional counterpart of L, by showing that $\vdash_M = \vdash_{(\cup)}$ and using 16.6. Since $\vdash_M \subset \vdash_{(\cup)}$ by the argument of 16.6, it is sufficient to suppose that $X \nvdash_M Y$ and to show in each case that $X \nvdash_{(\cup)} Y$.

If M is two-valued, let v be a valuation that invalidates $\langle X,Y\rangle$. Let q and r be distinct variables and let s be the substitution such that $s(p)=q$ if $v(p)$ is designated and $s(p)=r$ if not. If M is non-singular, $v'(s(A))=v(A)$ in each of the various valuations $v'$ such that $v'(q)$ is designated and $v'(r)$ undesignated. Hence $q,s(A) \vdash_M r$ for each A in Y, but $q,s(X) \nvdash_M r$. By 5.5 $s(X) \nvdash_\cup s(Y)$, and so $X \nvdash_{(\cup)} Y$.

If M is Kleene's matrix, let v be a valuation invalidating $\langle X,Y\rangle$. Let $p_0$, $p_{\frac{1}{2}}$ and $p_1$ be distinct variables and let s be the substitution such that $s(p)=p_i$ iff $v(p)=i$. Since only the three variables $p_i$ occur in the formulae of $s(V)$, and since there are only 27 ways of assigning values in M to them, any valuation in M partitions $s(V)$ in one of a finite number of ways. Let B be $p_1$ if Y is empty, and otherwise let B be a formula of the form $A_1 \vee \ldots \vee A_n$ such that each of $A_1,\ldots,A_n$ is in $s(Y)$ and such that no other formula of this kind is satisfied by a greater number of valuation-induced partitions of $s(V)$. In each case $A \vdash_M B$ for every A in $s(Y)$ either vacuously or because otherwise $A \vee B$ would be satisfied by more cases than B. But if $v'$ is any valuation such that $v'(p_i)=i$, we have $v'(s(A))=v(A)$ for every A, so that $v'(B)$

is undesignated for either choice of B, and we have $s(X) \nvdash_M B$. Thus $s(X) \nvdash_U s(Y)$ by 5.5, and so $X \nvdash_{(U)} Y$. A similar argument could be applied to the finite Łukasiewicz or Gödel matrices.

If M is a Gödel matrix, let p be any variable and s a substitution that maps the set of all variables one-to-one into the set of variables other than p. Then $s(X) \nvdash_M s(Y)$, and so let v be a valuation that invalidates $\langle s(X), s(Y) \rangle$. If M is well-ordered (for example, if M is finite), there exists an undesignated value a such that $a \leq v(A)$ for every A in s(Y). Assigning a as value to p it follows that $s(X), s(Y) \supset p \nvdash_M p$, but by modus ponens $A, s(Y) \supset p \vdash_M p$ for every A in s(Y). Hence $s(X) \nvdash_U s(Y)$ by 5.5, and so $X \nvdash_{(U)} Y$. If M is rational- or real-valued, it is isomorphic to the submatrix M′ obtained by omitting all values x for which $0 < x \leq \frac{1}{2}$, and so there is a valuation v′ which invalidates $\langle s(X), s(Y) \rangle$ in M′ and hence in M. Assigning $\frac{1}{2}$ as value to p it follows that $s(X), s(Y) \supset p \nvdash_M p$, and from here we argue as before.

If M is a Łukasiewicz matrix, let p, s and v be chosen as in the preceding case. Then for each A in s(Y) we have $v(A) > 0$ and so there exists an integer n such that $n.v(A) \geq 1$. Let Z comprise $A \supset^n p$ for each such A and n; then assigning 1 as value to p it follows that $s(X), Z \nvdash_M p$. But by modus ponens $A, Z \vdash p$ for each A in s(Y), so by 5.5 $s(X) \nvdash_U s(Y)$ and so $X \nvdash_{(U)} Y$.

The proof of the theorem shows that any fragment of the Kleene, Łukasiewicz or Gödel matrices is principal provided it contains $\vee$ or $\supset$ as the case may be, but in general a fragment of a principal matrix need not be principal. For example, the purely negational fragments of the Łukasiewicz matrices are never principal: they are all equivalent to one another and to the Cartesian powers of the corresponding classical fragment (see Section 17.3), and so by Theorem 17.17 they characterise the same single-conclusion calculus as the classical fragment does; but $\Lambda \vdash A, {\sim}A$ in the classical case and not in the Łukasiewicz one.

Not every finite matrix is therefore principal, and other examples of this are the nonempty singular matrices that characterise an inconsistent single-conclusion calculus. In each of the cases cited a principal matrix does however exist, namely the classical negational fragment and the empty matrix respectively. Whether this is true in general, or whether there exist single-conclusion calculi which are characterised by a finite matrix but which have no maximum many-valued counterpart, is an open question.

# 17 · Categoricity

## 17.1 *Multiple-conclusion calculi*

We begin by establishing a criterion for two matrices to characterise the same calculus. We write $M_1 \prec M_2$ to mean that every countable submatrix of $M_1$ is equivalent to a submatrix of $M_2$.

*Theorem 17.1* $M_1$ and $M_2$ characterise the same calculus iff $M_1 \prec M_2$ and $M_2 \prec M_1$.

The theorem is proved by two applications of the following lemma:

*Lemma 17.2* $\vdash_{M_2} \subset \vdash_{M_1}$ iff $M_1 \prec M_2$.

*Proof* Suppose that $M_1 \prec M_2$. If $X \nvdash_{M_1} Y$ then some valuation $v_1$ invalidates $\langle X,Y \rangle$ in $M_1$, and the truth-values involved in $v_1$ form a countable submatrix $M_1'$ of $M_1$. By hypothesis there is a structure-preserving correspondence between $M_1'$ and some submatrix $M_2'$ of $M_2$, and so let $v_2$ be a valuation in $M_2$ determined by assigning as value to each variable p any of the values of $M_2'$ that correspond to $v_1(p)$. It follows that $v_2$ also invalidates $\langle X,Y \rangle$ and hence $X \nvdash_{M_2} Y$. Conversely, suppose that $\vdash_{M_2} \subset \vdash_{M_1}$ and let $M_1'$ be any countable submatrix of $M_1$ with values $a_1, a_2, \ldots$ . Let $p_1, p_2, \ldots$ be a matching sequence of variables (empty, finite or infinite); let Z be the set of formulae involving only these variables; let $v_1$ be a valuation in $M_1'$ such that $v_1(p_i)=a_i$ for each i; and let $\langle Z_1,Z_2\rangle$ be the partition of Z that is invalidated by $v_1$. Then $Z_1 \nvdash_{M_1} Z_2$, whence by hypothesis $Z_1 \nvdash_{M_2} Z_2$ and consequently there is a valuation $v_2$ in $M_2$ that invalidates $\langle Z_1,Z_2\rangle$. Now the set $\{v_2(A): A\in Z\}$ is closed under the basic functions and so forms the set of values of a submatrix $M_2'$ of $M_2$, and it is easy to see that the correspondence $v_1(A)\leftrightarrow v_2(A)$ for A in Z is a structure-preserving correspondence between $M_1'$ and $M_2'$. Thus $M_1'$ is equivalent to $M_2'$ as required.

The definition of $\prec$ reflects the stipulation that a propositional calculus shall only have countably many variables. If no such stipulation were made a matrix would characterise each of a whole family of calculi based on the same connectives but with different-sized vocabularies of propositional variables. Theorem 17.1 could then be adapted to provide a criterion for two matrices to characterise the same calculi, simply by omitting the reference to countability in the definition of $\prec$ (whence by Theorem 14.19, $M_1 \prec M_2$ iff $M_1$ is equivalent to a submatrix of $M_2$). For countable matrices the new criterion is equivalent to that of the existing Theorem 17.1, but in general it is stronger. For example, taking $M_1$ and $M_2$ to be the real- and rational-valued Gödel matrices, every countable submatrix of $M_1$ is equivalent to a submatrix of $M_2$, but $M_1$ itself is not. Thus these two matrices characterise the same calculus if a countable vocabulary of variables is assumed for it (as it was in Theorem 13.3), but never do so if there are uncountably many variables.

*Theorem 17.3* Equivalent matrices characterise the same calculus. In particular a matrix and its quotient characterise the same calculus.

The theorem is a corollary of Theorems 17.1 and 14.19. Its

converse does not hold in general: that is to say, two non-equivalent matrices may characterise the same calculus. The rational- and real-valued Gödel matrices are an example, but it can happen even when both matrices are countable. In the latter case we have by Theorem 14.19 that $M_1 \prec M_2$ iff $M_1$ is equivalent to a submatrix of $M_2$, but it is possible for $M_1$ to be equivalent to a submatrix of $M_2$ and vice versa without $M_1$ and $M_2$ being equivalent. For example, let $M_1$ be the implicational fragment of the rational-valued Łukasiewicz matrix and let $M_2$ be the submatrix obtained by omitting the value 1. Then $M_1$ is equivalent to the submatrix of $M_2$ with values in the closed interval $[0,\frac{1}{2}]$, and $M_2$ is already a submatrix of $M_1$; but $M_1$ is not equivalent to $M_2$.

If all the matrices which characterise a many-valued calculus are equivalent, we say that it is *categorical*. By Theorems 14.18 and 17.3 a calculus is categorical iff all its simple matrices are isomorphic, so that (to within isomorphism) a categorical calculus is characterised by a unique simple matrix.

*Theorem 17.4* Every calculus characterised by a countable matrix not equivalent to any of its proper submatrices is categorical.

*Theorem 17.5* Every calculus characterised by a finite matrix is categorical.

*Theorem 17.6* Every calculus characterised by a monadic matrix is categorical.

*Theorem 17.7* Every many-valued calculus with no non-singulary connectives is categorical.

*Proofs* For 17.4, suppose that the calculus characterised by a countable matrix M is not categorical, so that it is characterised by a simple matrix M′ not equivalent to M. By 17.1 there is a structure-preserving correspondence between M and a proper submatrix of M′. Choosing one value corresponding to each value of M, we can therefore generate a countable proper submatrix $M_1'$ of M′, which is also equivalent to M by a restriction of the original correspondence. Let b be a value of M′ not in $M_1'$. We need

*Lemma 17.8* If S is a countable set of incongruent values of a matrix that characterises a propositional calculus, there is a countable submatrix in which the members of S are incongruent.

To prove the lemma, we note that for any distinct members a and a′ of S there is by hypothesis a finite sequence of truth-values such that $a, b_1, \ldots, b_n \neq a', b_1, \ldots, b_n$. The members of S are incongruent in the submatrix generated by S and these various sequences. This submatrix is therefore the required one, being countable since the assumption that the matrix characterises a propositional calculus ensures that the number of basic functions is finite.

Applying the lemma to b and the members of $M_1'$, there is a countable submatrix $M_2'$ of M′ which contains b and has $M_1'$ as a submatrix and in which b is not congruent to any value of $M_1'$. By 17.1 there is a structure-preserving correspondence between $M_2'$ and some submatrix $M_2$ of M, and by 14.19 this induces a similar correspondence between $M_1'$ and a submatrix $M_1$ of M. Moreover, since b is not congruent in $M_2'$ to any value of $M_1'$, it follows by 14.11 that $M_1$ is a proper submatrix of $M_2$ and hence of M. Since M and $M_1$ are both equivalent to $M_1'$ they are equivalent to each other. Thus M is equivalent to a proper submatrix, establishing the theorem.

For 17.5, a calculus characterised by a finite matrix is characterised by a finite simple one which cannot by 14.14 be equivalent to any matrix with fewer values. In particular it cannot be equivalent to any proper submatrix, and the result follows by 17.4.

For 17.6 and 17.7, suppose that M and M′ characterise the same calculus, and consider the (many-many) correspondence ↔ such that x↔x′ iff there exist countable submatrices of M and M′ with a structure-preserving correspondence between them in which x and x′ correspond. Since by 17.8 each value belongs to a countable submatrix, ↔ is a correspondence between M and M′ by 17.1. If a↔a′ then evidently a is designated iff a′ is, and F(a)↔F(a′) for each singulary connective F; and this alone is sufficient to establish 17.7. For 17.6 we note that for each singulary matrix function g′ of M′ there is a matrix function g of M, constructed in the same way from the corresponding basic functions, such that g(a) is designated iff g′(a′) is designated, so that values of M′ which correspond to the same value of M are inseparable. If now the calculus is characterised by a monadic matrix it is also characterised by its quotient, by 17.3; and we choose this latter to be M′. By 14.22 M′ is simple and monadic, so that values of M′ which correspond to the same value of M are identical; i.e. the correspondence ↔ is a mapping h of M onto M′. Moreover, h preserves structure since the submatrix $M_i$ of M generated by $a_1,\ldots,a_k$ is equivalent by 17.1 to a submatrix $M_i'$ of M′ which must contain $h(a_1),\ldots,h(a_k)$, whence $F(a_1,\ldots,a_k) \leftrightarrow F(h(a_1),\ldots,h(a_k))$ for each F. Thus every matrix M that characterises the calculus is equivalent to M′ as required.

We may call a calculus *n-valued* or *finite-valued* or *infinite-valued* if it is characterised by a simple matrix with n values or a finite or an infinite number, as the case may be; and *exclusively* n-valued if all its simple matrices have n values. By Theorems 17.5 and 14.18, no calculus is both finite-valued and infinite-valued, and every finite-valued calculus is exclusively n-valued for some n. An infinite-valued calculus, however, need not be exclusively n-valued for any n, as is shown by the calculus characterised by the rational- and real-valued Gödel matrices. Indeed no calculus can be exclusively n-valued for $n > 2^{\aleph_0}$, for

*Theorem 17.9* Every many-valued calculus is characterised by a matrix with at most $2^{\aleph_0}$ values.

*Proof* Let M characterise L. There are at most $2^{\aleph_0}$ pairs $\langle X,Y \rangle$ invalid in M, and it requires at most $2^{\aleph_0}$ valuations to invalidate them all. The values involved in these generate a submatrix of M with at most $2^{\aleph_0}$ values, and this evidently characterises L. (A similar argument establishes the result for single-conclusion calculi.)

17.2 *Examples*

Calculi characterised by Kleene's matrix or by one of the finite Łukasiewicz or Gödel matrices are categorical by Theorem 17.5, and similarly all many-valued singular calculi are categorical, being 0- or 1-valued. Minimum calculi containing no non-singulary connectives are categorical by Theorem 17.7: indeed the minimum calculus with no connectives at all is two-valued. (No other minimum calculus is characterised by a matrix with fewer than $2^{\aleph_0}$ values, since a different value must be available to invalidate each partition of the infinite set of formulae constructed out of a single variable.) On the other hand no minimum calculus containing non-singulary connectives is categorical, since

*Theorem 17.10* Every minimum calculus containing a non-singulary connective is characterised by arbitrarily large simple matrices.

*Proof* By 17.9 any minimum calculus L is characterised by a matrix M with at most $2^{\aleph_0}$ values. M is non-singular and so has a designated value, 0, say. Let S be a set of cardinal n greater than $2^{\aleph_0}$, disjoint from the set of values of M. We construct a matrix M′ by adding the members of S as undesignated values to those of M and extending each basic function F by defining, where any of $a_1,\ldots,a_k$ are in S, $F(a_1,\ldots,a_k) = 0$ if $a_1 = a_k$ and $F(a_1,\ldots,a_k) = a_1$ otherwise. M′ characterises L, for if $X \vdash Y$ then X and Y overlap and so $X \vdash_{M'} Y$, and conversely if $X \nvdash Y$ then $\langle X,Y\rangle$ can be invalidated in M and hence in M′. Moreover any k-ary basic function (k>1) separates $a_1,a_2,\ldots,a_k$ and $a_k,a_2,\ldots,a_k$ whenever $a_1$ and $a_k$ are distinct elements of S, so that the added values are all incongruent in M′. Thus the quotient of M′ has n values and is the required simple matrix.

The real- and rational-valued Łukasiewicz matrices are monadic and so characterise categorical calculi by Theorem 17.6. The same is true for any of their fragments that contain both ⊃ and ~, and fragments without ⊃ also characterise categorical calculi since their quotients are finite. On the other hand we have seen that a fragment containing ⊃ but not ~ will characterise the same calculus as the non-equivalent submatrix obtained by omitting the value 1, and the calculus is therefore not categorical.

The real- and rational-valued Gödel matrices characterise the same calculus (Theorem 13.3 or 17.1) but are not equivalent by Theorem 14.18, being simple but not isomorphic. The calculus in question is therefore not categorical. For other Gödel matrices we have

*Theorem 17.11* Every countable well-ordered Gödel matrix characterises a categorical calculus, different for each ordinal.

*Theorem 17.12* All uncountable well-ordered Gödel matrices characterise the same non-categorical calculus.

The first of these theorems shows incidentally that the sufficient conditions of Theorems 17.4 and 17.6 are not necessary for categoricity. For none of the matrices that characterise an infinite-valued calculus here is monadic (by Theorem 14.22), and each is equivalent to the submatrix obtained by omitting the least of its undesignated congruence classes. To prove the theorems we need the following lemmas:

*Lemma 17.13* If a calculus is characterised by a matrix with just one designated (undesignated) value, every simple matrix that characterises it has just one designated (undesignated) value.

*Lemma 17.14* If a calculus is characterised by a Gödel matrix, every simple matrix that characterises it is a Gödel matrix.

*Lemma 17.15* Two Gödel matrices characterise the same calculus iff every countable subset of values of either is order-isomorphic to a subset of values of the other.

*Proofs* For 17.13 let M have just one designated (undesignated value a, and let M′ be any simple matrix characterising the same calculus. If b and c are distinct designated (undesignated) values of M′, there exists by 17.8 a countable submatrix $M_1'$ of $M'$ in which b and c are incongruent values. By 17.1 $M_1'$ is equivalent to a submatrix of M, so that both b and c correspond to a in some structure-preserving correspondence. By 14.11 it follows that b and c are congruent in $M_1'$, contrary to hypothesis.

For 17.14, let M be any simple matrix that characterises the same calculus as a Gödel matrix $M'$ with designated value $0'$. By 17.13 M has a unique designated value 0, and we write $a \geq b$ if $a \supset b = 0$. If a and b are distinct values of M there exists by 17.8 a countable submatrix $M_1$ of M in which a and b are incongruent. Every submatrix of $M'$ is simple, so by 17.1 and 14.14 there is a nomomorphism h of $M_1$ onto a submatrix $M_1'$ of $M'$. If $a \geq b$ and $b \geq a$ then $h(a) \supset h(b) = h(b) \supset h(a) = h(0) = 0'$. It follows that $h(a) = h(b)$ and so by 14.11 that a and b are congruent in $M_1$, whence $a = b$. Thus $\geq$ is anti-symmetric. We can show similarly that $\geq$ is connected and transitive and so orders the values of M; that 0 is the least and 1 (i.e. $\sim 0$) the greatest value in this ordering; that there is at least one other value; and that the basic functions follow the required Gödel pattern.

For 17.15, let M and $M'$ be Gödel matrices characterising the same calculus and let 0 and $0'$ be their least values and 1 and $1'$ their greatest. The submatrix $M_1$ generated by any countable subset S of values of M is isomorphic to a submatrix $M_1'$ of $M'$, by 17.1 and 14.14. Using the fact that $a \geq b$ iff $a \supset b = 0$ (or $0'$), it follows that the values of $M_1$ are order-isomorphic to those of $M_1'$. Hence S is order-isomorphic to a subset of values of $M_1'$ and so of $M'$. Conversely, if M and $M'$ characterise distinct calculi some countable submatrix $M_1$ of (say) M is isomorphic to no submatrix of $M'$, by 17.1. Hence the set S of values of $M_1$ is order-isomorphic to no set of values of $M'$ that contains $0'$ and $1'$. Since S has a least and a greatest member (viz.

0 and 1) it follows that it is not order-isomorphic to any subset of values of M′.

For 17.11, suppose that a calculus is characterised by a countable Gödel matrix M whose values are well-ordered with ordinal $\alpha$, and also by a simple matrix M′. By 17.14 M′ is a Gödel matrix, and since an ordered set is well-ordered iff it has no subset of order type $\omega^*$, it follows by 17.15 that the values of M′ are well-ordered, say with ordinal $\alpha'$. Also by 17.15, since M is countable $\alpha \leq \alpha'$, and similarly M′ has no subset of ordinal $\alpha+1$. Hence $\alpha=\alpha'$, and M and M′ are isomorphic as required. It also follows from 17.15 that countable Gödel matrices M and M′ of different ordinals $\alpha$ and $\alpha'$ (with, say, $\alpha<\alpha'$) characterise distinct calculi, since no subset of values of M has ordinal $\alpha'$. 17.12 follows similarly from 17.15, since every uncountable well-ordered set has subsets of every countable ordinal.

## 17.3 *Single-conclusion calculi*

A many-valued single-conclusion calculus is *categorical* if all the matrices that characterise it are equivalent, i.e. if all the simple matrices that characterise it are isomorphic. Though categoricity is common among multiple-conclusion calculi it is rare among single-conclusion ones, as the following theorem shows.

*Theorem 17.16* In a categorical single-conclusion calculus every theorem is a substitution instance of one in which no variable occurs more than once.

To prove this we need to construct the *Cartesian power* $M^S$ of a matrix M indexed by a nonempty set S. $M^S$ is the matrix whose values are the various families of values of M indexed by S, a family $\{a_i\}$ being designated iff $a_i$ is designated in M for each i in S, and whose basic functions are of the form $\lambda\{a_i\}\ldots\{b_i\}.\{F(a_i,\ldots,b_i)\}$ for each basic function F of M. Since index sets of the same cardinality give rise to isomorphic Cartesian powers, we write $M^n$ with systematic ambiguity to stand for any Cartesian power of M indexed by a set of cardinal n. (We have defined $M^S$ only for nonempty S, and hence $M^n$ only for non-zero n. We could have defined $M^\Lambda$ and so $M^o$, but $M^\Lambda$ would always be singular and the case n=0 would be an exception to Theorem 17.17.)

*Theorem 17.17* M and $M^n$ characterise the same single-conclusion calculus.

*Proofs* For 17.17, the families $\{a_i\}$ such that $a_i=a_j$ for all i and j in S form a submatrix of $M^S$ isomorphic to M, so if $X \nvdash_M B$ then $\langle X,B\rangle$ can be invalidated in this submatrix and consequently in $M^S$. For the converse we note that each valuation in $M^S$ is in effect a family $\{v_i\}$ of valuations in M, in which each formula A takes the value $\{v_i(A)\}$. Now if $\langle X,B\rangle$ is invalidated by some valuation $\{v_i\}$ in $M^S$ then $v_i(A)$ must be designated in M for each i in S and each A in X, while $v_i(B)$ is undesignated for some i in S. Thus for some i, $v_i$ invalidates $\langle X,B\rangle$ in M, so that $X \nvdash_M B$ as required.

For 17.16, let M be a simple matrix characterising a categorical single-conclusion calculus L. Since, as we shall see, inconsistent calculi are not categorical, M is nonempty and we may form the Cartesian power $M^M$ of M indexed by its own values. The values of $M^M$ are thus families of the form $\{b_x\}$ with x and $b_x$ both ranging over the values of M. We want to show that if A is not a theorem of L neither is any formula obtained from A by identifying propositional variables, and by 17.17 it is sufficient to show that for each non-theorem A there is a valuation in $M^M$ in which A takes an undesignated value and in which the variables are all assigned congruent values. Since L is categorical, M and $M^M$ are equivalent, and so by 14.14 and 14.15 there is an isomorphism $a \leftrightarrow S_a$ between M and $|M^M|$. Suppose that for each a and y in M there were to exist $a_y$ in M such that $a_y \neq b_y$ for each family $\{b_x\}$ in $S_a$. Then taking a to be the value for which $\{x_x\} \in S_a$, and taking y to be a, we would obtain the contradiction $a_a \neq a_a$. So for some a and y, there is for each c in M a value $\{b_x\}$ in $S_a$ such that $b_y = c$. In particular there exists a valuation $\{v_x\}$ in $M^M$ in which each variable p is assigned a value in $S_a$ such that $v_y(p) = v(p)$ where v is a valuation in M such that v(A) is undesignated. But then $v_y = v$, so $\{v_x(A)\}$ is undesignated as required.

Theorem 17.16 means that scarcely any single-conclusion calculus that has any theorems is categorical. For example, neither the classical calculus nor any calculus characterised by a Łukasiewicz or a Gödel matrix can be categorical, since it has $p \supset p$ but not $p \supset q$ as a theorem. The calculus characterised by Kleene's matrix is not caught by this result since it has no theorems, but nevertheless it too is not categorical. For by Theorem 16.5 every many-valued single-conclusion calculus is characterised by a matrix in which no genuinely multiple-conclusion inference is valid (i.e. in which $X \vdash B_1, \ldots, B_n$ for n≥ only if $X \vdash B_i$ for some i). It follows that whenever a matrix permits genuinely multiple-conclusion inferences the relevant single-conclusion calculus is not categorical. Kleene's matrix is a case in point, for in it $A \vee B \vdash A, B$ although $A \vee B \nvdash A$ and $A \vee B \nvdash B$. Similarly no fragment of the classical propositional calculus can be categorical if it contains $\supset$ or $\vee$ or $\sim$.

Inconsistent single-conclusion calculi are not categorical since they are characterised both by empty and nonempty matrices, but the other singular ones are categorical: the simple matrices that characterise them have a single (undesignated) value. Minimum single-conclusion calculi are never categorical, apart from the exceptional case where no connectives are present at all. For each is characterised by any simple matrix M that characterises the corresponding minimum multiple-conclusion calculus, and we saw in Section 17.2 that M has at least $2^{\aleph_0}$ values if any connectives are present. But at most $\aleph_0$ valuations in M suffice to invalidate the various pairs of the form ⟨V-B,B⟩, and the values involved in these generate a countable submatrix which evidently also characterises the single-conclusion calculus.

Distinctions based on the cardinality of the matrices that characterise a calculus, which can be sharply drawn for multiple-conclusion calculi, become blurred for single-conclusion ones. Thus a single-conclusion calculus can be both finite-valued and infinite-valued, and even one which is exclusively finite-valued (i.e. such that every simple matrix that characterises it is finite) need not be exclusively n-valued for any finite n. The classical calculus exemplifies the first possibility, and its fragment in ~ exemplifies the second. Whether multiple- or single-conclusion, if a calculus is categorical it will of course be exclusively n-valued for some n, but in neither case have we any reason to expect the converse to be true, though we have no counterexample to offer.

The construction of Cartesian powers can also be used to provide matrices characterising the minimum many-valued counterpart of any given many-valued single-conclusion calculus.

*Theorem 17.18* If a single-conclusion calculus is characterised by a nonempty matrix M, its minimum many-valued counterpart is characterised by $M^n$ for any infinite n.

*Proof* Let S be any set of infinite cardinal n. By 17.17 $M^S$ characterises a many-valued counterpart of the calculus in question, and it is sufficient to show that it is contained in the minimum many-valued counterpart. By 16.5, 16.3, and 5.4 this amounts to showing that if $X \nvdash_M B$ for each B in Y, where Y is nonempty, then $\langle X,Y \rangle$ is invalid in $M^S$. Since there are not more than n formulae in Y there is a many-one mapping $i \to B_i$ of S onto Y. For each i, if $X \nvdash_M B_i$, let $v_i$ be a valuation which invalidates $\langle X,B_i \rangle$ in M. Then $\{v_i\}$ invalidates $\langle X,Y \rangle$ as required.

From this result we can derive a criterion for two matrices to characterise the same single-conclusion calculus. We know that if either matrix is empty it is necessary and sufficient that the other should have no undesignated values. For the general case we have:

*Theorem 17.19* Two nonempty matrices $M_1$ and $M_2$ characterise the same single-conclusion calculus iff $M_1 \prec M_2^{\aleph_0}$ and $M_2 \prec M_1^{\aleph_0}$.

*Proof* If $M_1$ and $M_2$ each characterise the single-conclusion calculus L then by 17.18 both $M_1^{\aleph_0}$ and $M_2^{\aleph_0}$ characterise the minimum many-valued counterpart of L. Hence by 17.2 $M_1 \prec M_2^{\aleph_0}$ and $M_2 \prec M_1^{\aleph_0}$ as required. Conversely, if $M_1 \prec M_2^{\aleph_0}$ then $X \vdash_{M_1} B$ whenever $X \vdash_{M_2} B$, by 17.2 and 17.17. Similarly, if $M_2 \prec M_1^{\aleph_0}$ then $X \vdash_{M_2} B$ whenever $X \vdash_{M_1} B$, whence the required result.

In the case where one of the matrices is principal we can sharpen this result to describe the spectrum of matrices that characterise the calculus in question.

*Theorem 17.20* If L is a consistent single-conclusion calculus characterised by a principal matrix M, then M′ characterises L iff $M \prec M' \prec M^{\aleph_0}$.

*Proof* Since L is consistent M is nonempty, and so by 17.18 $M^{\aleph_0}$ characterises the minimum many-valued counterpart of L. As M is principal it characterises the maximum many-valued counterpart, whence by 17.2 M′ characterises L iff $M \prec M' \prec M^{\aleph_0}$.

Another corollary is a partial analogue of Lemma 17.13.

*Theorem 17.21* If a consistent single-conclusion calculus is characterised by a matrix with only one designated value, every simple matrix that characterises it has only one designated value.

*Proof* If $M_1$ has just one designated value, so also does $M_1^{\aleph_0}$. If $M_2$ is any nonempty simple matrix characterising the same single-conclusion calculus, $M_2 \prec M_1^{\aleph_0}$ by 17.19, and $M_2$ has just one designated value by the argument of 17.13. The corresponding result for undesignated values does not carry over to the single-conclusion case, as Theorem 18.5 illustrates.

# 18 · Two-valued logic

## 18.1 *Axiomatisation*

The classical matrix characterises the calculus PC for which rules of inference were given in Section 2.3. We show this as a corollary of

*Theorem 18.1* Any matrix whose values are t (designated) and f (undesignated) characterises the same calculus as the rule 'From $\{A_i : a_i = t\}$ infer $FA_1 \ldots A_k$, $\{A_i : a_i = f\}$ if $F(a_1, \ldots, a_k) = t$, and from $\{A_i : a_i = t\}$, $FA_1 \ldots A_k$ infer $\{A_i : a_i = f\}$ if $F(a_1, \ldots, a_k) = f$, for each k-ary F and k-tuple of values $a_1, \ldots, a_k$'.

*Proof* Let M be the matrix and R the corresponding rule. It is evident from the construction of R that $\vdash_R \subset \vdash_M$. Conversely, suppose that $X \nvdash_R Y$, so that by cut there is a partition $\langle T, U \rangle$ such that $X \subset T$ and $Y \subset U$ and $T \nvdash_R U$. Let v be the valuation in M determined by assigning t as value to every variable belonging to T and f to every variable belonging to U. Then for every A we have $T \vdash_R A, U$ if $v(A) = t$ and $T, A \vdash_R U$ if $v(A) = f$. This is proved by induction on the complexity of A, the basis (in which A is a variable) being established by appeal to overlap, and the induction step (in which A is of the form $FA_1 \ldots A_k$) by appeal to the appropriate instance of R for which each $a_i = v(A_i)$, together with cut for the set $\{A_1, \ldots, A_k\}$. It follows that $A \in T$ iff $v(A) = t$, so that v invalidates $\langle T, U \rangle$ and hence $X \nvdash_M Y$ as required.

Theorem 18.1 applies to any non-singular two-valued matrix, including fragments of the classical one and variants based on an unorthodox choice of truth-functions. In the case of the classical matrix itself, based on ⊃, &, ∨ and ~, the corresponding rules are

| | | |
|---|---|---|
| From A,B infer A⊃B | From A,B infer A&B | From A,B infer A∨B |
| From A,A⊃B infer B | From A,A&B infer B | From A infer A∨B,B |
| From B infer A⊃B,A | From B,A&B infer A | From B infer A∨B,A |
| From Λ infer A⊃B,A,B | From A&B infer A,B | From A∨B infer A,B |

From A,~A infer Λ    From Λ infer ~A,A

The rules for PC represent a simplification of this list, for each of them either appears in it or is derivable by cut from two of its members, while conversely each of the present rules either appears in the list in Section 2.3 or is derivable from one of them by dilution. Like the present ones, the rules for PC can be shown to be independent by using matrices obtained by altering just one cell of the relevant classical truth-table.

A corollary of Theorem 18.1 is a subformula theorem for the rules for PC (discussed further in Section 20.1).

*Theorem 18.2* If $X \vdash_{PC} Y$ then $X \vdash_{PC'} Y$, where PC′ comprises just those instances of the rules for PC each of whose premisses and conclusions is a subformula of a member of X or Y.

*Proof* Let M be the classical matrix and R the corresponding rule as described in 18.1, and let R′ comprise just those instances of R whose premisses and conclusions are all subformulae of members of X and Y. Since each instance of R′ is either one of PC′ or derivable from one of them by dilution, if $X \nvdash_{PC'} Y$ then $X \nvdash_{R'} Y$. By the argument of 18.1 there exists a partition ⟨T,U⟩ such that

> $T \nvdash_{R'} U$ and $X \subset T$ and $Y \subset U$, and a valuation v such that $A \in T$ iff $v(A)=t$ for all subformulae A of members of X and Y. In particular v invalidates $\langle X,Y \rangle$, so that $X \nvdash_M Y$ and by 18.1 $X \nvdash_{PC} Y$.

Despite the definition of PC′, deducibility by PC′ is not quite the same thing as deducibility using only subformulae of members of X and Y, since the latter relation is generally not closed under overlap (Section 3.3). It is however easy to verify from the discussion of $\vdash_R^Z$ in Section 3.3 (using the fact that X and Y are subsets of the relevant Z) that for any adequate variety of proof Y is deducible from X by the rules for PC, using only subformulae of members of X and Y, iff $X \vdash_{PC'} Y$. Thus a corollary of Theorem 18.2 is a proof-theoretic version of the subformula theorem: if Y is deducible from X by the rules for PC it is deducible using only subformulae of members of X and Y.

## 18.2 *Duality*

The *converse* of a matrix M is defined as the matrix M* having the same truth-values and functions as M but such that a value is designated in M* iff it is undesignated in M. It follows that $X \vdash_M Y$ iff $Y \vdash_{M*} X$: converse matrices characterise converse calculi.

When M has the same number of designated values as undesignated ones, it is possible to set up a one-one correspondence $x \leftrightarrow x*$ between M and M* such that x is designated in M iff x* is designated in M*. Every valuation v in M can then be matched with a valuation v* in M* defined by setting $v*(p)=(v(p))*$ for each variable p. It does not follow that $v*(A)=(v(A))*$ for formulae other than the variables, but if conditions are favourable it is possible to define for every formula A a *dual formula* A* such that v*(A*)=(v(A))* for every v. By Lemma 16.11 this condition need not determine the dual formula uniquely, but it is convenient to define a particular dual as follows.

For each k-ary connective F let F* be the truth-function such that $F^*(a_1^*,\ldots,a_k^*) = (F(a_1,\ldots,a_k))^*$ for all $a_1^*,\ldots,a_k^*$. If F* is not a matrix function it is easy to show that no dual formula exists for $Fp_1\ldots p_k$. Otherwise (and in particular if M is functionally complete) let $F^*p_1\ldots p_k$ be any formula, in which $p_1,\ldots,p_k$ are the only variables to appear, associated with F* in the manner of Section 14.1. (If F* is a basic function the corresponding connective can be used to construct $F^*p_1\ldots p_k$, and is said to be the *dual* of the connective F.) We define $F^*A_1\ldots A_k$ by substitution in $F^*p_1\ldots p_k$, and then define A* inductively, by setting p*=p for each variable p, and setting $(FA_1\ldots A_k)^* = F^*A_1^*\ldots A_k^*$.

It is easily shown by induction on the complexity of A that this definition of A* satisfies the condition that v*(A*)=(v(A))*. It follows that v(A) is designated in M iff v*(A*) is designated in M*, and hence, defining X* as {A*: A ∈ X} and Y* similarly, that $X \vdash_M Y$ iff $X^* \vdash_{M^*} Y^*$. Putting this together with the earlier result about converse matrices in general, we have the *duality principle*: $X \vdash_M Y$ iff $Y^* \vdash_M X^*$.

The classical matrix satisfies the conditions for the formulation of a duality principle, for there is a unique one-one correspondence of the required kind, namely that in which t*=f and f*=t, and in addition the matrix is functionally complete, so that every formula has a dual. (Indeed & and ∨ are dual connectives, and ~ is self-dual.) We therefore have

*Theorem 18.3* $X \vdash_{PC} Y$ iff $Y^* \vdash_{PC} X^*$.

Taking the special case in which X and Y each consist of a single formula, we obtain a limited form of duality principle which can be carried over to the single-conclusion case, namely $A \vdash B$ iff $B^* \vdash A^*$. To derive the principle of duality as enunciated by logicians occupied with theoremhood (cf Church, 1956, *161), we need to take a different special case, in which Y consists of a single formula but X is empty. This gives $\Lambda \vdash A$ iff $A^* \vdash \Lambda$, and since $A^* \vdash \Lambda$ iff $\Lambda \vdash \sim A^*$ by the truth-table for negation, the duality principle for theorems follows in the form: $\vdash A$ iff $\vdash \sim A^*$.

## 18.3 *Counterparts*

PC is the maximum many-valued counterpart of the classical calculus, for by Theorem 16.12 the classical matrix is principal. The proof of that theorem shows that PC is also the maximum propositional counterpart, but we can go further and show that

*Theorem 18.4* PC is the maximum counterpart of the classical propositional calculus.

*Proof* Let L be the classical calculus and M the classical matrix. Since PC is a counterpart of L it is sufficient to show that if $X \vdash_U Y$ then $X \vdash_{PC} Y$. Suppose then that $X \nvdash_{PC} Y$, so that some valuation v invalidates $\langle X,Y \rangle$ in M. There exist formulae B such that v(B)=f (viz., either A or ~A for any A), and v shows that $X, Y{\supset}B \nvdash B$ for any such B. Hence $X \nvdash_U Y$ by 5.14.

PC is categorical and so exclusively two-valued. This is a corollary of Theorem 17.5 and of 17.6, and was also proved by Carnap (1943) by an argument special to the two-valued case, namely by showing in effect that in any matrix in which the rules for PC are valid the basic functions must obey the classical tables with respect to the designated/undesignated dichotomy. By contrast the single-conclusion part of PC, the classical propositional calculus, is not categorical by Theorem 17.16, and although it is two-valued it is not exclusively so, being $2^n$-valued for every n as well as $\aleph_0$-valued. To describe its characteristic matrices let us call a matrix *Boolean* if it forms a Boolean algebra with respect to &, ∨, and ~, the unit of this algebra being the only designated value. (It is assumed that ⊃ is related to &, ∨, and ~ in the standard way, i.e. that $x \supset y = {\sim}x \vee y$.)

*Theorem 18.5* The simple matrices that characterise the classical propositional calculus are precisely the Boolean matrices.

*Proof* Using 16.11 it is easy to verify that the axioms of Boolean algebra hold good in any simple matrix that characterises the classical calculus. For example, $p \vee q$ and $q \vee p$ are synonymous and so by 16.11 $a \vee b$ and $b \vee a$ are congruent for every a and b, and since the matrix is simple it follows that $a \vee b = b \vee a$ as required. Moreover since $\vdash A \vee {\sim}A$ the unit of the algebra is designated, and is the only designated value by 17.21. This half of the proof is due in essence to Church, 1953. For the converse one could try starting from Church's observation that any Boolean matrix M′ characterises classical theoremhood, but to extend this to cover consequence via the deduction theorem would require a demonstration that $\vdash_{M'}$ is compact. Instead we argue as follows. The Boolean algebra formed by any Boolean matrix M′ is isomorphic to an algebra of subsets of a nonempty set S, with S as unit (Stone, 1936), and hence is isomorphic to a subalgebra of the Boolean algebra constituted by the set of all subsets of S. But the latter is isomorphic to the Cartesian power $M^S$ of the classical matrix M, each subset S′ corresponding to the family $\{a_i\}$ such that $a_i$=t or f according as $i \in S'$ or not. Hence M′ is isomorphic to a submatrix of $M^S$. Since also M is isomorphic to the submatrix of M′ generated by its unit 1, it follows by 17.17 that M′ characterises the classical calculus. Moreover M′ is simple. For if a and b are congruent $\sim a \vee b = \sim b \vee a = 1$, since each of these is congruent to $\sim a \vee a$, which is the only designated value 1; and hence $a = a\&(\sim a \vee b) = a\&b = (\sim b \vee a)\&b = b$.

The Cartesian powers of the classical matrix have a special place among the matrices for the classical calculus, as we shall see, and we begin by providing rules of inference appropriate to them. Let R be the singleton-conclusion rules corresponding to any set of rules for the classical calculus, together with 'from A,~A infer Λ'. For each positive integer n let $R_n$ be obtained by adding to R the rule 'from $\bigcup A_i \vee A_j$ infer $A_0, A_1, \ldots, A_n$', the union being taken over all i and j such that $0 \le i < j \le n$.

*Theorem 18.6* Each Cartesian power $M^n$ of the classical matrix M characterises the same calculus as the above rules $R_n$ (for finite n) or R (for infinite n).

*Proof* Let L be the classical calculus. If n is infinite $M^n$ characterises the negative variant of the minimum counterpart of L, by 16.5 and 17.18; and so does R, by 5.15 and 5.27. If n is finite, to show that $\vdash_{R_n} \subset \vdash_{M^n}$ it is sufficient to establish the validity of the distinctive rule of $R_n$. Suppose that for some valuation $v = \{v_i\}$ in $M^n$, $v(A_i)$ is undesignated for each i, i.e. for each i ($0 \le i \le n$) there is a j ($1 \le j \le n$) such that $v_j(A_i)$=f. Since the range of i outnumbers that of j it follows that some $i_1$ and $i_2$ (say with $i_1 < i_2$) must correspond to the same j, so that $v_j(A_{i_1} \vee A_{i_2}) = f$ and $v(A_{i_1} \vee A_{i_2})$ is undesignated as required.

To show conversely that $\vdash_{M^n} \subset \vdash_{R_n}$, suppose that $X \nvdash_{R_n} Y$. Then by cut there is a partition $\langle T,U \rangle$ such that $X \subset T$ and $Y \subset U$ and $T \nvdash_{R_n} U$, and we call a subset of U *critical* if $A \vee B \in T$ for all distinct A and B in the subset. By the special rule of $R_n$ no critical set can have more than n members, and there exist nonempty critical sets, e.g. $\{A\&{\sim}A\}$; so let $Z = \{C_1,\ldots,C_m\}$ be a critical set of greatest possible cardinal m ($1 \le m \le n$). Let $v = \{v_i\}$ be the assignment of truth-values in $M^m$ such that $v_i(A) = t$ iff $A \vee C_i \in T$. To show that v is a valuation it is therefore sufficient to show that (i) $A \vee C_i \in T$ iff ${\sim}A \vee C_i \notin T$, (ii) $(A\&B) \vee C_i \in T$ iff $A \vee C_i \in T$ and $B \vee C_i \in T$, (iii) $v_i(A \vee B) = v_i({\sim}({\sim}A\&{\sim}B))$, and (iv) $v_i(A \supset B) = v_i({\sim}(A\&{\sim}B))$. For (i), since $C_i \in U$ we have $T \nvdash C_i$, and since $A \vee C_i, {\sim}A \vee C_i \vdash C_i$ it follows that if $A \vee C_i \in T$ then ${\sim}A \vee C_i \notin T$. The converse implication is established by the informal tree proof below. For (ii) it is sufficient to note that $(A\&B) \vee C_i \vdash A \vee C_i$ and $(A\&B) \vee C_i \vdash B \vee C_i$ and $A \vee C_i, B \vee C_i \vdash (A\&B) \vee C_i$; for (iii) that $(A \vee B) \vee C_i \vdash {\sim}({\sim}A\&{\sim}B) \vee C_i$ and vice versa; and similarly for (iv).

$${\sim}A \vee C_i \in U$$

| $A \vee C_i \in U$ | $A \vee C_i \in T$ |
|---|---|

| $B \vee C_i \in Z$ for some B in $\{A,{\sim}A\}$ | $Z \cup \{A \vee C_i, {\sim}A \vee C_i\} - \{C_i\}$ is not critical |
|---|---|
| $C_i \vee B \vee C_i \in T$ | |
| $B \vee C_i \in T$ | |

| $A \vee C_i \vee {\sim}A \vee C_i \in U$ | $C_j \vee B \vee C_i \in U$ for some B in $\{A,{\sim}A\}$ and some $j \ne i$ |
|---|---|
| $A \vee {\sim}A \in U$ | $C_i \vee C_j \in U$ |

Now for any A in T, since $A \vdash A \vee C_i$ it follows that $A \vee C_i \in T$ and so $v_i(A) = t$ for each i. But if $A \in U$ then either $A = C_i$ for some i, in which case since $C_i \vee C_i \vdash C_i$ we have $A \vee C_i \notin T$ and so $v_i(A)=f$; or Z,A has more members than Z and so is not critical, in which case since Z itself is critical it follows that $A \vee C_i \notin T$ for some i and so $v_i(A)=f$. Thus v invalidates $\langle T,U \rangle$, and since $M^m \prec M^n$ it follows by 17.2 that $X \nvdash_{M^n} Y$ as required.

Since the classical calculus is negative all its many-valued counterparts are negative propositional ones, by Theorem 16.3. In the compact case the converse is also true:

*Theorem 18.7* Every compact negative propositional counterpart of the classical calculus is many-valued and characterised by some Cartesian power of the classical matrix.

*Proof* Let L′ be any compact negative propositional counterpart of the classical calculus L, and let a set Y be called *trivial* in L′ if (for every X) $X \vdash' Y$ iff $X \vdash B$ for some B in Y. If every nonempty finite set is trivial so is every nonempty set, since L′ is compact; and so in this case L′ is the negative variant of the minimum counterpart of L and by 16.5 and 17.18 it is characterised by every infinite Cartesian power $M^n$ of the classical matrix M. Otherwise let n be the greatest integer such that every nonempty set with not more than n members is trivial. Since all unit sets are trivial $n \geq 1$, and we show that $M^n$ characterises L′.

If $X \vdash' Y$ then $X' \vdash' Y'$ for some finite subsets X′ and Y′ of X and Y. Let $v=\{v_i\}$ be any valuation in $M^n$, and for each i ($1 \leq i \leq n$) let $A_i$ be the disjunction of those members A of Y′ such that $v_i(A)=f$ (if $v_i(A)=t$ for all A in Y′, let $A_i$ be $p \& {\sim}p$). Then $v_i(A_i)=f$, so that $v(A_i)$ is undesignated. If v(A) is designated for some A in Y′ then v satisfies $\langle X,Y \rangle$. Otherwise each A in Y′ appears as a disjunct in some $A_i$ or other, so that $A \vdash A_i$ and hence by cut for Y′ we have $X' \vdash' A_1,\ldots,A_n$. But $\{A_1,\ldots,A_n\}$ is trivial, so $X' \vdash A_i$ for some i; and since each $v(A_i)$ is undesignated it follows by 17.17 that v does not satisfy X′ and so again satisfies $\langle X,Y\rangle$. Thus $X \vdash_{Mn} Y$ as required.

For the converse it is sufficient by 18.6 to show that $R_n \subset \vdash'$ and since $R \subset \vdash'$ by the hypothesis that L′ is a negative counterpart of L, we need only show that $\bigcup p_i \vee p_j \vdash' p_o,\ldots,p_n$, where the union is taken over all i and j such that $0 \leq i < j \leq n$. Let $Y = \{B_o,\ldots,B_n\}$ be a non-trivial set and let X be such that $X \vdash' Y$ but $X \nvdash B_i$ for any i, so that for each i there is a valuation $v_i$ in M which invalidates $\langle X,B_i\rangle$. We may assume that $v_o(p)=t$ for every variable p, for if not we can redefine X and Y to make it so, by substituting $\sim p$ for each p for which $v_o(p)=f$. Let $C_i$ be $p_o \& \ldots \& p_{i-1} \& p_{i+1} \& \ldots \& p_n$, and let s be the substitution such that s(p) is the disjunction of those $C_i$ for which $v_i(p)=t$. We show that (i) $\bigcup p_i \vee p_j \vdash A$ for each A in s(X), and (ii) $\bigcup p_i \vee p_j, s(B_i) \vdash p_i$ for each i. For this it is sufficient to show that every valuation v in M which satisfies $\bigcup p_i \vee p_j$ satisfies s(X) and each $\langle s(B_i),p_i\rangle$. If v satisfies $\bigcup p_i \vee p_j$, either $v(p_i)=t$ for every i or $v(p_j)=f$ for some j but $v(p_i)=t$ for all $i \neq j$. In the former case v evidently satisfies $\langle s(B_i),p_i\rangle$ for every i. Also $v(s(p))=t=v_o(p)$ for every variable p and hence $v(s(A))=v_o(A)$ for every formula A; and since $v_o$ satisfies X it follows that v satisfies s(X). In the second of the two cases $v(C_j)=t$ while $v(C_i)=f$ for all $i \neq j$, whence $v(s(p))=v_j(p)$ for every variable p and so $v(s(A))=v_j(A)$ for every A. Thus v satisfies s(X) since $v_j$ satisfies X. Similarly $v(s(B_j))=v_j(B_j)=f$, and so v satisfies $\langle s(B_j),p_j\rangle$; while v satisfies $\langle s(B_i),p_i\rangle$ for all $i \neq j$ since $v(p_i)=t$. This establishes (i) and (ii), and since $s(X) \vdash' s(Y)$ it follows by cut for s(X) and s(Y) that $\bigcup p_i \vee p_j \vdash' p_o,\ldots,p_n$ as required.

*Theorem 18.8* Every compact propositional counterpart of the classical calculus is axiomatisable.

*Proof* By 4.3 it is sufficient to show that any compact propositional counterpart L of the classical calculus is characterised by recursive rules, and for this we may assume that R and $R_n$ are based on recursive rules for the classical calculus and so are themselves recursive. If L is negative then by 18.6 and 18.7 it is characterised by R or $R_n$ as the case may be. If L is positive its negative variant is also a compact propositional counterpart, since the classical calculus is both compact and negative. The same argument as before shows that this variant is characterised by R or $R_n$, and it is clear from 5.27 that a calculus is characterised by recursive rules iff its variant is.

Not all many-valued counterparts of the classical calculus are compact. For example, if M′ is the Boolean matrix whose values are the finite sets of natural numbers and their complements, and if $\ldots,p_{-2},p_{-1},p_o,p_1,p_2,\ldots$ are distinct variables, we have $\bigcup p_{i+1} \supset p_i \vdash_{M'} \bigcup p_i \supset p_{i+1}$ when the unions are taken over all integers i, but not when either union is taken over a finite set. If we relax the condition of compactness there is no result corresponding to Theorem 18.7, for

*Theorem 18.9* Not every negative propositional counterpart of the classical calculus is many-valued.

*Proof* Let $R_\omega$ be the infinite analogue of the rules $R_n$, obtained by adding to R the infinite rule 'from $\bigcup A_i \vee A_j$ infer $A_o,A_1,A_2,\ldots$' where the union is taken over all integers i and j such that $0 \leq i < j$. The propositional calculus characterised by $R_\omega$ is a negative counterpart of the classical one, since $\vdash_R \subset \vdash_{R_\omega} \subset \vdash_{R_1}$ and $A, \sim A \vdash_{R_\omega} \Lambda$. But it is not many-valued, for by 18.5 any simple matrix $M_o$ that characterised it would have to be Boolean, and we can show that this is impossible. For if $M_o$ is finite and Boolean it is isomorphic to $M^n$ for some finite n, and so by 18.6 $\vdash_{R_\omega} = \vdash_{R_n}$. But in any case $\vdash_{R_\omega} \subset \vdash_{R_{n+1}} \subset \vdash_{R_n}$, so that $\vdash_{R_{n+1}} = \vdash_{R_n}$. Hence by 18.6 the same calculus is characterised by two simple finite matrices - $M^n$ and $M^{n+1}$ - with different numbers of values, contrary to 17.5. On the other hand, if $M_o$ is an infinite Boolean matrix, there exists an infinite sequence of distinct values $a_o, a_1, \ldots$ such that $\ldots a_{i+1} \subset a_i \subset \ldots \subset a_o$, and the assignment of $\sim a_i \vee a_{i+1}$ as value to $p_i$ invalidates the instance $\langle \bigcup p_i \vee p_j, \{p_o, p_1, \ldots\} \rangle$ of $R_\omega$.

The results of this section do not necessarily apply to fragments of the classical calculus and the corresponding fragmentary counterparts. For example, we have seen that PC is the maximum counterpart of the classical calculus (Theorem 18.4). By way of contrast consider the fragment of the classical calculus based on & alone. We note that in this fragment $X \vdash B$ iff each variable in B occurs in a member of X, and we can see from this by Theorem 5.5 that $\Lambda \vdash_U A, B$ whenever A and B have no variable in common (e.g. $\Lambda \vdash_U p, q$ for each pair of distinct variables). Thus the maximum counterpart is not only stronger than the corresponding fragment of PC, but is not even a propositional calculus.

Another example is the theorem that the simple characteristic matrices for the classical calculus are precisely the Boolean ones (Theorem 18.5). Consider instead the fragmentary calculus based on ~ alone. The corresponding fragment of a Boolean matrix with more than two values is never simple, since all the values other than the unit 1 and its complement are congruent to each other. The quotient of every such matrix is the negational fragment of Kleene's matrix and this is a simple characteristic matrix for the fragmentary calculus, but is not a fragment of any Boolean matrix.

As a final example, since by Theorems 18.5 and 17.5 each finite Boolean matrix characterises a distinct counterpart, the classical calculus has an infinite number of many-valued counterparts. In contrast to this the fragment based on ~ has only two (characterised by the two-valued and three-valued matrices respectively), and the fragment based on & has only one (for the same two-valued matrix characterises the maximum many-valued counterpart by Theorem 16.12 and the minimum one by Theorems 18.1 and 5.15; indeed the same argument shows that the fragment has a unique propositional counterpart).

# 19 · Axiomatisation

## 19.1 *Finite axiomatisation*

Although in principle a calculus may be axiomatised by supplying recursive rules of inference for it, in practice we expect the number of rules to be finite. For example the classical propositional calculus, characterised by the classical matrix M, can in principle be axiomatised by the rule 'from $A_1,\ldots,A_m$ infer B if $A_1,\ldots,A_m \vdash_M B$'. As an exercise in axiomatisation this is not trivial, since it calls for a demonstration that $\vdash_M$ is compact; and if it is unsatisfactory it is because the rule is felt to be a portmanteau containing infinitely many distinct rules. But to make this feeling coherent we must be able to say what is to count as a single rule. For propositional calculi, and for other calculi too if a suitable definition of substitution can be devised, the appropriate idea is that of a scheme. The familiar example is the *axiom scheme*, a set of axioms comprising every substitution instance s(A) of a given formula A. By analogy we define a *rule scheme* to be a set of rules comprising 'from s(X) infer s(Y)' for every substitution instance of given X and Y. A single-conclusion rule scheme will similarly comprise 'from s(X) infer s(B)' for every substitution instance of given X and B. Axiom schemes are covered by this definition as rule schemes with zero premisses. Most of the familiar rules - modus ponens, adjunction, etc. - are rule schemes in this sense. An exception is the rule of substitution 'from

A infer s(A)', but it is ordinarily only used for developing systems of theorems, not for drawing consequences from premisses. It would in any case be too strong for such use, since as it stands it licences p ⊢ q and so cannot be a rule for a non-singular propositional calculus. We may therefore take the rule scheme to be the unit of rule, and say that a propositional calculus is *finitely axiomatisable* if it can be characterised by a finite number of rule schemes.

We shall show that every finite-valued calculus is finitely axiomatisable (Theorem 19.12). Our method falls into two stages. Using unspecified sets of formulae as parameters, we first describe a set of rules R which are at least strong enough for a matrix M in the sense that $\vdash_M \subset \vdash_R$. The second stage is to specify particular parameters for which the converse holds, and for which moreover R is equivalent to a finite set of rule schemes. This method is inspired by the well-known work of Rosser and Turquette, who made use of individual formulae as parameters in describing rules for many-valued single-conclusion calculi (cf. Section 19.8). But with single conclusions there is no guarantee that the second stage can be carried out, so that their method yields an axiomatisation only in those favourable cases for which suitable parametric formulae happen to be available; whereas with multiple conclusions we show that suitable parameters can be devised for any finite matrix. As a corollary we are able to supply finite axiomatisations for any finite-valued single-conclusion calculus that contains either a disjunction or else an implication satisfying the deduction theorem.

## 19.2 *Monadic matrices*

We begin with the case of monadic matrices, which by Theorem

14.24 includes all functionally complete ones, since it gives rise to a simple pattern of rules which is interesting in its own right as well as providing an introduction to the general case. With each formula A and each truth-value a of a matrix M let there be associated two sets of formulae $T_a(A)$ and $U_a(A)$. We call each set a *parameter*, and use $t_a(A)$ and $u_a(A)$ respectively to stand for any member of it. The notation $T_a(A)$ and $U_a(A)$ is not governed by the convention that requires T and U to stand for complementary sets: the choice of parameters is entirely arbitrary at this stage. In particular we do not need to assume that they are finite, though if they are not the rules R1 and R2 below will have instances with infinitely many premisses or conclusions and so will only be rules in the extended sense discussed in Chapter 6. Similarly we do not at this stage need to assume that M is monadic or finite, though if it is not finite R3 will generally not be a finite rule. For each set S of truth-values we use $t_S(A)$ to stand for any set consisting of one representative member $t_a(A)$ of $T_a(A)$ for every a in S, and we use $u_S(A)$ similarly. If $T_a(A)$ is empty for any a in S it follows that no sets $t_S(A)$ exist; but if S is empty we merely have $t_S(A) = \Lambda$. Let R be the following rules:

R1 From $T_a(A)$ infer $A, U_a(A)$ if a is designated, and from $A, T_a(A)$ infer $U_a(A)$ if a is undesignated.

R2 From $\bigcup T_{a_i}(A_i)$ infer $t_{F(a_1,\ldots,a_k)}(FA_1\ldots A_k)$, $\bigcup U_{a_i}(A_i)$, and from $\bigcup T_{a_i}(A_i)$, $u_{F(a_1,\ldots,a_k)}(FA_1\ldots A_k)$ infer $\bigcup U_{a_i}(A_i)$, for each k-ary F and each choice of $t_{F(a_1,\ldots,a_k)}(FA_1\ldots A_k)$ or $u_{F(a_1,\ldots,a_k)}(FA_1\ldots A_k)$ as the case may be, the unions being taken over $1 \le i \le k$.

R3 From $u_{S_1}(A)$ infer $t_{S_2}(A)$, for each partition $\langle S_1, S_2\rangle$ of the truth-values and each choice of $u_{S_1}(A)$ and $t_{S_2}(A)$.

*Lemma 19.1* If v is any valuation, $\bigcup T_{v(p)}(p) \vdash_R t_{v(A)}(A)$,

$\bigcup U_{v(p)}(p)$ and $\bigcup T_{v(p)}(p), u_{v(A)}(A) \vdash_R \bigcup U_{v(p)}(p)$, the unions being taken over all the variables p in A.

*Lemma 19.2* If v is any valuation, $\bigcup T_{v(p)}(p) \vdash_R A, \bigcup U_{v(p)}(p)$ if v(A) is designated and $\bigcup T_{v(p)}(p), A \vdash_R \bigcup U_{v(p)}(p)$ if v(A) is undesignated, the unions being taken over all the variables p in A.

*Proofs* For 19.1 we argue by induction on the complexity of A, appealing to overlap for the basis, when A is a variable p, and to R2 and cut for $\bigcup(T_{v(A_j)}(A_j) \cup U_{v(A_j)}(A_j))$ in the induction step, when $A = FA_1 \ldots A_k$. 19.2 follows by R1 and cut for $T_{v(A)}(A) \cup U_{v(A)}(A)$.

*Lemma 19.3* $\vdash_M \subset \vdash_R$.

*Proof* Suppose $X \nvdash_R Y$. Then by cut there is a partition $\langle T,U \rangle$ such that $X \subset T$ and $Y \subset U$ and $T \nvdash_R U$. If for some variable p and every a there exists either (i) some $t_a(p)$ in U or (ii) some $u_a(p)$ in T, then by taking $\langle S_1, S_2 \rangle$ to be the partition of the truth-values in which $S_2$ is the set of a for which (i) holds, we have by dilution from R3 that $T \vdash_R U$, contrary to hypothesis. Therefore there exists a valuation v such that $T_{v(p)}(p) \subset T$ and $U_{v(p)}(p) \subset U$ for every p. It follows from 19.2 that $A \notin U$ if v(A) is designated and $A \notin T$ if v(A) is undesignated. Hence v invalidates $\langle T,U \rangle$ and so $X \nvdash_M Y$ as required.

We want to choose parameters so that the rules R are valid in the matrix M and equivalent to a finite set of rule schemes. With this double aim in mind we say that a choice of parameters is *standard* if for some variable p and every a, (i) each valuation v invalidates $\langle T_a(p), U_a(p) \rangle$ iff v(p) is congruent to a, and (ii) $T_a(p)$ and $U_a(p)$ contain no variable other than p, and

$T_a(A)$ and $U_a(A)$ are obtained for each A by substituting A for p in $T_a(p)$ and $U_a(p)$. It evidently follows that v invalidates $\langle T_a(A), U_a(A)\rangle$ iff v(A) is congruent to a.

*Lemma 19.4* If the parameters are standard, $\vdash_R \subset \vdash_M$.

*Proof* We show that, given standard parameters, every valuation v in M satisfies R. If v satisfies $\langle T_a(A), U_a(A)\rangle$ it will evidently satisfy R1. If not, then as remarked above v(A) is congruent to a and so is designated iff a is designated, whence again R1 is satisfied. Similarly R2 will certainly be satisfied unless v invalidates $\langle T_{a_i}(A_i), U_{a_i}(A_i)\rangle$ for each i; and in this case each $v(A_i)$ is congruent to $a_i$ and so $v(FA_1 \ldots A_k)$ is congruent to $F(a_1,\ldots,a_k)$, whence v invalidates $\langle T_{F(a_1,\ldots,a_k)}(FA_1\ldots A_k), U_{F(a_1,\ldots,a_k)}(FA_1\ldots A_k)\rangle$ and hence the rule is satisfied. For R3, since v invalidates $\langle T_{v(A)}(A), U_{v(A)}(A)\rangle$, any instance of the rule must either contain an undesignated premiss $u_{v(A)}(A)$ or a designated conclusion $t_{v(A)}(A)$, and the rule is therefore satisfied.

*Lemma 19.5* For every monadic matrix there exist standard parameters which are finite if the matrix is finite.

*Proof* Let $f_1,\ldots,f_j,\ldots$ be the singular matrix functions of a monadic matrix M. Let Z be a set of formulae $A_1,\ldots,A_j,\ldots$ such that p is the only variable to appear in any $A_j$, and each $A_j$ is associated with $f_j$ in the way described in Section 14.1, i.e. $f_j(v(p)) = v(A_j)$ for every valuation v. For each a let $\langle T_a(p), U_a(p)\rangle$ be the partition of Z which is invalidated when p is assigned the value a. Since M is monadic v(p) is congruent to a iff for each j both $f_j(v(p))$ and $f_j(a)$ are designated or both are undesignated; i.e. iff for each j either

$v(A_j)$ is designated and $A_j \in T_a(p)$ or $v(A_j)$ is undesignated and $A_j \in U_a(p)$; i.e. iff v invalidates $\langle T_a(p), U_a(p)\rangle$. Condition (i) of standardness is therefore satisfied, and so is condition (ii) if we define $T_a(A)$ and $U_a(A)$ by substitution from $T_a(p)$ and $U_a(p)$. Moreover if M is finite so is the number of distinct singulary matrix functions, whence Z and consequently every parameter is finite.

*Theorem 19.6* Every calculus characterised by a finite monadic matrix is finitely axiomatisable.

*Proof* It is readily verified that for finite standard parameters each of the rules R1-3 is equivalent to a finite set of (finite) rule schemes, and the theorem follows from 19.3, 19.4 and 19.5.

This result not only ensures the existence of a finite axiomatisation but provides an algorithm for generating it, since it is an effective matter to choose parameters by the method of Lemma 19.5 for a finite matrix. It may be noted too that if the algorithm, when applied to M, produces rules R based on parameters $T_a(A)$ and $U_a(A)$, then applied to the converse matrix M* it produces parameters obtained by interchanging $T_a(A)$ and $U_a(A)$ throughout, and the resulting set of rules is the converse of R. Both remarks apply equally to the general case considered in Section 19.5 below.

## 19.3 *Examples*

A number of simplifications can be exploited when applying Theorem 19.6. We have seen that suitable parameters for a finite monadic matrix can always be obtained from partitions of the set Z of formulae $A_1,\dots,A_n$ associated with the various singulary matrix functions $f_1,\ldots,f_n$, but it is usually possible to simplify the rules by using for each a some partition of a proper subset $Z_a$ of Z.

In a non-singular matrix it is always possible, for example, to ensure that every instance of R1 and R3 is a case of overlap, so that the required axiomatisation can be obtained from R2 alone. To do so, we take $f_1$ to be the identity function and $A_1$ to be p, and each $A_i$ is to be included in $Z_a$ iff, for some b, $f_i$ is the first function in the list to separate a from b. We then as before take $\langle T_a(p),U_a(p)\rangle$ to be the partition of $Z_a$ that is invalidated when v(p)=a. Since the matrix is non-singular, $p \in Z_a$ for every a, and so each instance of R1 is a case of overlap. To show the same for R3, suppose that $\langle\{A_{i_1},\ldots,A_{i_k}\},\{A_{i_{k+1}},\ldots,A_{i_m}\}\rangle$ is an instance of R3 (with p as A), in which no formula appears both as a premiss and as a conclusion, and in which without loss of generality we may take $i_m$ to be the largest index. We show that $\langle\{A_{i_1},\ldots,A_{i_k}\},\{A_{i_{k+1}},\ldots,A_{i_{m-1}}\}\rangle$ is a similar instance of R3; and by repeating the argument it follows that $\langle\Lambda,\Lambda\rangle$ is an instance of R3 and hence that the matrix is singular, contrary to hypothesis. The presence of $A_{i_m}$ as a conclusion indicates that it is a member of $T_a(p)$ for at least one a, so that $f_{i_m}$ is the first function to separate a from, say, b. Thus $A_{i_m} \in U_b(p)$, and so $T_b(p)$ or $U_b(p)$ must be represented in this instance of R3 by some other formula $A_{i_j}$. Then $f_{i_j}$ fails to separate a and b (since $i_j < i_m$) but must be the first to separate b and, say, c. But then $f_{i_j}$ is also the first to separate a and c, and we may therefore use $A_{i_j}$ as a representative of $T_a(p)$ and dispense with $A_{i_m}$ as required.

A second possibility is to omit p from $Z_a$ in a case where the consequent need for R1 is compensated by economies in the other rules. And a quite different shortcut is possible in connection with R2 whenever the truth-table for a connective has a row or column filled with the same value. For when the value of $F(a_1,\ldots,a_k)$ is independent of $a_1$, say, we can omit $T_{a_1}(A_1)$ and $U_{a_1}(A_1)$ from the corresponding instances of R2 and still carry through the proof of Lemma 19.1.

The rules for PC can be seen as illustrating these ideas. Theorem 18.1 represents in retrospect an application of the general method to the two-valued case, exploiting the first simplification by taking $Z_t = Z_f = \{p\}$, so that R1 and R3 become redundant and R2 assumes the form displayed in the theorem. And if we make the further simplifications possible when the truth-tables have a row or column of ts or fs, the argument leads directly to the rules for PC.

Kleene's matrix provides another illustration. Consider for simplicity's sake the version in which $\vee$ and $\sim$ alone are primitive, & and $\supset$ being defined from them. The sequence $A_1,\ldots,A_n$ can be taken to be $p, \sim p, p\vee\sim p, \sim(p\vee\sim p)$, and the first procedure described above makes $Z_0 = \{p\}$ and $Z_{\frac{1}{2}} = Z_1 = \{p, \sim p\}$. We may however omit p from $Z_1$, since $\sim$ separates 1 from every other value, and thereby simplify R2 at the expense of including a scheme under R1. Thus $T_0(A) = \{A\}$, $T_1(A) = \{\sim A\}$, $U_{\frac{1}{2}}(A) = \{A, \sim A\}$, and $U_0(A) = U_1(A) = T_{\frac{1}{2}}(A) = \Lambda$. Finally we can simplify R2 by postulating single schemes to correspond to the row $0\vee x=0$ and the column $x\vee 0=0$. The resulting rules are

| | |
|---|---|
| R1 From $A, \sim A$ infer $\Lambda$ | R2.7 From $\sim A, \sim B$ infer $\sim(A\vee B)$ |
| R2.1 From $A$ infer $\sim\sim A$ | R2.8 From $A\vee B$ infer $A, \sim A, B, \sim B$ |
| R2.2 From $\sim A$ infer $A, \sim A$ | R2.9 From $\sim(A\vee B)$ infer $A, \sim A, B, \sim B$ |
| R2.3 From $\sim\sim A$ infer $A, \sim A$ | R2.10 From $\sim B, A\vee B$ infer $A, \sim A$ |
| R2.4 From $\sim A$ infer $\sim A$ | R2.11 From $\sim B, \sim(A\vee B)$ infer $A, \sim A$ |
| R2.5 From $A$ infer $A\vee B$ | R2.12 From $\sim A, A\vee B$ infer $B, \sim B$ |
| R2.6 From $B$ infer $A\vee B$ | R2.13 From $\sim A, \sim(A\vee B)$ infer $B, \sim B$ |

Eliminating redundancies, these are equivalent to K1-9 below.

K1 From A,~A infer Λ
K2 From A infer ~~A
K3 From ~~A infer A
K4 From A infer A∨B
K5 From B infer A∨B
K6 From A∨B infer A,B
K7 From ~A,~B infer ~(A∨B)
K8 From ~(A∨B) infer ~A
K9 From ~(A∨B) infer ~B.

In particular R2.10 and R2.12 are diluted forms of modus ponens, which is derivable by cut from K1 and K6. Modus ponens is however not strong enough to take the place of any single one of K1-9, since it is valid in each of the matrices which serve as independence examples for them. These matrices differ from Kleene's only in that (for K1) all values are designated, (K2) $\sim 0=\frac{1}{2}$, (K3) $\sim\frac{1}{2}=1$, (K4) $0\vee\frac{1}{2}=\frac{1}{2}$, (K5) $\frac{1}{2}\vee 0=\frac{1}{2}$, (K6) $\frac{1}{2}\vee\frac{1}{2}=0$, (K7) $1\vee 1=\frac{1}{2}$, (K8) $\frac{1}{2}\vee 1=1$, (K9) $1\vee\frac{1}{2}=1$.

Unlike its analogue for the non-monadic case (Lemma 19.8 below), Lemma 19.3 holds whether or not the matrix is finite, and even in the infinite case it is possible by Lemma 19.5 to find parameters for which the rules are valid and therefore characterise the calculus in question. In general this will not serve any useful purpose, since the resulting rules are not only liable to have instances with infinitely many premisses or conclusions, but they may also not be equivalent to any finite number of rule schemes. It is however sometimes possible to arrive in this way at a finite axiomatisation or (if the calculus in question is not compact) the next best thing, namely a finite set of infinite rule schemes. As examples consider the two infinite matrices of possibility described at the end of Section 16.1. The sequence $A_1,A_2,\ldots$ can be taken to be $p, \Diamond p, \Diamond^2 p, \ldots$, and following the first of the simplified methods we have $T_o(A) = \{A\}$ and $U_o(A) = \Lambda$; $T_{n+1}(A) = \{\Diamond^{n+1}A\}$ and $U_{n+1}(A) = \{A, \Diamond A, \ldots, \Diamond^n A\}$; and $T_\omega(A) = \Lambda$ and $U_\omega(A) = \{A, \Diamond A, \Diamond^2 A, \ldots\}$. Then all instances of the rules for M′

reduce to cases of overlap, except for one scheme under R2, namely 'from A infer ◇A'. Thus we obtain a finite axiomatisation despite the fact that some of the parameters, as well as the matrix, are infinite. All the rules for M are similarly redundant, except for 'from A infer ◇A' under R2 and an infinite rule scheme 'from Λ infer A, ◇A, $◇^2$A,...' under R3.

## 19.4 *Limitations*

We have seen that standard parameters can always be found for a monadic matrix but, conversely, they can be found only in the monadic case. For if the parameters are standard and a and b are any incongruent values, v invalidates $\langle T_a(p), U_a(p)\rangle$ if v(p)=a but not if v(p)=b. There must therefore be a formula in $T_a(p)$ or $U_a(p)$ whose associated matrix function separates a and b, whence a≠b. The matrix is therefore monadic.

| ≡ | 0 | 1 | 2 | 3 |
|---|---|---|---|---|
| *0 | 0 | 2 | 0 | 2 |
| *1 | 3 | 1 | 3 | 1 |
| 2 | 0 | 2 | 0 | 2 |
| 3 | 3 | 1 | 3 | 1 |

The method of axiomatisation of Section 19.2 is nevertheless applicable to some non-monadic matrices, for which suitable non-standard parameters can be found. For example, for each value a in the matrix above let $\langle T_a(p), U_a(p)\rangle$ be the partition of {p,p≡q} that is invalidated by assigning a as value to p and 0 to q, and let $T_a(A)$ and $U_a(A)$ be obtained by substitution for p. Thus $T_o(A) = U_3(A)$ = {A,A≡q} and $U_o(A) = T_3(A)$ = Λ and $T_1(A) = U_2(A)$ = {A} and $U_1(A) = T_2(A)$ = {A≡q}. Then

R1 and R3 reduce to overlap, while R2 is derivable from the axiom scheme $A\equiv(A\equiv B)$ and the rule schemes 'from $A\equiv B, B\equiv C$ infer $C\equiv A$' and 'from $\Lambda$ infer $A\equiv B, B\equiv C, C\equiv A$', the derivation being straightforward once it is seen that $\Lambda \vdash B\equiv B$ by the second rule scheme and hence $B\equiv C \vdash C\equiv B$ by the first. Since these schemes are valid in the matrix they therefore provide the required finite axiomatisation by Lemma 19.3. The parameters in this example actually satisfy neither of the two conditions for standardness, and this is not accidental, for

*Theorem 19.7* For no choice of parameters satisfying either of the two standardness conditions are the rules R of Section 19.2 valid in a non-monadic matrix.

*Proof* If the rules are valid in the matrix M then by 19.3 R and M characterise the same calculus L, so that $\vdash_R = \vdash_M = \vdash$. If the parameters satisfy condition (i) then for some p and each a, each valuation v invalidates $\langle T_a(p), U_a(p)\rangle$ iff v(p) is congruent to a. Hence the parameters obtained by substituting A for every variable in $T_a(p)$ and $U_a(p)$ are standard; but this we have seen to be impossible for a non-monadic matrix. If the parameters satisfy (ii) then $T_a(p)$ and $U_a(p)$ contain no variable other than p, and $T_a(A)$ and $U_a(A)$ are obtained by substitution for p. Let $a_1,\ldots,a_k$ be any values of M and let $p, p_1,\ldots,p_k$ be distinct. By R2 and cut, if $T_{F(a_1,\ldots,a_k)}(Fp_1\ldots p_k) \vdash U_{F(a_1,\ldots,a_k)}(Fp_1\ldots p_k)$ then $\bigcup T_{a_i}(p_i) \vdash \bigcup U_{a_i}(p_i)$, and by 15.1 $T_{a_i}(p_i) \vdash U_{a_i}(p_i)$ for some i. It follows that $\{a\colon T_a(p) \nvdash U_a(p)\}$ is closed under the basic functions and so forms the set of values of a submatrix M′ of M. Let R′ be the rules of Section 19.2 corresponding to M′, using the relevant selection of the parameters given for M. Then $R_1'$ and $R_2'$ are subrules of R1 and R2, while each instance of $R_3'$ is derivable by cut from all the corresponding instances of R3 together with the fact that $T_a(A) \vdash U_a(A)$ for each a not in M′. Thus $\vdash_{R'} \subset \vdash$, and since $\vdash_{M'} \subset \vdash_{R'}$ by 19.3 and $\vdash \subset \vdash_{M'}$ because M′ is a submatrix of M, it follows that $\vdash = \vdash_{M'}$. Hence by 14.22 and 17.6, to show that M is monadic it is sufficient to show that M′ is. So let a and b be incongruent values of M′, so that there exist values $a_1,\ldots,a_n$ in M′ such that $a,a_1,\ldots,a_n \neq b,a_1,\ldots,a_n$. By 19.2 there is a formula B, whose distinct variables are $p,p_1,\ldots,p_n$, such that $T_a(p), \bigcup T_{a_i}(p_i) \vdash B, U_a(p), \bigcup U_{a_i}(p_i)$, say, but $T_b(p), \bigcup T_{a_i}(p_i), B \vdash U_b(p), \bigcup U_{a_i}(p_i)$. By cut, $T_a(p), T_b(p), \bigcup T_{a_i}(p_i) \vdash U_a(p), U_b(p), \bigcup U_{a_i}(p_i)$, and since each $a_i$ is in M′ it follows by 15.1 that $T_a(p), T_b(p) \vdash U_a(p), U_b(p)$. Then, say, some formula A in $U_a(p) \cup U_b(p)$ is such that v(A) is designated when v(p)=a. But $A \notin U_a(p)$, since otherwise $T_a(p) \vdash U_a(p)$ by 19.2. So $A \in U_b(p)$, and by the same argument v(A) is undesignated when v(p)=b. Hence a≠b in M′, so that M′ is monadic.

The method of Section 19.2 is however not applicable to non-monadic matrices in general. For example, take a four-valued Boolean matrix and let M be the fragment based on ∨ alone, as shown below. The only binary matrix functions of M are the two projective functions and ∨ itself, associated respectively with the formulae p, q and p∨q. If parameters $T_a(p)$ and $U_a(p)$ could make the rules R valid, so also would the parameters $T'_a(p)$ and $U'_a(p)$ obtained from them by substituting q for every variable other than p; and we may assume that $T'_a(p)$ and $U'_a(p)$ are subsets of {p,q,p∨q}. It is then a straightforward matter to show by enumerating the possibilities that the rules cannot after all be valid, and hence that the method fails for M.

| ∨ | 1 | 2 | 3 | 4 |
|---|---|---|---|---|
| *1 | 1 | 1 | 1 | 1 |
| 2 | 1 | 2 | 1 | 2 |
| 3 | 1 | 1 | 3 | 3 |
| 4 | 1 | 2 | 3 | 4 |

## 19.5 *The general case*

To deal with the general case we need to associate parameters with finite sequences or n-tuples of truth-values instead of individual values. To simplify the presentation we use a vector notation, writing (for arbitrary finite n) n-tuples of truth-values, variables and formulae as **a**, **p**, **A**, etc. (We write **a**,a for the n+1-tuple obtained by appending a to the n-tuple **a**, etc.)

Suppose that for each $n \geq 0$, sets of formulae $T_{\mathbf{a}}(\mathbf{A})$ and $U_{\mathbf{a}}(\mathbf{A})$ are associated with each n-tuple **a** of truth-values and each n-tuple **A** of formulae; and suppose in particular that $T_{\mathbf{a}}(\mathbf{A}) = U_{\mathbf{a}}(\mathbf{A}) = \Lambda$ if n=0. As before, we call these sets *parameters*, and use $t_{\mathbf{a}}(\mathbf{A})$ or $u_{\mathbf{a}}(\mathbf{A})$ to stand for any member of them; and if S is a set of n-tuples of values we use $t_S(\mathbf{A})$ to stand for any set consisting of one member from $T_{\mathbf{a}}(\mathbf{A})$ for each **a** in S, and $u_S(\mathbf{A})$ similarly. Let R consist of the following rules, each being posited for every n:

R1 From $T_{\mathbf{a},a}(\mathbf{A},A)$ infer A, $U_{\mathbf{a},a}(\mathbf{A},A)$ if a is designated, and from A, $T_{\mathbf{a},a}(\mathbf{A},A)$ infer $U_{\mathbf{a},a}(\mathbf{A},A)$ if a is undesignated.

R2 From $\bigcup T_{\mathbf{a},a_i}(\mathbf{A},A_i)$, $u_{S_1}(\mathbf{A},A_1,\ldots,A_k)$ infer $\bigcup U_{\mathbf{a},a_i}(\mathbf{A},A_i)$, $t_{S_2}(\mathbf{A},A_1,\ldots,A_k)$, $t_{\mathbf{a},F(a_1,\ldots,a_k)}(\mathbf{A},FA_1\ldots A_k)$ and from $\bigcup T_{\mathbf{a},a_i}(\mathbf{A},A_i)$, $u_{S_1}(\mathbf{A},A_1,\ldots,A_k)$, $u_{\mathbf{a},F(a_1,\ldots,a_k)}(\mathbf{A},FA_1\ldots A_k)$ infer $t_{S_2}(\mathbf{A},A_1,\ldots,A_k)$, $\bigcup U_{\mathbf{a},a_i}(\mathbf{A},A_i)$, for each k-ary F, each $a_1,\dots,a_k$ in the submatrix generated by the values in $\mathbf{a}$, each partition $\langle S_1,S_2\rangle$ of the set of n+k-tuples of greater dimension than $\mathbf{a}$, and each choice of $u_{S_1}$, $t_{S_2}$ etc; the unions being taken over $1\le i\le k$.

R3 From $T_{\mathbf{a}}(\mathbf{A})$, $u_{S_1}(\mathbf{A},A)$, $u_{S_1'}(\mathbf{A},A)$ infer $t_{S_2'}(\mathbf{A},A)$, $t_{S_2}(\mathbf{A},A)$, $U_{\mathbf{a}}(\mathbf{A})$ for each partition $\langle S_1,S_2\rangle$ of the set of n+1-tuples of greater dimension than $\mathbf{a}$, each partition $\langle S_1',S_2'\rangle$ of the set of n+1-tuples $\mathbf{a},a$ for a in $\mathbf{a}$, and each choice of $u_{S_1}$ etc.

*Lemma 19.8* If M is finite, $\vdash_M \subset \vdash_R$.

*Proof* Suppose that $X \nvdash_R Y$, so that by cut there is a partition $\langle T,U\rangle$ such that $X \subset T$ and $Y \subset U$ and $T \nvdash_R U$. Consider the n-tuples $\mathbf{a}$ such that for some $\mathbf{A}$, $T_{\mathbf{a}}(\mathbf{A}) \subset T$ and $U_{\mathbf{a}}(\mathbf{A}) \subset U$. At least one such exists, namely the 0-tuple, and the dimensions of these n-tuples are bounded above by the number of truth-values. We may therefore take $\mathbf{a}$ (with its corresponding $\mathbf{A}$) to be such an n-tuple with the largest possible dimension. Then for each n, and each n-tuple $\mathbf{B}$, there is a partition $\langle S_1,S_2\rangle$ of the set of n-tuples of greater dimension than $\mathbf{a}$, and a pair of sets $u_{S_1}(\mathbf{B})$ and $t_{S_2}(\mathbf{B})$ which are contained in T and U respectively. Let $Z_1 = T_{\mathbf{a}}(\mathbf{A}) \cup \bigcup u_{S_1}(\mathbf{B})$ and $Z_2 = U_{\mathbf{a}}(\mathbf{A}) \cup \bigcup t_{S_2}(\mathbf{B})$, the unions being taken over all n and $\mathbf{B}$. Then $Z_1 \subset T$ and $Z_2 \subset U$. If for each X′ and Y′ we write $X' \vdash Y'$ when $X',Z_1 \vdash_R Z_2,Y'$, it is easy to verify that $\vdash$ is a consequence relation. Then $T \nvdash U$ and hence $X \nvdash Y$. Taking M′ to be the submatrix generated by the values in $\mathbf{a}$, and for each a in M′ taking $T_a(A)$ and $U_a(A)$ to be $T_{\mathbf{a},a}(\mathbf{A},A)$ and $U_{\mathbf{a},a}(\mathbf{A},A)$, let R′ be the rules of Section 19.2. It is readily verified that for each instance $\langle X',Y'\rangle$ of each rule in R′, $Y' \cup Z_2$ is an immediate consequence of $X' \cup Z_1$ by R, and so $X' \vdash Y'$. Hence $\vdash_{R'} \subset \vdash$, and so by 19.3 $\vdash_M \subset \vdash_{M'} \subset \vdash_{R'} \subset \vdash$, so that $X \nvdash_M Y$ as required.

The lemma may fail without the assumption that the parameters are empty for 0-tuples, for if $T_{\mathbf{a}}(\mathbf{A})$ and $U_{\mathbf{a}}(\mathbf{A})$ overlap for every $\mathbf{a}$ and $\mathbf{A}$, every instance of R reduces to a case of overlap.

We write $v(\mathbf{A})$ for the n-tuple of values assigned by a valuation v to the n-tuple $\mathbf{A}$. Following the pattern of Section 19.2, we say that the parameters are *standard* if, for each n, for some n-tuple $\mathbf{p}$ of distinct variables and each n-tuple $\mathbf{a}$, (i) each valuation v invalidates $\langle T_{\mathbf{a}}(\mathbf{p}),U_{\mathbf{a}}(\mathbf{p})\rangle$ iff $v(\mathbf{p})\approx\mathbf{a}$, and (ii) $T_{\mathbf{a}}(\mathbf{p})$ and $U_{\mathbf{a}}(\mathbf{p})$ contain no variables other than those in $\mathbf{p}$, and $T_{\mathbf{a}}(\mathbf{A})$ and $U_{\mathbf{a}}(\mathbf{A})$ are obtained by substitution for $\mathbf{p}$ in $T_{\mathbf{a}}(\mathbf{p})$ and $U_{\mathbf{a}}(\mathbf{p})$. Evidently for standard parameters, v invalidates $\langle T_{\mathbf{a}}(\mathbf{A}),U_{\mathbf{a}}(\mathbf{A})\rangle$ iff $v(\mathbf{A})\approx\mathbf{a}$.

*Lemma 19.9* If the parameters are standard, $\vdash_R \subset \vdash_M$.

*Proof* We show that for standard parameters, every valuation v in M satisfies R. If v satisfies $\langle T_{\mathbf{a},a}(\mathbf{A},A),\ U_{\mathbf{a},a}(\mathbf{A},A)\rangle$ it will evidently satisfy R1; if not, then since the parameters are standard $v(\mathbf{A}),v(A) \approx \mathbf{a},a$, and by 14.3 $v(A)$ is designated iff a is designated, whence again the rule is satisfied. Using the notation of R2, that rule is satisfied if (i) $v(\mathbf{A}),v(A_i) \not\approx \mathbf{a},a_i$ for some i, or (ii) $v(\mathbf{A}),v(FA_1\ldots A_k) \approx \mathbf{a},F(a_1,\ldots,a_k)$, or (iii) $\dim(v(\mathbf{A}),v(A_1),\ldots,v(A_k)) > \dim(\mathbf{a})$, and we show that these exhaust the possibilities. For if (i) fails $v(\mathbf{A})\approx\mathbf{a}$, and so if (iii) also fails then by 14.8 there exist $b_1,\ldots,b_k$ in $\mathbf{a}$ such that $v(\mathbf{A}),v(A_1),\ldots,v(A_k) \approx \mathbf{a},b_1,\ldots,b_k$. Hence for each i we have $v(\mathbf{A}),v(A_i) \approx \mathbf{a},b_i$; and so $\mathbf{a},a_i \approx \mathbf{a},b_i$ by the failure of (i). By 14.12, $a_i$ and $b_i$ are congruent in the submatrix generated by $\mathbf{a}$, so that by 14.9 $v(\mathbf{A}),v(A_1),\ldots,v(A_k) \approx \mathbf{a},a_1,\ldots,a_k$, and hence (ii) holds. Similarly R3 is satisfied if (i) $v(\mathbf{A})\not\approx\mathbf{a}$, or (ii) $v(\mathbf{A}),v(A) \approx \mathbf{a},a$ for some a in

$\mathbf{a}$, or (iii) dim(v(A),v(A)) > dim($\mathbf{a}$), and we show that these cover all the possibilities. For if (i) and (iii) fail, then as for R2 there exists a in $\mathbf{a}$ such that v(A),v(A) ≃ $\mathbf{a}$,a; i.e. (ii) holds.

Although the assumption of standardness thus succeeds in its first objective of validating the rules, it is not in general sufficient for the second objective of providing a finite set of rule schemes. For although for each n the rules R are equivalent to a finite set of schemes, we need to posit them for every n. We may however obviate this need by choosing parameters which are *nested* in the sense that whenever j,...,k are drawn (possibly with repetitions) from 1,...,n, we have $T_{a_j,\ldots,a_k}(A_j,\ldots,A_k) \subset T_{a_1,\ldots,a_n}(A_1,\ldots,A_n)$ and similarly $U_{a_j,\ldots,a_k}(A_j,\ldots,A_k) \subset U_{a_1,\ldots,a_n}(A_1,\ldots,A_n)$.

*Lemma 19.10* If the parameters are nested then $\vdash_M \subset \vdash_{R'}$, where R′ is the subset of R corresponding to n-tuples $\mathbf{a}$ of distinct values only.

*Proof* If the parameters are nested, the n-tuple $\mathbf{a}$ in the proof of 19.8 may be chosen so that its values are all distinct, and the proof then goes through as before.

*Lemma 19.11* For any finite matrix there exist finite, standard, nested parameters.

*Proof* Let the matrix have m values and let $\mathbf{p}$ be an m-tuple of distinct variables. Let $f_1$,... be the distinct m-ary matrix functions, and let Z be a set of formulae $A_1$,..., in which only the variables in $\mathbf{p}$ appear, associated with them in the manner of Section 14.1. Let $\langle T'_{\mathbf{a}}(\mathbf{p}), U'_{\mathbf{a}}(\mathbf{p})\rangle$ be for each m-tuple $\mathbf{a}$ the partition of Z that is invalidated by a valuation v for which v($\mathbf{p}$)=$\mathbf{a}$, and let $T'_{\mathbf{a}}(A)$ and $U'_{\mathbf{a}}(A)$ be obtained by substitution for p in $T'_a(p)$ and $U'_a(p)$. Then v invalidates $\langle T'_a(A), U'_a(A)\rangle$ iff $v(A) \approx a$. As parameters we take $T_{a_1,\ldots,a_n}(A_1,\ldots,A_n) = \bigcup T'_{a_j,\ldots,a_k}(A_j,\ldots,A_k)$ and $U_{a_1,\ldots,a_n}(A_1,\ldots,A_n) = \bigcup U'_{a_j,\ldots,a_k}(A_j,\ldots,A_k)$, where the unions are taken over all m-tuples j,...,k whose elements are drawn (possibly with repetitions) from 1,...,n. The parameters are therefore nested, they are finite because Z is finite, and they satisfy condition (ii) of standardness. For condition (i), by 14.4 v invalidates $\langle T_a(A), U_a(A)\rangle$ if $v(A) \approx a$. Conversely, if $v(A_1),\ldots,v(A_n) \neq a_1,\ldots,a_n$ then by 14.27 there exists an m-tuple j,...,k whose elements are drawn (possibly with repetitions) from 1,...,n, such that $v(A_j),\ldots,v(A_k) \neq a_j,\ldots,a_k$. It follows that v satisfies $\langle T'_{a_j,\ldots,a_k}(A_j,\ldots,A_k),\ U'_{a_j,\ldots,a_k}(A_j,\ldots,A_k)\rangle$, and hence satisfies $\langle T_{a_1,\ldots,a_n}(A_1,\ldots,A_n),\ U_{a_1,\ldots,a_n}(A_1,\ldots,A_n)\rangle$ as required.

*Theorem 19.12* Every finite-valued calculus is finitely axiomatisable.

*Proof* Given an m-valued matrix and finite parameters satisfying condition (ii) of standardness, the rules R′ of 19.10 have no instances for n>m, and for each n≤m are easily seen to be equivalent to a finite set of rule schemes. The theorem then follows by 19.9-11. The compactness of finite-valued calculi, proved otherwise as 13.1, is a corollary.

## 19.6 *Further examples*

In the case of the finite Gödel matrices we may take standard nested parameters for $A_1,\ldots,A_n$ by appropriately partitioning the various $A_i$, $\sim A_i$ and $A_i \supset A_j$. A literal application of the method of Section 19.5 will produce many cumbersome rule schemes, but we can obtain quite elegant axiomatisations by following its spirit rather than its letter. Similarly, although the method proper is confined to finite matrices, we can obtain finite axiomatisations for the calculi characterised by certain infinite matrices, and we take the rational- and real-valued Gödel matrices as an example. We also provide rules for the finite-valued Łukasiewicz calculi (but not of course for the infinite-valued ones, since they are not even compact). Most of the rules required are of the form 'from $\Lambda$ infer A' and are represented in the list below by the corresponding axiom A. The notation $C_m(A)$ in ($13_m$) is explained as follows: by appropriate application of Lemma 13.5 there exists for each m a formula $C_m(p)$, in which p is the only variable and $\supset$ and $\sim$ the only connectives, such that for every valuation v in the rational-valued Łukasiewicz matrix, $v(C_m(p)) = 0$ iff $v(p) = \frac{k}{m-1}$ for some integer k; and $C_m(A)$ is obtained from $C_m(p)$ by substitution.

| | | | |
|---|---|---|---|
| (1) | $A\supset.B\supset A$ | (9) | $(A\supset\sim B)\supset.B\supset\sim A$ |
| (2) | $(A\supset B)\supset.(B\supset C)\supset.A\supset C$ | (10) | $\sim A\supset.A\supset B$ |
| (3) | $A\supset(A\vee B)$ | (11) | from A, $\sim$A infer $\Lambda$ |
| (4) | $B\supset(A\vee B)$ | (12) | from A, $A\supset B$ infer B |
| (5) | $(A\supset C)\supset.(B\supset C)\supset.(A\vee B)\supset C$ | ($13_m$) | $C_m(A)$ |
| (6) | $(A\&B)\supset A$ | (14) | $((A\supset B)\supset B)\supset.(B\supset A)\supset A$ |
| (7) | $(A\&B)\supset B$ | (15) | $(A\supset.A\supset B)\supset.A\supset B$ |
| (8) | $(C\supset A)\supset.(C\supset B)\supset.C\supset(A\&B)$ | (16) | from $\Lambda$ infer $A\supset B$, $B\supset A$ |

($17_m$) from $\Lambda$ infer $A_1$, $A_1\supset A_2$,..., $A_{m-1}\supset A_m$

*Theorem 19.13* The calculus characterised by the real- and rational-valued Gödel matrices is characterised by (1)-(12), (15), (16); and any fragment that contains $\supset$ is characterised by the schemes in which the relevant connectives appear.

*Theorem 19.14* The calculus characterised by an m-valued Gödel matrix is characterised by (1)-(12), (15), $(17_m)$; and any fragment that contains $\supset$ is characterised by the schemes in which the relevant connectives appear.

*Theorem 19.15* The calculus characterised by an m-valued Łukasiewicz matrix is characterised by (1)-(12), $(13_m)$, (14), $(17_m)$; and any fragment that contains $\supset$ is characterised by the schemes in which the relevant connectives appear.

*Proofs* Let L be the calculus characterised by the rules described for any matrix M covered by the theorems. It is easy to check that the rules are valid in M, and hence that $X \vdash_M Y$ whenever $X \vdash Y$. For the converse we need the following

*Lemma 19.16* Let $R_1$ be the rules (1), (2), (12) plus either (16) or $(17_m)$, let $\vdash_1$ be consequence by $R_1$, let $\vdash_2$ be consequence by $R_1$ plus (9)-(10), let $\vdash_3$ be consequence by $R_1$ plus (14), let $\vdash_4$ be consequence by $R_1$ plus (15), and let $\vdash_5$ be consequence by $R_1$ plus (9)-(10) and either (14) or (15). Then

| | | | |
|---|---|---|---|
| (i) | $A \vdash_1 B\supset A$ | (xi) | $A\supset B \vdash_2 A\supset{\sim}{\sim}B$ |
| (ii) | $A\supset B \vdash_1 (B\supset C)\supset.A\supset C$ | (xii) | ${\sim}B, A\supset B \vdash_2 {\sim}A$ |
| (iii) | $A\supset B, B\supset C \vdash_1 A\supset C$ | (xiii) | $A\supset.(C\supset B)\supset B \vdash_3 A\supset.(B\supset C)\supset C$ |
| (iv) | $\Lambda \vdash_1 A\supset B, B\supset A$ | (xiv) | $A\supset.B\supset C \vdash_3 B\supset.A\supset C$ |
| (v) | $\Lambda \vdash_1 A\supset A$ | (xv) | $B\supset C, (C\supset B)\supset.A\supset B \vdash_3 A\supset C$ |
| (vi) | $A\supset{\sim}B \vdash_2 B\supset{\sim}A$ | (xvi) | $A\supset.A\supset B \vdash_4 A\supset B$ |
| (vii) | $A\supset{\sim}B, B \vdash_2 {\sim}A$ | (xvii) | $\Lambda \vdash_4 A\supset B, (A\supset B)\supset B$ |
| (viii) | ${\sim}A \vdash_2 A\supset B$ | (xviii) | $A\supset.B\supset C \vdash_4 B\supset.A\supset C$ |
| (ix) | $\Lambda \vdash_2 B\supset{\sim}{\sim}B$ | (xix) | $\Lambda \vdash_5 (A\supset{\sim}(B\supset B))\supset{\sim}A$ |
| (x) | $\Lambda \vdash_2 {\sim}{\sim}(B\supset B)$ | | |

Proof of lemma: (i)-(iii) are obtained from (1) and (2) using (12). (iv) follows at once from (16), while from $(17_m)$, taking each $A_n$ to be A or B according as n is odd or even, we have $\Lambda \vdash_1 A, A\supset B, B\supset A$, whence the result by (i). (v) follows from (iv), taking B to be A. (vi)-(viii) are obtained from (9) and (10) using (12). (ix) follows from (v) and (vi), taking A to be $\sim B$. By (ix) and (12), $B\supset B \vdash_2 \sim\sim(B\supset B)$, and (x) follows using (v). (xi) follows from (iii) and (ix), taking C to be $\sim\sim B$. (xii) follows from (vii), substituting $\sim B$ for B and using (xi). (xiii) follows from (iii), since by (14) $((C\supset B)\supset B)\supset.(B\supset C)\supset C$ is an axiom. For (xiv) we use a proof adapted from McCall and Meyer, 1966: taking A to be B in (xiii) and using (1), we have $\Lambda \vdash_3 B\supset.(B\supset C)\supset C$; by (ii) $A\supset.B\supset C \vdash_3 ((B\supset C)\supset C)\supset.A\supset C$; and by (iii) $B\supset.(B\supset C)\supset C, ((B\supset C)\supset C)\supset.A\supset C \vdash_3 B\supset.A\supset C$. For (xv) we apply (xiv) both to the premiss and to the conclusion of (xiii), and use (12). (xvi) is obtained from (15) using (12). For (xvii) we use a proof suggested by Dummett, 1959: $(A\supset B)\supset A \vdash_4 (A\supset B)\supset.(A\supset B)\supset B$ by (ii), and $\Lambda \vdash_4 (A\supset B)\supset A, A\supset.A\supset B$ by (iv), so by cut $\Lambda \vdash_4 A\supset.A\supset B, (A\supset B)\supset.(A\supset B)\supset B$, and the result follows by (xvi). For (xviii) $A\supset.B\supset C, (B\supset C)\supset C \vdash_4 A\supset C$ by (iii) and $B\supset C, C\supset.A\supset C \vdash_4 B\supset.A\supset C$ likewise, and the result follows by cut, since $\Lambda \vdash_4 C\supset.A\supset C$ by (1), $A\supset C \vdash_4 B\supset.A\supset C$ by (i) and $\Lambda \vdash_4 B\supset C, (B\supset C)\supset C$ by (xvii). Finally $(A\supset\sim(B\supset B))\supset.(B\supset B)\supset\sim A \vdash_5 (B\supset B)\supset.(A\supset\sim(B\supset B))\supset\sim A$ by (xiv) and (xviii), and (xix) follows by (v), (9) and (12).

For 19.13 and 19.14, suppose that $X \nvdash Y$, so that by cut there exists a partition $\langle T,U\rangle$ such that $X \subset T$ and $Y \subset U$ and $T \nvdash U$. We define a valuation in M which invalidates $\langle T,U\rangle$ and hence invalidates $\langle X,Y\rangle$, as required. We write $A \geq B$ if $A\supset B \in T$. If $A\supset B \in T$ and $B\supset C \in T$ then since $T \nvdash U$ it follows by (iii) of 19.16 that $A\supset C \in T$; i.e. $\geq$ is transitive. Likewise $\geq$ is reflexive by (v). The relation that holds between A and B if both $A \geq B$ and $B \geq A$ is thus an equivalence relation, and we write $|A|$ for the class of formulae equivalent to A. If $A \geq B$ then $A' \geq B'$ for every $A'$ in

$|A|$ and $B'$ in $|B|$, and accordingly we can write $|A| \geq |B|$. The equivalence classes are then partially ordered by $\geq$, and since by (iv) either $A \geq B$ or $B \geq A$, this ordering is a simple ordering. By a classical result there therefore exists an order-preserving mapping, v say, of the equivalence classes of formulae into the rationals in the interval [0,1]. In the m-valued case we note that there can be at most m equivalence classes. For if $|A_1|,\ldots,|A_{m+1}|$ were distinct classes and, say, $|A_{m+1}| \geq |A_m| \geq \ldots \geq |A_2| \geq |A_1|$, then it could not be that $A_1 \geq A_2$, so it must be that $A_1 \supset A_2 \in U$ and so $A_2 \in U$ by (i). Similarly $A_2 \supset A_3 \in U,\ldots,$ $A_m \supset A_{m+1} \in U$, whence by $(17_m)$ $T \vdash U$, contrary to hypothesis. In the m-valued case too, there is therefore an order-preserving mapping of the equivalence classes into the values of the appropriate Gödel matrix.

Now $A \supset A \in T$ by (v), and if $A \in T$ then by (i) and (12) $A \geq B$ iff $B \in T$. It follows that $T = |A \supset A|$ and that it is the least in the ordering of the equivalence classes. We may therefore choose the mapping v so that $v(T)=0$. Where $\sim$ is included among the connectives, let $\bar{T} = \{A: \sim A \in T\}$. Now $\sim(B \supset B) \in \bar{T}$ by (x), and when $B \in \bar{T}$ then by (viii) and (xii) $A \geq B$ iff $A \in \bar{T}$. It follows as before that $\bar{T} = |\sim(B \supset B)|$ and that it is the greatest in the ordering of the equivalence classes; and $\bar{T} \neq T$ by (11). We may therefore choose v so that $v(\bar{T})$ is 1.

We now write v also for the induced mapping $A \to v(|A|)$ of the formulae into the truth-values. Since $v(T)=0$, $v(A)$ is designated iff $A \in T$. It therefore only remains to show that v is a valuation, that is (a) $v(A \supset B) = 0$ or $v(B)$ according as $v(A) \geq v(B)$ or not, and (where the relevant connectives are present) (b) $v(\sim A) = v(A) \supset 1$, (c) $v(A \vee B) = \min(v(A),v(B))$, and (d) $v(A \& B) = \max(v(A),v(B))$. For (a), if $v(A) \geq v(B)$ then $A \geq B$, so that $A \supset B \in T$ and hence

$v(A\supset B) = 0$; otherwise $A\supset B \in U$, so that $(A\supset B)\supset B \in T$ by (xvii), and since $B\supset .A\supset B \in T$ by (1) it follows that $v(A\supset B) = v(B)$ as required. For (b), $\sim A$ and $A\supset\sim(B\supset B)$ are in the same equivalence class by (10) and (xix), so $v(\sim A) = v(A\supset\sim(B\supset B)) = v(A)\supset v(\sim(B\supset B)) = v(A)\supset 1$. For (c), we choose C from A and B so that $v(C) = \min(v(A),v(B))$. Then $v(C) \geq v(A\vee B)$ since by (3) and (4) $v(A) \geq v(A\vee B)$ and $v(B) \geq v(A\vee B)$; and $v(A\vee B) \geq v(C)$ by (5) and (12). (d) is proved similarly from (6)-(8).

For 19.15 we proceed as for 19.14, except that if there are n equivalence classes of formulae we take v in the first instance to be a mapping onto the values $0, \frac{1}{n-1}, \ldots, 1$ of an n-valued Łukasiewicz matrix M′. If n=2 the classical matrix serves as a two-valued Łukasiewicz matrix for this purpose and if n=1 (which by (11) can only happen in the absence of negation) a matrix with a single designated value serves similarly as a degenerate one-valued case. We establish that v is a valuation as before, except that if $v(A) < v(B)$ we now need to show that $v(A\supset B)$ is not $v(B)$ but $v(B)-v(A)$. Let $v(B) = \frac{k}{n-1}$, and let $A_o, \ldots, A_k$ be arbitrary formulae such that $v(A_i) = \frac{i}{n-1}$, including A as the appropriate $A_i$. If $i<j\leq k$ then $A_i\supset A_j \in U$ and $B\supset A_j \in T$, whence $(A_j\supset B)\supset .A_i\supset B \in U$ by (xv); i.e. $v(A_i\supset B) > v(A_j\supset B)$. By (1) $v(B) \geq v(A_o\supset B)$, so $\frac{k}{n-1} \geq v(A_o\supset B) > v(A_1\supset B) > \ldots > v(A_k\supset B) \geq 0$. Accordingly $v(A_i\supset B) = v(B)-v(A_i)$ for each i, and so in particular $v(A\supset B) = v(B)-v(A)$.

This shows that v invalidates $\langle T,U\rangle$ in M′: to conclude that $\langle T,U\rangle$ is invalid in M we need to show that M′ is isomorphic to a submatrix of M. For fragments lacking ~ this is immediate from the fact that $n\leq m$, while otherwise it is sufficient to note that v assigns a value in M to each formula A, since by $(13_m)$ $C_m(A) \in T$ and hence $v(C_m(A)) = 0$.

The proof of 19.15 can easily be adapted to allow

$C_m(p)$ to be any formula such that $v(C_m(p))$ is designated whenever $v(p)$ is $\frac{k}{m-1}$ for integral k, but such that for each n less than m and not a factor of m-1 there is an integer k for which $v(C_m(p))$ is undesignated when $v(p)$ is $\frac{k}{n}$. So for example $C_3(p)$ could be taken to be $p \supset p$ and $C_4(p)$ to be $(p \supset {\sim}p) \supset .({\sim}p \supset p) \supset .({\sim}p \supset p) \supset p$.

## 19.7 *Single-conclusion calculi*

We do not know whether every single-conclusion calculus characterised by a finite matrix is finitely axiomatisable or not, but there are two ways in which the results of this chapter can be applied to certain classes of single-conclusion calculi.

The first way is to approach the single-conclusion calculus L characterised by a finite matrix M by applying Theorem 19.6 or 19.12 to the quotient of $M^{\aleph_0}$. For by Theorems 17.18, 17.3 and 16.5, $|M^{\aleph_0}|$ characterises the minimum counterpart of L or its negative variant, and from a finite axiomatisation of either we can always derive a finite axiomatisation of L. To do so we replace each rule scheme 'from s(X) infer s(Y)' by 'from s(X) infer s(A)', where A is chosen as follows: if Y is empty we take A to be any variable that does not appear in any member of X (so that by appropriate substitution s(A) can be any formula); otherwise we know from Theorem 5.4 that $X \vdash B$ for some B in Y, and we take A to be B.

The utility of this approach is limited by the fact that $|M^{\aleph_0}|$ is generally not finite. The limitation is not absolute, since the methods of Section 19.2 can be applied to certain infinite matrices, as we saw in Section 19.3. Taking for example the two infinite matrices for possibility discussed there, it is easily verified that M′ is isomorphic to $|M^{\aleph_0}|$, and consequently from the finite axiomatisation of the calculus characterised by M′ we obtain one for the single-conclusion calculus characterised by M, namely 'from A infer ◇A'. On the other hand the method used to obtain the initial axiomatisation of M′ depended on its being monadic, and this is exceptional, since in general $|M^{\aleph_0}|$ will not be monadic even when M is. There is however one class of matrices for which $|M^{\aleph_0}|$ is always finite and monadic, and for which the present approach is therefore always suitable:

*Lemma 19.17* If a finite matrix has only singulary basic functions, every Cartesian power of it is equivalent to a finite matrix.

*Theorem 19.18* Every single-conclusion calculus which is characterised by a finite matrix and has only singulary connectives, is finitely axiomatisable.

*Proofs* For 19.17 we establish the stronger result that if M has m values, every Cartesian power $M^n$ for $n \geq m$ is equivalent to every other and to a matrix of fewer than $2^m$ values. Let M′ be the matrix whose values are the various nonempty subsets of values of M, a set S being designated in M′ iff all its members are designated in M, and whose basic functions are of the form $\lambda S.\{F(a): a \in S\}$ for each basic function F of M. The values of $M^n$ are families $\{a_i\}$ indexed by I, say, and we let h be the mapping of $M^n$ into M′ such that $h(\{a_i\}) = \{a_i: i \in I\}$. Evidently h is structure-preserving, and if $n \geq m$ it is a homomorphism of $M^n$ onto M′, whence $M^n$ is equivalent to M′. 19.18 follows by the argument set out in the text, from 19.17, 14.26 and 19.6.

The second way is to apply Theorem 19.6 or 19.12 directly to M to obtain a finite axiomatisation of the corresponding multiple-conclusion calculus, and then if possible manipulate the resulting rules to obtain suitable rules for its single-conclusion part L. We shall show how this can be done if L contains a disjunction or the deduction theorem holds in it. For this purpose it is only necessary that disjunction and implication should behave schematically; i.e. the following theorem holds even if $\vee$ is not a (primitive) connective provided $s(A\vee B) = s(A)\vee s(B)$, and the same applies to $\supset$ in Theorem 19.21 and the results of Section 19.8.

*Theorem 19.19* Every single-conclusion calculus which is characterised by a finite matrix and contains a disjunction as a connective, is finitely axiomatisable.

*Proof* Let L be the calculus in question. By 19.12 there exists a finite set of rule schemes R′ which characterise a multiple-conclusion counterpart of L, and so by 5.37 there exist corresponding rules R1-3 for L. We show that R1-3 are equivalent to a finite set of rule schemes. Since $s(A\vee B) = s(A)\vee s(B)$, we may take R1-3 to be a set of schemes, e.g. the rules R2 corresponding to a scheme 'from $s(X)$ infer $s(Y)$' of R′ can be taken to consist of distinct schemes 'from $s(X\vee q)$ infer $s(B_1\vee\ldots\vee B_n\vee q)$' for each sequence $B_1,\ldots,B_n$ such that $\{B_1,\ldots,B_n\} = Y$ and each way of construing the notation $B_1\vee\ldots\vee B_n\vee q$; where q may be any variable not occurring in any member of X or Y, so that by appropriate substitution any formula C may be taken as $s(q)$. There are infinitely many formulae of the required form $B_1\vee\ldots\vee B_n\vee q$, but by R3 they are all interdeducible, and so it is sufficient to choose one of the relevant schemes to represent them all. Finally R3 is equivalent to the schemes (1) from A infer $A\vee B$, (2) from $A\vee B$ infer

B∨A, (3) from A∨A infer A, (4) from (A∨B)∨C infer A∨(B∨C), (5) from A∨(B∨C) infer (A∨B)∨C, (6) from A∨C infer (A∨B)∨C (7) from (A∨B)∨C infer (B∨A)∨C, (8) from (A∨A)∨C infer A∨C, (9) from ((A∨B)∨C)∨D infer (A∨(B∨C))∨D, (10) from (A∨(B∨C))∨D infer ((A∨B)∨C)∨D. For, arguing as in 5.38 (using schemes 6-10 to deal with 1-5 respectively and 4-5 to deal with 6-10), if X ⊢ B in the calculus L characterised by 1-10 then X∨C ⊢ B∨C. So if Z,A ⊢ C and Z,B ⊢ C then Z∨B,A∨B ⊢ C∨B and Z∨C,B∨C ⊢ C∨C, whence Z,A∨B ⊢ B∨C and Z,B∨C ⊢ C by 1-3, and hence Z,A∨B ⊢ C by cut. Conversely if Z,A∨B ⊢ C then Z,A ⊢ C and Z,B ⊢ C by 1, so that ∨ is a disjunction in L and hence R3 ⊂ ⊢ by 5.34, as required. (In fact 1, 3, 7 and 10 alone are sufficient, the other six schemes being derivable from them.)

To illustrate this result, consider the version of the classical calculus based on ∨ and ~. Theorem 18.1 provides a finite axiomatisation for a counterpart of it, comprising 'from A,~A infer Λ', 'from Λ infer A,~A', and four schemes for disjunction. Applying Theorem 5.37, the first rule gives rise under R1 to 'from A∨B,~A∨B infer B'. (As observed in Section 5.5, the simpler form 'from A,~A infer B' is insufficient as can be seen by considering the validity or otherwise of the relevant rules in the 3-valued matrix differing from Kleene's only in that $\sim\frac{1}{2}$, $\frac{1}{2}\vee 1$ and $1\vee\frac{1}{2}$ are all 0.) The second rule gives rise to an infinite set of axiom schemes under R2 of which A∨~A∨C is a sufficient representative, and this can be simplified to A∨~A by virtue of R3; and the rules for disjunction give rise to rules under R2 all of which are instances of R3 and so dispensable. We therefore arrive at the following finite axiomatisation for the classical calculus:

| | |
|---|---|
| A∨~A | From A∨A infer A |
| From A∨B, ~A∨B infer B | From (A∨B)∨C infer (B∨A)∨C |
| From A infer A∨B | From (A∨(B∨C))∨D infer ((A∨B)∨C)∨D |

Another illustration is the single-conclusion calculus characterised by Kleene's matrix. We obtained in Section 19.3 a finite axiomatisation of a counterpart of it, taking $\vee$ and $\sim$ as basic. Of the relevant rule schemes, K1 gives rise to 'from A$\vee$C, $\sim$A$\vee$C infer C' under R1, and K2-3 and K7-9 each give rise to rule schemes under R2 of which a single representative suffices, while the rules under R2 corresponding to the purely disjunctive schemes K4-6 are all covered by R3. We therefore have the following finite axiomatisation:

| | |
|---|---|
| From A$\vee$C, $\sim$A$\vee$C infer C | From $\sim$A$\vee$C, $\sim$B$\vee$C infer $\sim$(A$\vee$B)$\vee$C |
| From A$\vee$C infer $\sim\sim$A$\vee$C | From A infer A$\vee$B |
| From $\sim\sim$A$\vee$C infer A$\vee$C | From A$\vee$A infer A |
| From $\sim$(A$\vee$B)$\vee$C infer $\sim$B$\vee$C | From (A$\vee$B)$\vee$C infer (B$\vee$A)$\vee$C |
| From $\sim$(A$\vee$B)$\vee$C infer $\sim$A$\vee$C | From (A$\vee$(B$\vee$C))$\vee$D infer ((A$\vee$B)$\vee$C)$\vee$D |

The above axiomatisation for the classical calculus is unusual in having several rule schemes and only one axiom scheme. An axiomatisation of the usual sort, with one rule and several axioms, can of course be derived by introducing modus ponens and converting each rule 'from $A_1,\dots,A_m$ infer B' into an axiom $\sim A_1\vee\dots\vee\sim A_m\vee B$. No such device is possible in the Kleene case, however, since the calculus does not have any theorems and consequently an axiomatisation of it cannot have any axioms.

*Lemma 19.20* If a calculus is characterised by rules R′, the deduction theorem holds for $\supset$ in its single-conclusion part L iff L is characterised by (R1) axioms and rules for intuitionist implication, and (R2) axioms $A_1\supset.\dots\supset.A_m\supset.(B_1\supset C)\supset.\dots\supset.(B_n\supset C)\supset C$ whenever $\{A_1,\dots,A_m\}$ R′ $\{B_1,\dots,B_n\}$.

*Theorem 19.21* Every single-conclusion calculus which is characterised by a finite matrix, and in which the deduction theorem holds for some connective, is finitely axiomatisable (being characterised by a finite set of axiom schemes with modus ponens).

*Proofs* For 19.20 we may take R1 to be the axiom schemes $A\supset.B\supset A$ and $(A\supset.B\supset C)\supset.(A\supset B)\supset.A\supset C$ with modus ponens, and let R be R1 plus R2. If L is characterised by R the deduction theorem can be proved in the standard way, given that R2 consists entirely of axioms. For the converse it is sufficient by 5.16 and 5.1 to show that if (i) the deduction theorem holds for $\supset$ in L then (ii) if X R B then $X \vdash B$ and (iii) if X R′ Y and $Z,A \vdash_R B$ for every A in Y then $Z,X \vdash_R B$. For (ii), the validity of modus ponens is presupposed by (i) and that of the axioms in R1 follows by applying (i) to $A,B \vdash A$ and $A,\ A\supset.B\supset C,\ A\supset B \vdash C$. Also each axiom in R2 corresponds to an instance of R′, and if $\{A_1,\ldots,A_m\}$ R′ $\{B_1,\ldots,B_n\}$ then by 5.14 $A_1,\ldots,A_m$, $B_1\supset C,\ldots,B_n\supset C \vdash C$, whence the result by (i). For (iii), if X R′ Y then by R2 and modus ponens $X,\ Y\supset C \vdash_R C$ for every C, whence the result by 5.14 and 5.5.

For 19.21, let L be the calculus in question and let the deduction theorem hold for $\supset$. By 19.12 there exists a finite set of schemes which characterises a counterpart of L, and so by 19.20 L is characterised by the corresponding R1-2. We may take R1 to be a finite set of schemes as for 19.20. As in the proof of 19.19, each scheme of R′ gives rise to an infinite number of axiom schemes under R2, but they are all interdeducible by R1 and the resulting deduction theorem, so that a single representative scheme suffices. This can moreover be simplified whenever the original scheme of R′ has a singleton conclusion; for by R1 we have $B \vdash (B\supset C)\supset C$ and $(B\supset B)\supset B \vdash B$, and so, using the deduction theorem, the scheme $A_1\supset.\ldots\supset.A_m\supset.(B\supset C)\supset C$ can be replaced by $A_1\supset.\ldots\supset.A_m\supset B$.

To illustrate this approach in the case of the Gödel and Łukasiewicz matrices, we need to add the following to the rules (1)-($17_m$) of Section 19.6:

(18) $((A\supset B)\supset C)\supset.((B\supset A)\supset C)\supset C$

($19_m$) $(A_1\supset C)\supset.((A_1\supset A_2)\supset C)\supset\ldots\supset.((A_{m-1}\supset A_m)\supset C)\supset C$

($20_m$) $(A\supset^m B)\supset.A\supset^{m-1}B$

($21_m$) $(A_1\supset^{m-1}C)\supset.((A_1\supset A_2)\supset^{m-1}C)\supset\ldots\supset.((A_{m-1}\supset A_m)\supset^{m-1}C)\supset C$

*Theorem 19.22* The single-conclusion calculus characterised by the rational- and real-valued Gödel matrices is characterised by (1)-(10), (12), (15), (18); and any fragment that contains $\supset$ is characterised by those schemes in which the relevant connectives appear.

*Theorem 19.23* The single-conclusion calculus characterised by an m-valued Gödel matrix is characterised by (1)-(10), (12), (15), ($19_m$); and any fragment that contains $\supset$ is characterised by those schemes in which the relevant connectives appear.

*Theorem 19.24* The single-conclusion calculus characterised by an m-valued Łukasiewicz matrix is characterised by (1)-(10), (12), ($13_m$), (14), ($20_m$), ($21_m$); and any fragment that contains $\supset$ is characterised by those schemes in which the relevant connectives appear.

*Proofs* For 19.22 and 19.23, the deduction theorem holds in all the relevant calculi by 14.1, so by 19.20 they are characterised by the rules R1-2 corresponding to the appropriate choice of R′ given by 19.13 or 19.14. R1 is represented by (1), (2), (15) and (12), which constitute an alternative axiomatisation of the positive implicational calculus. Using for R2 the simplification suggested in the proof of 19.21, (1)-(10) and (15) correspond to themselves and (10) also corresponds to (11), and similarly

(18) corresponds to (16), and ($19_m$) to ($17_m$). To (12) there corresponds $(A \supset B) \supset . A \supset B$, but this is derivable from R1.

For 19.24, we cannot appeal to the deduction theorem for $\supset$ since it does not hold; but it does hold for $\supset^{m-1}$. We can therefore apply 19.20 to the rules R′ supplied by 19.15, provided we use $\supset^{m-1}$ in the formation of R1 and R2. (1), (2), (12), (14) and ($20_m$) are together stronger than R1, but to show that they will do duty for it we need only show that they are valid in the matrix, that $A, A \supset^{m-1} B \vdash B$ in the calculus L which they characterise, and that the deduction theorem holds for $\supset^{m-1}$ in L. The first two points are straightforward but in order to establish the third by induction in the classical way we need to show that (i) $B \vdash A \supset^{m-1} B$, (ii) $\vdash A \supset^{m-1} A$ and (iii) $A \supset^{m-1} (B \supset C), A \supset^{m-1} B \vdash A \supset^{m-1} C$. By (1) and (12) $C \vdash A \supset C$ and (i) is obtained by repeated application of this. (ii) follows similarly, since $\vdash A \supset^{2} A$ by (1). For (iii) we note first from (2), (12) and (iv) $\vdash (A \supset . B \supset C) \supset . B \supset . A \supset C$ (proved like (xiv) of 19.16) that (v) $B \supset C \vdash (A \supset B) \supset (A \supset C)$. It follows that (vi) if $B_1, \ldots, B_n$ is a permutation of $A_1, \ldots, A_n$ then $A_1 \supset \ldots \supset . A_n \supset D \vdash B_1 \supset \ldots \supset . B_n \supset D$; the proof is by iteration of the special case in which $A_j = B_{j+1}$, $A_{j+1} = B_j$, and otherwise $A_i = B_i$, the special case following from (iv) with $A = A_j$, $B = B_j$ and $C = A_{j+2} \supset \ldots \supset D$, by repeated application of (v). So $A \supset^{m-1} . B \supset C \vdash B \supset . A \supset^{m-1} C$, and by repeated application of (v), $B \supset . A \supset^{m-1} C \vdash (A \supset^{m-1} B) \supset (A \supset^{2m-2} C)$. Hence $A \supset^{m-1} . B \supset C$, $A \supset^{m-1} B \vdash A \supset^{2m-2} C$, and we may use ($20_m$) to reduce the conclusion of this to $A \supset^{m-1} C$ as required. For R2, simplifying as before, (1)-(10), ($13_m$) and (14) correspond to themselves. To (11) straightforwardly corresponds $A \supset^{m-1} . \sim A \supset^{m-1} B$, but by (vi) this is equivalent to $A \supset^{m-2} . \sim A \supset^{m-2} (\sim A \supset . A \supset B)$, and so is derivable from (10) and (1). Similarly, given that ($21_m$) is valid in the matrix, it can be used to correspond to ($17_m$).

To (12) would correspond $(A\supset B)\supset^{m-1}.A\supset^{m-1}B$, but this is derivable from (1) using (vi).

By Theorem 13.6 the calculus which forms the subject of Theorem 19.22 is Dummett's LC. His axioms and rules are essentially those devised by Kleene (1952) for the intuitionist calculus, plus $(A\supset B)\vee(B\supset A)$. This last corresponds to our (18), a purely implicative version being needed for our result about fragments (see also Bull, 1962, and for a variant of (19) see Thomas, 1962). The other differences reflect our attempt to draw on a common stock of axioms for the Gödel and Łukasiewicz cases. Thus we use (2) and (15) instead of Dummett's $(A\supset.B\supset C)\supset.(A\supset B)\supset.A\supset C$, since unlike the latter (2) is valid in the Łukasiewicz case too. Similarly we use (8) instead of Dummett's $A\supset.B\supset(A\&B)$ since the latter, surprisingly enough, is not equivalent to (8) in the context of the other Łukasiewicz axioms; and we use (9) instead of Dummett's $(A\supset B)\supset.(A\supset\sim B)\supset\sim A$ since the latter is invalid in the Łukasiewicz case.

## 19.8 *Rosser and Turquette*

Finally we explore the relation between the results of Section 19.2 and those of Rosser and Turquette, 1952, insofar as they are concerned with propositional calculi. Following their notation, let M be a matrix whose values are the integers 1,...,m, and with each value a and each A let there be associated a formula $J_a(A)$. Let $\supset$ be a binary connective and consider the following schemes:

RT1 $J_a(A)\supset.A\supset B$ if a is undesignated.

RT2 $J_a(A)\supset A$ if a is designated.

RT3 $J_{a_1}(A_1)\supset\ldots\supset.J_{a_k}(A_k)\supset J_{F(a_1,\ldots,a_k)}(FA_1\ldots A_k)$ for each k-ary F.

RT4 $(J_1(A)\supset B)\supset.\ldots\supset.(J_m(A)\supset B)\supset B$

RT5 $A\supset.B\supset A$

RT6 $(J_a(A)\supset.B\supset C)\supset.(J_a(A)\supset B)\supset.J_a(A)\supset C$

RT7 From $A, A\supset B$ infer $B$.

*Lemma 19.25* If RT1-7 are valid in a finite matrix M, they characterise the same single-conclusion calculus as M.

*Proof* Defining $T_a(A) = \{J_a(A)\}$ and $U_a(A) = \Lambda$, the rules R of Section 19.2 take the form

R1 From $J_a(A)$ infer A if a is designated, and from $J_a(A)$,A infer $\Lambda$ if a is undesignated.

R2 From $J_{a_1}(A_1),\ldots,J_{a_k}(A_k)$ infer $J_{F(a_1,\ldots,a_k)}(FA_1\ldots A_k)$ for each k-ary F.

R3 From $\Lambda$ infer $J_1(A),\ldots,J_m(A)$.

If L is the single-conclusion calculus characterised by RT1-7 we begin by showing that $R \subset \vdash_U$, and for this it is sufficient by 5.5 to show (i) $J_a(A) \vdash A$ if a is designated, (ii) $J_a(A), A \vdash B$ if a is undesignated, (iii) $J_{a_1}(A_1),\ldots, J_{a_k}(A_k) \vdash J_{F(a_1,\ldots,a_k)}(FA_1\ldots A_k)$ and (iv) if $Z, J_a(A) \vdash B$ for every a then $Z \vdash B$. (i)-(iii) are immediate from RT1-3 and RT7. For (iv) we first use RT5-6 to establish this restricted form of deduction theorem: if $Z, J_a(A) \vdash B$ then $Z \vdash J_a(A)\supset B$ (cf. Turquette, 1958). The result then follows by RT4 (with RT7) and cut for $\{J_1(A)\supset B,\ldots,J_m(A)\supset B\}$.

Since $\vdash_U$ is a consequence relation by 5.11, it follows that $\vdash_R \subset \vdash_U$. Moreover $\vdash_M \subset \vdash_R$ by 19.3, and $\vdash_\cap \subset \vdash_M$ if RT1-7 are valid in M. Thus $\vdash_\cap \subset \vdash_M \subset \vdash_U$, whence the result follows by 5.3.

The argument of this lemma can be used to establish Rosser and Turquette's result concerning the set of valid formulae of a finite matrix. For RT2-7 are the axioms and rules presented by Turquette (1958) in his refinement of their original result, which had used three slightly stronger axiom schemes in place of RT6, and we have

*Theorem 19.26* If RT2-7 are valid in a finite matrix M they generate as theorems precisely the formulae valid in M.

*Proof* RT1 is only needed in the proof of 19.25 to match R1 for undesignated a, and this in turn is only needed to establish the second part of 19.2 in the course of proving 19.3. In 19.3 we show that if $X \nvdash_R Y$ then there is a valuation v which invalidates ⟨X,Y⟩ in M, using the first part of 19.2 to show that v(A) is undesignated for each A in Y, and the second part to show that v(A) is designated for each A in X. If X is empty this latter demonstration is unnecessary, and so the theorem follows by the argument of 19.25.

It is not surprising that Lemma 19.25 should appeal to a stronger set of axioms and rules than Theorem 19.26, for the former is concerned with consequence in general and the latter only with the special case of valid formulae. Nor is it surprising that the difference should take the form of an extra axiom rather than an extra rule with premisses, for the results depend on a version of the deduction theorem, whose proof is unaffected by the addition of axioms but not of rules in general. The fact that the additional ingredient is an axiom does however make it possible to refine Lemma 19.25 in the light of the other result. For whenever RT1-7 are valid in the matrix then by Theorem 19.26 RT1 must be deducible from the remainder, and so we have

*Theorem 19.27* If RT1-7 are valid in a finite matrix M then RT2-7 characterise the same single-conclusion calculus as M.

| ⊃ | 1 | 2 | 3 |
|---|---|---|---|
| *1 | 1 | 2 | 3 |
| 2 | 1 | 1 | 3 |
| 3 | 1 | 1 | 1 |

| | $J_1$ |
|---|---|
| 1 | 1 |
| 2 | 3 |
| 3 | 3 |

| | $J_2$ |
|---|---|
| 1 | 3 |
| 2 | 1 |
| 3 | 3 |

| | $J_3$ |
|---|---|
| 1 | 3 |
| 2 | 3 |
| 3 | 1 |

**For an example to separate the two theorems, it is easy to verify that RT2-7 are valid in the matrix M above, but that RT1 is not. RT2-7 therefore generate as theorems precisely the formulae valid in M; but they do not generate all the inferences valid in M. For $J_2(A), A \vdash_M B$; but $J_2(A), A \nvdash_{RT2-7} B$ since RT2-7 are valid in the matrix M′ obtained by designating 2 as well as 1, while $J_2(A), A \nvdash_{M'} B$.**

# Part IV · Natural deduction

# 20 · Natural deduction

## 20.1 *Proof by cases*

It may be possible to use a set of multiple-conclusion rules R to provide a novel proof technique for a single-conclusion calculus L, by treating a proof from X to {B} by R as a proof of B from X in L. In particular, direct proofs by multiple-conclusion rules may be an alternative to the typically indirect methods of systems of natural deduction; indeed it was for just this purpose that Kneale introduced them.

We use Gentzen's notation for natural deduction, with square brackets to indicate a premiss of a subordinate proof represented by a column of dots. Thus Figure 20.1 shows the pattern of proof by cases, where it is assumed that $B_1,\ldots,B_n$ exhaust the various alternative cases which might arise on the basis of the premisses $A_1,\ldots,A_m$, so that when C is shown to follow from each of the $B_j$ this is taken as proving it from $A_1,\ldots,A_m$.

$$\dfrac{\begin{matrix}[B_1] & \ldots & [B_n] & & \\ \vdots & & \vdots & & \\ C & & C & A_1 \ldots A_m\end{matrix}}{C}$$

Figure 20.1

$$\begin{matrix}\dfrac{A_1 \quad \ldots \quad A_m}{B_1 \quad \ldots \quad B_n} \\ \begin{matrix}\vdots & \quad\quad\quad & \vdots \\ C & & C\end{matrix}\end{matrix}$$

Figure 20.2

Figure 20.2 shows how the same effect is achieved by a development using the rule 'from $A_1,\ldots,A_m$ infer $B_1,\ldots,B_n$'. The problem therefore becomes one of reducing the various natural-deduction rules to the form of Figure 20.1. The next figures show how this is done for the classical connectives, with the relevant natural-deduction rule to the left, its simulation as a proof by cases in the centre and the corresponding development to the right. Thus Figure 20.3 shows how, assuming the presence of 'from B infer A⊃B', the rule of ⊃ introduction can be replaced by 'from Λ infer A,A⊃B'. Similarly Figure 20.4 shows how ∨ elimination is replaced by 'from A∨B infer A,B'; there is no centre diagram here, since the rule is of the required form as it stands. Figure 20.5 shows how, assuming the presence of 'from B,~B infer A', the rule of ~ introduction is replaced by 'from Λ infer A,~A', and the rule of ~ elimination is dealt with in exactly the same way by interchanging A and ~A in the figure. There is no figure for conjunction since it, alone of the connectives, is adequately characterised without using indirect rules.

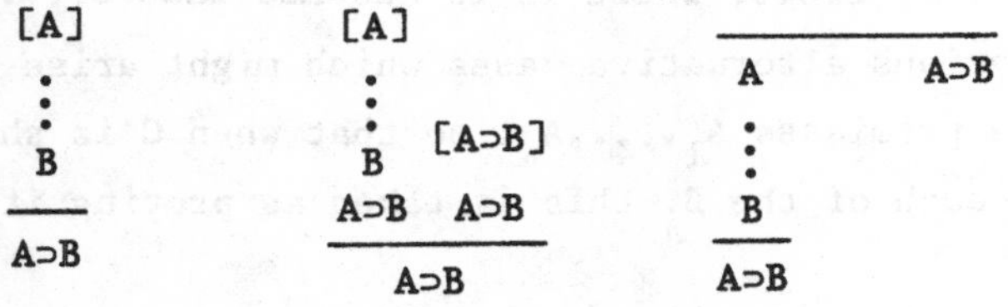

$$\dfrac{\begin{matrix}[A]\\ \vdots\\ B\end{matrix}}{A\supset B} \qquad \dfrac{\begin{matrix}[A] & \\ \vdots & [A\supset B]\\ B & \\ A\supset B & A\supset B\end{matrix}}{A\supset B} \qquad \begin{matrix}\overline{\begin{matrix}A & \quad A\supset B\end{matrix}}\\ \vdots\qquad\quad\\ \dfrac{B}{A\supset B}\qquad\quad\end{matrix}$$

Figure 20.3

$$\dfrac{\begin{matrix}[A] & [B] & \\ \vdots & \vdots & \\ C & C & A\vee B\end{matrix}}{C} \qquad\qquad \begin{matrix}\dfrac{A\vee B}{\begin{matrix}A & \quad & B\end{matrix}}\\ \begin{matrix}\vdots & \quad & \vdots\\ C & \quad & C\end{matrix}\end{matrix}$$

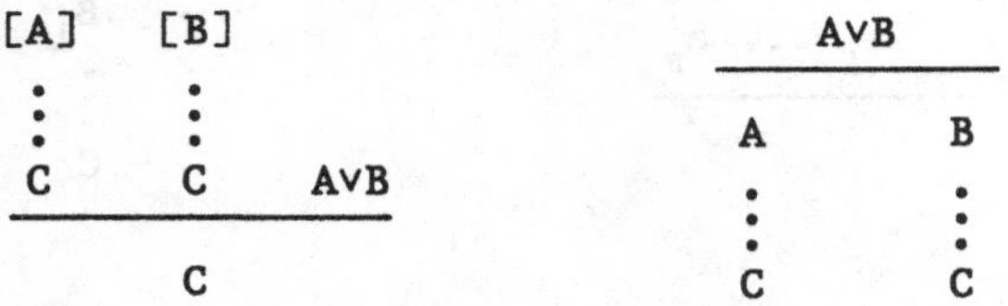

Figure 20.4

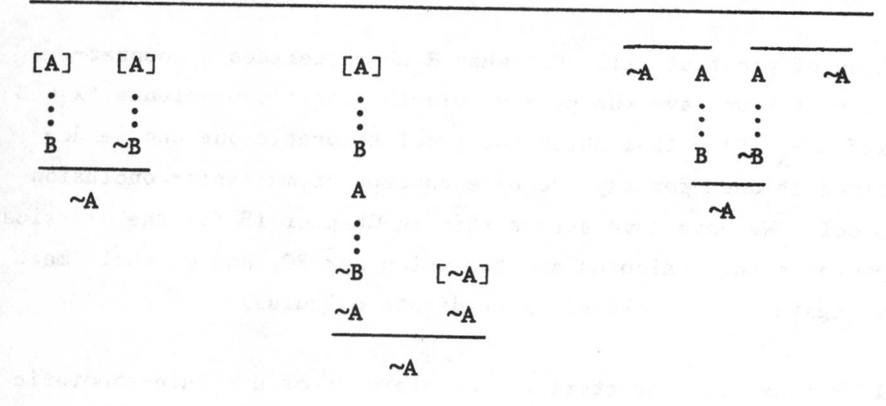


Figure 20.5

Success in such a project will be marked by a result of the form '$X \vdash B$ iff there is an articulated natural-deduction proof from X to {B} by R', but this is complicated by there being no unique idea of an articulated natural-deduction proof. For natural-deduction rules are not autonomous: their very formulation involves a reference to proofs, so that on pain of circularity they can only be understood as clauses in an inductive definition of proof; and different systems may use different frameworks for the definition. For example, Fitch (1952) keeps a principal proof disjoint from those subordinate to it, reiterating formulae into them as required; while at the other extreme Lemmon (1965) allows the same occurrence of A in a proof by ⊃ introduction to be both a premiss of the main proof and the operative premiss of the subordinate one (so that presumably a user would have to assert and suppose A in the same breath). Because of the rule of reiteration an articulated Fitch proof is liable to be non-standard and non-abstract, whereas an articulated Lemmon proof will be standard and abstract but may break our bounds for graph arguments by allowing A to occur as a non-initial premiss.

It may not be necessary, however, to invoke any particular

idea of proof at all. For when R characterises a counterpart of L we have the purely rule-theoretic equivalence '$X \vdash B$ iff $X \vdash_R \{B\}$', from which the proof-theoretic one can be derived at once for any adequate concept of multiple-conclusion proof. We have come across this in Chapter 18 for the classical propositional calculus and the rules for PC, and we shall meet it again for the classical predicate calculus.

If R does not characterise a counterpart of L a rule-theoretic equivalence is impossible, but a proof-theoretic one may obtain nonetheless. Let us say that R is *sound* with respect to L if $R \subset \vdash_U$ (cf. Theorems 5.3 and 5.5). Then even if R is unsound we may be able, by restricting the way the unsound instances are used in proofs, to define a class of 'sound' proofs for which the desired equivalence holds: $X \vdash B$ iff there is a sound proof from X to {B} by R. We shall illustrate this for the intuitionist propositional calculus in Section 20.3.

One of the most interesting results that may distinguish a system of natural deduction is the 'subformula' theorem that if there exists a proof of B from X there exists one consisting entirely of subformulae of B and subformulae of members of X. When the theorem holds - it fails for the usual natural deduction textbook versions of classical logic - it may be obtained as a corollary of a normal form theorem for proofs, the subformula property being obvious for proofs of a suitable normal form. (See Prawitz, 1965, on his intuitionist system. The subformula theorem fails for his classical system, as his substitute result on p.42 attests.) It is inevitable that the subformula theorem for a system of natural deduction should be expressed in proof-theoretic terms since, as was noted above, the existence of an appropriate proof is the only thing that consequence by natural-deduction rules can mean. By contrast rules of inference as we have been expounding them do not require any conception of proof either to enunciate them or to define consequence by them. One can therefore envisage a subformula theorem which is intrinsic to a set of such rules and does not refer to proofs at all, namely (in the multiple-conclusion case): if $X \vdash_R Y$ then $X \vdash_{R'} Y$, where R′ comprises just those instances of R whose premisses and conclusions are subformulae of members of X and Y. We have already established such a result for PC (Theorem 18.2), and in the next section we do so for a counterpart of the classical predicate calculus, with a subformula theorem for the latter as a corollary.

For any adequate concept of proof a proof-theoretic version of the subformula theorem will be a corollary of the rule-theoretic one (cf. the discussion following Theorem 18.2), whether or not a normal form theorem for proofs has been established or is even true. On the other hand Kneale proofs are not adequate: indeed it was precisely in order to demonstrate their inadequacy that we wanted a subformula theorem in Section 8.3. We thus had no choice there but to state a proof-theoretic version of the result and derive it via a normal form theorem for Kneale proofs; and a similar situation arises in Section 20.3 in connection with intuitionistically sound proofs. A normal form theorem in this context says that whenever there is a proof $\pi$ of the appropriate kind from X to Y there is a normal one as defined in Section 8.3. One would expect to establish this by induction on some feature of $\pi$, and in this connection it may be possible to regard a proof in a sequent calculus as if it were the history of a construction of $\pi$, the antecedents and succedents of each sequent representing the premisses and conclusions of the proof at that stage of its construction. (Prawitz, 1965, suggests a similar link between a sequent $A_1,\ldots,A_m \rightarrow B_1,\ldots,B_n$ and a natural-deduction proof of $B_n$ from $A_1,\ldots,A_m,{\sim}B_1,\ldots,{\sim}B_{n-1}$.)

In particular Gentzen's 'cut' rule would correspond to junction, 'mix' to multiple junction for cross-referenced-circuit proofs, 'contraction' to identification of premisses and conclusions for cornered-circuit proofs and to election for cross-referenced Kneale ones, and so on. One might then hope to equate normal proofs with those for which there exists a construction represented by a cut-free proof in the sequent calculus, in which case the desired theorem would become a corollary of a Gentzen-style proof of cut-elimination in the sequent calculus. It may however be possible to establish the normal form theorem by an inductive argument of lesser complexity than the proof of cut-elimination. In the case of Kneale proofs, for example, it was unnecessary to consider the manner of their construction and sufficient to argue in terms of a straightforward feature of the proofs themselves, viz. the number of their edges (Theorem 8.9). We leave it to the reader to assess the proof of our other normal form result (Theorem 20.8) in this light.

## 20.2 *Classical predicate calculus*

Expositions of the predicate calculus which agree in their treatment of sentences, i.e. of formulae without free variables, may differ in their treatment of free variables, some adopting the 'conditional' interpretation, others the 'generality' one and the natural-deduction versions neither. By 'the classical predicate calculus', then, we shall mean what is common to the various expositions, a calculus whose universe V consists entirely of sentences. We shall see that the orthodox semantics for the classical predicate calculus also characterises its maximum counterpart, and we present rules which (though they make use of free variables) in effect characterise the latter and thereby provide a proof technique for the classical calculus on the lines of the preceding section. We postpone comment on the rules until these results have been established.

We assume a countable vocabulary of predicates together with the connectives ⊃, &, ∨ and ~, the existential quantifier (∃), a countably infinite stock of *variables* and a separate countably infinite stock $b_1, b_2, \ldots$ of *parameters* (cf. Prawitz, 1965). The intention is to reserve the variables exclusively for use as bound variables (vacuous quantification being allowed or not as desired), and to use the parameters as free variables. With this explanation the definition of 'formula' is routine. If desired the notion of propositional function (as in Lemmon, 1965; cf. Prawitz's 'pseudo-formulae') can be used as an auxiliary to cover such expressions as A(x) in (∃x)A(x), and our formulae sieved out as the propositional functions all of whose variables are bound. A *sentence* is a formula containing no parameters.

For any parameter b and any formula of the form (∃x)A(x), let A(b) be obtained from A(x) by substituting b for all occurrences of x which are not bound by a quantifier in A(x). The formulae of the form (∃x)A(x) can be enumerated as $A_1, A_2, \ldots$ in such a way that the parameter $b_m$ does not occur in $A_n$ if $n \leq m$ (e.g. by sieving them from an arbitrary enumeration, taking $A_n$ to be the first available one that satisfies the condition). We can then enunciate the required rules of existential generalisation and existential instantiation:

EG From A(b) infer (∃x)A(x).

EI From (∃x)A(x) infer $A(b_n)$ if (∃x)A(x) is $A_n$.

The semantics is based on the standard idea of a *valuation* of the formulae, defined by the choice of a (nonempty) domain of individuals, the interpretation of each k-ary predicate as a k-ary relation between individuals, the assignment of an individual as value to each parameter, and the usual inductive clauses determining the truth-value of compound formulae. The clause for quantified formulae, '$(\exists x)A(x)$ is true iff $A(x)$ is true for some value i of x', can be taken at face value if propositional functions have been introduced and values assigned to variables as well as to parameters; otherwise '$A(x)$ is true for value i of x' is to be read as short for '$A(b)$ is true for value i of b', where b is any parameter not occurring in $A(x)$. Each valuation determines a partition $\langle T,U \rangle$ of the formulae into true and untrue, and we say that it *satisfies* $\langle X,Y \rangle$ if $\langle T,U \rangle$ does so. Similarly we say that $\langle X,Y \rangle$ is *valid in a domain* if it is satisfied by every valuation in that domain.

*Theorem 20.1* $X \vdash Y$ in the maximum counterpart of the classical predicate calculus iff $X \vdash_R Y$, where X and Y are any sets of sentences and R comprises EG, EI and the rules for PC. In particular, if X is any set of sentences and B any sentence, $X \vdash B$ in the classical predicate calculus iff $X \vdash_R \{B\}$.

*Proof* Let X and Y be any sets of sentences and let L be the classical predicate calculus, so that for any sentence B we have $X \vdash B$ iff $\langle X,B \rangle$ is valid in every domain. Then exactly the argument of 18.4 shows that $X \vdash_U Y$ iff $\langle X,Y \rangle$ is valid in every domain. Suppose therefore that there is a valuation which invalidates $\langle X,Y \rangle$. We reassign values to the successive parameters as follows. Suppose that values have been reassigned to $b_1,\ldots,b_{n-1}$ and let $A_n$ be $(\exists x)A(x)$. If at this stage $A(x)$ is true for some value i of x we assign i as value to $b_n$; otherwise we assign $b_n$ its original value (or an arbitrary one). We note that the truth-values of $A_n$ and $A(b_n)$ are unaffected by any reassignments after that of $b_n$, since no subsequent parameter occurs in either formula, and the truth-values of members of X and

Y are unaffected by the whole process, since they contain no parameters at all. The result of the process is therefore a valuation which invalidates ⟨X,Y⟩ but satisfies EI, and since it obviously satisfies EG and the rules for PC we have $X \nvdash_R Y$.

Conversely, suppose that $X \nvdash_R Y$, so that by cut for the set of all formulae there is a partition ⟨T,U⟩ such that $T \nvdash_R U$ and $X \subset T$ and $Y \subset U$. We define a valuation by taking the parameters as individuals and assigning each as value to itself, and interpreting each k-ary predicate F as the relation which holds between $b_{i_1},\ldots,b_{i_k}$ iff $Fb_{i_1}\ldots b_{i_k} \in T$. We show that each formula C is true in this valuation iff $C \in T$, arguing by induction on the number of connectives and quantifiers. The basis is immediate from the definition. For the induction step C may be (i) A⊃B, (ii) A&B, (iii) A∨B, (iv) ~A or (v) (∃x)A(x). Cases (i)-(iv) are dealt with as in the corresponding proof for propositional calculus (18.1); for example, from $T \nvdash_R U$ and the rules 'from Λ infer A, ~A' and 'from A, ~A infer Λ' it is immediate that $\sim A \in T$ iff $A \in U$, whence the result by the induction hypothesis. For (v), if $(\exists x)A(x) \in T$ then by EI $A(b_n) \in T$ for the appropriate n, and so by the induction hypothesis $A(b_n)$ is true. Hence A(x) is true for the value $b_n$ of x, and so (∃x)A(x) is true. Conversely if (∃x)A(x) is true then A(x) is true for some value b of x, whence A(b) is true. By the induction hypothesis it follows that $A(b) \in T$, whence $(\exists x)A(x) \in T$ by EG.

The argument actually establishes that if $X \nvdash_R Y$ then ⟨X,Y⟩ is invalid in a countably infinite domain, and hence that ⟨X,Y⟩ is valid in every domain iff it is valid in a countably infinite one.

We define the subformula relation to be the ancestral of that holding between A and each of A⊃B, B⊃A etc. and between A(b) and $(\exists x)A(x)$.

*Theorem 20.2* $X \vdash B$ in the classical predicate calculus iff $X \vdash_{R'} \{B\}$, where B is any sentence, X is any set of sentences and R′ comprises just those instances of EG, EI and the rules for PC each of whose premisses and conclusions is a subformula of B or of a member of X.

*Proof* By 20.1 it is sufficient to show that if $X \vdash_R Y$ then $X \vdash_{R'} Y$, where X and Y are any sets of sentences and R is as in 20.1. This is established from 20.1 in the same way as 18.2 was established from 18.1, by applying the second half of the argument to R′ and to those C that are subformulae of members of X or Y.

The calculus we have presented has a number of special features. Some are purely for expository convenience, like the omission of proper names and the inclusion of all four connectives as primitive. Universal quantification can be introduced by defining $(x)$ as $\sim(\exists x)\sim$ and versions of universal instantiation (UI) and universal generalisation (UG) can be derived from EG and EI as in Figure 20.6, UI being 'from $(x)A(x)$ infer $A(b)$' and UG 'from $A(b_n)$ infer $(x)A(x)$ when $(\exists x)\sim A(x)$ is $A_n$'. Test cases involving all four rules are shown in Figures 20.7 and 20.8. The former is a sound pattern of deduction, since m and n can always be chosen so that $(\exists x)(y)Rxy$ is $A_m$ and $(\exists y)\sim(\exists x)Rxy$ is $A_n$; but the latter is unsound, since the

$A(b)$ $\sim A(b)$ $(\exists x)\sim A(x)$ $\sim(\exists x)\sim A(x)$

UI

$\sim(\exists x)\sim A(x)$ $(\exists x)\sim A(x)$ $\sim A(b_n)$ $A(b_n)$

UG

Figure 20.6

step by EI requires that $(\exists y)Rb_m y$ is $A_n$ and hence that $m<n$, while the step by UG requires that $(\exists x)\sim Rxb_n$ be $A_m$ and hence that $n<m$.

| | | | | |
|---|---|---|---|---|
| $(\exists x)(y)Rxy$ | | | $(x)(\exists y)Rxy$ | |
| $(y)Rb_m y$ | by EI | | $(\exists y)Rb_m y$ | by UI |
| $Rb_m b_n$ | by UI | | $Rb_m b_n$ | by EI ? |
| $(\exists x)Rxb_n$ | by EG | | $(x)Rxb_n$ | by UG ? |
| $(y)(\exists x)Rxy$ | by UG | | $(\exists y)(x)Rxy$ | by EG |

Figure 20.7 Figure 20.8

The separation of parameters from bound variables has some distinguished precedents outside the context of natural deduction. It is known too that the subformula theorem is liable to fail for a sequent version of the predicate calculus when the same variable occurs free and bound in the same sequent (cf. Kleene, 1952, p.450). If one regards the subformula theorem as having only a technical significance one can dismiss this, as Kleene does, as a local restriction which 'does not detract from the usefulness of the theorem'; but otherwise a separation based on principle seems to be called for. The behaviour of parameters closely resembles that of extra-logical constants (Lemmon rightly calls them 'arbitrary names' or 'dummy names'), and our use of them in Theorem 20.1 is comparable to the use of constants in Henkin's (1949) completeness theorem. It was with this in mind that we mentioned the possibility of using propositional functions as an auxiliary in defining the formulae, for the parameter-free propositional functions would correspond to the formulae of Henkin's actual calculus before it is extended by the addition of his constants.

The crucial feature of our version of existential instantiation is its independence of any concept of proof. The usual natural-deduction version, pictured below, requires that b should not occur in the other premisses of the subordinate proof or in its conclusion; this restriction still needs to be expressed when the rule is translated into 'from (∃x)A(x) infer A(b)' as in Figures 20.1 and 20.2, but to do so requires a reference to proofs. To formulate the restriction one would need to proceed in the spirit of Section 20.3, by (1) taking, say, standard cornered-circuit proofs as the norm; (2) defining a 'sound' proof $\pi$ as one in which b does not occur in any initial or final vertex of $\pi^c$ other than c, where c is the relevant occurrence of A(b) and $\pi^c$ is as in Section 20.3; and possibly (3) requiring that the various $\pi^c$ be nested, analogously to the nesting of subordinate natural-deduction proofs.

$$\dfrac{\begin{matrix}[A(b)] \\ \vdots \\ B\end{matrix} \qquad (\exists x)A(x)}{B}$$

In our calculus, as in the usual expositions of the predicate calculus, formulae containing parameters are thought of as Hilbertian 'ideal' formulae - auxiliaries introduced to facilitate inferences between the 'real' formulae, i.e. the sentences. But it is evident that the parameter in our EI functions like a shorthand for an ε-term of Hilbert and Bernays, 1939, and it should be possible to adapt the rule to accommodate ε-terms explicitly, so that the deduction would take place through 'real' formulae.

In Chapter 6 we cited Carnap and Kneale as developing a predicate calculus, not with an essentially single-conclusion rule like our EI but with one on the lines of 'from (∃x)Fx infer $Fx_1, Fx_2, \ldots$', where the conclusions are intended to represent 'all propositions that are values of the function expressed by Fx' (Kneale, 1962, p.547). Carnap (1943) gives effect to this intention by postulating a fixed, countably infinite domain of individuals and assuming that the calculus contains a proper name $b_i$ for each individual. He thereby justifies his rule (call it $EI_{\aleph_0}$) 'from (∃x)A(x) infer $A(b_1), A(b_2), \ldots$', but naturally obtains a non-compact calculus whose single-conclusion part is not the orthodox predicate calculus. Kneale is less explicit, but it seems that he has in mind the orthodox semantics of truth in every domain, and that he must therefore first envisage a distinct rule $EI_n$ appropriate to each possible cardinality n of the domain of individuals, and then identify the intersection of the resulting calculi. That he has something like this in mind may perhaps be inferred from his answer to another question - if n is infinite how can one use $EI_n$ when the proofs at one's disposal are finite? His answer seems to be that a finite proof can sometimes be read as a shorthand for an infinite one, like $\pi$ and $\pi_{\aleph_0}$ below. If $\pi$ can be so read for one choice of n it can obviously be read in a similar way for any n, finite or infinite; and with this we are beginning to move towards the goal of representing just the inferences that are valid for every choice of n. If this is Kneale's plan, one might be able to carry it out by (1) taking a class of proofs known to be adequate for infinite rules, say standard cornered-circuit abstract proofs; (2) showing how a finite proof using a suitable version of existential instantiation can be expanded, concertina-fashion, into a possibly infinite one using $EI_n$, for any n; and (3) showing how a proof using $EI_{\aleph_0}$ can be converted into a finite one using existential instantiation, by choosing a distinct $b_i$ for each instance of $EI_{\aleph_0}$, deleting the edges from each stroke b to all but the relevant member of $Y_b$, appealing to the infinite analogue of Theorem 7.8 followed by Theorems 12.1 and 7.14, and relettering the result.

$$\begin{array}{ccc} & (\exists x)A(x) & & \multicolumn{2}{c}{\underline{\quad(\exists x)A(x)\quad}} \\ \pi & A(b) & \pi_{\aleph_0} & A(b_1) & A(b_2)\ \ldots \\ & \vdots & & \vdots & \vdots \\ & B & & B & B \end{array}$$

## 20.3 *Intuitionist propositional calculus*

We plan to utilise the rules for PC to provide a proof technique for Heyting's intuitionist propositional calculus. For the reason given below we shall not call on any zero-conclusion rules, so we consider the following *rules for the positive variant of PC*, as prescribed by Theorem 5.27.

| | | |
|---|---|---|
| From A,A⊃B infer B | From B infer A⊃B | From Λ infer A,A⊃B |
| From A&B infer A | From A&B infer B | From A,B infer A&B |
| From A infer A∨B | From B infer A∨B | From A∨B infer A,B |
| From A,~A infer B | From Λ infer A,~A | |

An inspection of these rules reveals that the only ones which are unsound with respect to Heyting's calculus are 'from Λ infer A,A⊃B' and 'from Λ infer A,~A'. Since they also are the only rules with zero premisses, we can restrict their application by restricting the way initial strokes can occur in proofs. We choose standard cornered-circuit proofs as the norm for this purpose, though other choices are doubtless possible. We therefore say that a standard cornered-circuit proof by the rules for the positive variant of PC is (intuitionistically) *sound* if there is a major edge (Section 8.3) from each initial stroke b to a vertex which *discharges* b, i.e. which lies on every directed path from b to a final vertex. For example, Figure 20.9 shows a pair of isomorphic proofs whose initial stroke is discharged by the conclusion, but only in $\pi_1$ is the edge joining the two a major one. Thus $\pi_1$ is a sound proof from A to ~~A but $\pi_2$ is not a sound proof from ~~A to A.

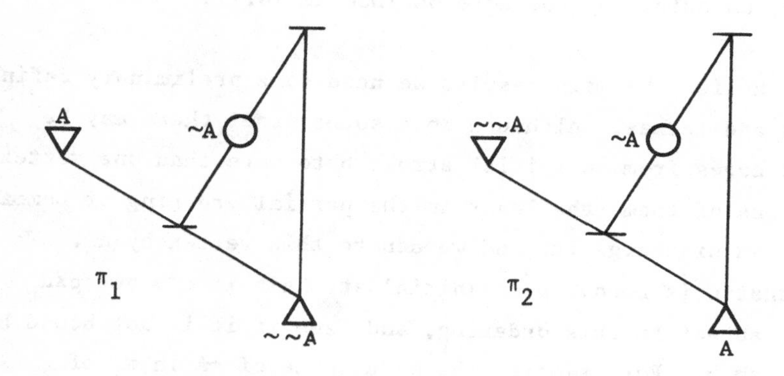


Figure 20.9

We shall vindicate this definition by showing that there is a sound proof from X to {B} iff X ⊢ B in Heyting's calculus, and derive a subformula property for the latter as a corollary of a normal form theorem for sound proofs. More generally, we show that deducibility by sound proofs is equivalent to consequence in the maximum positive compact counterpart of the Heyting calculus, the restriction on sign being needed because we have left out zero-conclusion rules. We have done so because it greatly simplifies the definition of discharge, an idea which is of interest in connection with the articulation of natural-deduction proofs in general. We note too that it is not necessary to appeal to zero-conclusion rules when transforming the natural-deduction ones in the manner of Section 20.1 (cf. Figure 20.5 where only 'from B, ~B infer A' is assumed). We leave the reader to see if our result can be extended to take in the full rules for PC on the one hand and the maximum compact counterpart of the Heyting calculus on the other. The latter is also the maximum propositional counterpart, as we noted when proving Theorem 16.7. The maximum counterpart does not come up for discussion in this context, being neither compact (Theorem 5.9) nor a propositional calculus (by the note on Theorem 16.7).

To establish the main results we need some preliminary definitions and lemmas. Although in a sound proof there may be major edges from an initial stroke b to more than one vertex, only one of them (the least in the partial ordering of Lemma 7.1) can discharge it, and we denote this vertex by $d_b$. We say that c is *bound by* an initial stroke b if $c<b$ but $c \nless d_b$ with respect to this ordering, and *free* if it is not bound by any such b. For example, the occurrence of ~A in $\pi_1$ of Figure 20.9 is bound by the initial stroke, but the other formula vertices, like the premisses and conclusions of every sound proof, are free.

Given any formula vertex c of an argument $\pi$, we write $\pi(c)$ for the standard argument whose graph is the maximum formal subgraph of $\pi$ that is a union of c and directed paths to c. Such a graph necessarily exists, since there is at least one formal subgraph of the kind described, namely {c}, and since the union of any family of such subgraphs is one of the same kind.

Again, given any formula vertex c of an argument $\pi$, we write $\pi^c$ (read '$\pi$ below c') for the argument whose graph is the union of c and those corner-free paths of $\pi$ in which c is an initial endpoint (so that c is initial in $\pi^c$), c being a premiss of $\pi^c$ but otherwise each vertex of $\pi^c$ being a premiss or conclusion iff it is one in $\pi$. Similarly we define $\pi_c$ (read '$\pi$ above c') as the union of c and all the corner-free paths of $\pi$ in which c is a final endpoint, with c as a conclusion but otherwise each vertex being a premiss or conclusion iff it is so in $\pi$. We have implicitly made use of these constructs in the proof of Theorem 9.5. So, for example, taking c to be the intermediate occurrence of A⊃A in $\pi$ of Figure 9.6, $\pi_1'$ and $\pi_2'$ are respectively $\pi_c$ and $\pi^c$; and similarly in Figure 10.2.

*Lemma 20.3* If a sound proof has any initial strokes, it has an initial stroke b for which $d_b$ is free.

*Lemma 20.4* If c is free in a sound proof $\pi$ then $\pi(c)$ is a sound proof with c as its sole conclusion, its premisses being free in $\pi$ and having no major edges to them in $\pi$.

*Lemma 20.5* If $\pi$ is a standard cornered-circuit proof, $\pi^c$ and $\pi_c$ are standard proofs whose graphs are abstract subgraphs of that of $\pi$.

*Lemma 20.6* If $\pi$ is a sound proof then $\pi^c$ is sound, and so is $\pi_c$ if c is free in $\pi$.

*Proofs* For 20.3, if the set of vertices of the form $d_b$ is nonempty, it has a minimal member in the partial ordering of 7.1, and this member is evidently free.

For 20.4, $\pi(c)$ is by definition a standard cornered-circuit proof with c as its sole final vertex. For each initial stroke b of $\pi$, if $\pi(c)$ and the union $\delta$ of the directed paths from b to $d_b$ have any vertex in common, then $\delta$ is a subgraph of $\pi(c)$. For if there is any common vertex, every directed path from it to c contains $d_b$ since c is free, and so $d_b$ is in $\pi(c)$; and hence the union of $\pi(c)$ with $\delta$ and the edges to strokes in $\delta$ and the vertices adjoining these edges is both a formal subgraph and the union of directed paths to c, and so must be a subgraph of $\pi(c)$. It follows that every initial stroke b of $\pi(c)$ is discharged by $d_b$ in $\pi(c)$ as in $\pi$, and the edge from b to $d_b$ is in $\pi(c)$. Hence $\pi(c)$ is sound. It also follows that no initial formula vertex of $\pi(c)$ is in $\delta$. Hence each such vertex a is free in $\pi$ and without an edge to it from any initial stroke, so if there is an edge E to a, it must be from an intermediate stroke b. But $Y_b$ cannot be a singleton since if it were, the union of $\pi(c)$, E, b, the edges to b and the vertices adjoining these edges would be both a formal subgraph and the union of directed paths to c, and hence a subgraph of $\pi(c)$; and then a would not after all be initial in $\pi(c)$. So $X_b = \{A \vee B\}$ and $Y_b = \{A,B\}$ for some A and B, and hence E is not major.

For 20.5, it is sufficient by 10.3 to show that $\pi^c$ is standard and that it contains all edges of $\pi$ which adjoin its strokes. (A similar argument serves for $\pi_c$.) Since c is evidently standard in $\pi^c$, we need only show that (i) $\pi^c$ contains all the edges adjoining each stroke of $\pi^c$, and (ii) each formula vertex of $\pi^c$ other than c is initial, final or intermediate in $\pi^c$ according as it is in $\pi$. If either of these conditions fails, there is a path $\alpha = \{a_1, E_1, \ldots, E_{k-1}, a_k\}$ in $\pi^c$ in which $a_1$ is c and is initial, and an edge $E_k$ joining $a_k$ and, say, $a_{k+1}$ such that $E_k \notin \pi^c$ and $\{E_{k-1}, a_k, E_k\}$ is not a corner. If $a_{k+1} \notin \alpha$, let $\beta$ be $\{a_1, \ldots, E_k, a_{k+1}\}$; if $a_{k+1} = a_i$ for some i, $1 \le i \le k$, then $\{a_i, E_i, \ldots, E_k, a_{k+1}\}$ is a circuit which can have its corner only at $a_i$, and we let $\beta$ be $\{a_1, E_k, a_k\}$ if i=1, and $\{a_1, \ldots, a_{i-1}, E_{i-1}, a_i, E_k, a_k\}$ otherwise. In each case $\beta$ is corner-free and $a_1$ is initial, so that $E_k \in \pi^c$, contrary to hypothesis.

For 20.6, $\pi^c$ is standard by 20.5, and every directed path $\alpha$ of $\pi^c$ from an initial stroke b to a conclusion is also a directed path from b to a conclusion in $\pi$, and so contains $d_b$. Hence $d_b$ discharges b in $\pi^c$ as well as in $\pi$, and the edge from b to $d_b$ is in $\pi^c$ by 20.5. Accordingly $\pi^c$ is sound. The proof for $\pi_c$ is similar, except that here α may be a path from b to c; but if c is not bound by b, α contains $d_b$ in this case also.

*Theorem 20.7* Deducibility by sound proofs is the same as consequence in the maximum positive compact counterpart of Heyting's propositional calculus; in particular X ⊢ B in Heyting's calculus iff there is a sound proof from X to {B}.

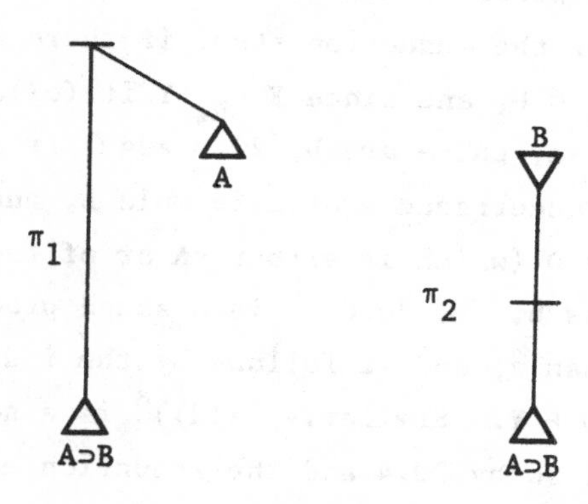


Figure 20.10

*Proof* Let L be the maximum positive compact counterpart of Heyting's calculus. By 5.36 it is characterised by adding 'from A∨B infer A,B' to the singleton-conclusion rules corresponding to modus ponens and the following axioms for the Heyting calculus (Kleene, 1952): A⊃.B⊃A, (A⊃B)⊃.(A⊃.B⊃C)⊃.A⊃C, A⊃.B⊃(A&B), (A&B)⊃A, (A&B)⊃B, A⊃(A∨B), B⊃(A∨B), (A⊃C)⊃.(B⊃C)⊃.(A∨B)⊃C, (A⊃B)⊃.(A⊃~B)⊃~A and ~A⊃.A⊃B. It is left to the reader to show that there are sound proofs corresponding to all these axioms. (The task is simplified by showing first that the deduction theorem holds for deducibility by sound proofs: given such a proof from X,A to B, we may obtain one from X to A⊃B by joining a replica of $\pi_1$ of Figure 20.10 at each premiss A, and a replica of $\pi_2$

at each conclusion B, and identifying the resulting conclusions.) Since soundness is preserved under junction and identification of premisses and conclusions, it follows as in 10.2 that if $X \vdash Y$ there is a sound proof from X to Y.

For the converse we show by induction on the length of $\pi$ (measured by the total number of vertices and edges) that if $\pi$ is a sound proof from X to Y then $X \vdash Y$. The basis is trivial. For the induction step, if there are no initial strokes then $R_\pi \subset \vdash$, and since $X \vdash_{R_\pi} Y$ it follows that $X \vdash Y$. Otherwise, there are by 20.3 edges from an initial stroke b to an occurrence a of a formula A, and to a free occurrence d of D (which is either $\sim A$ or of the form $A \supset B$) which discharges b. By 20.6 $\pi^d$ is a sound proof from X,D to Y shorter than $\pi$, and it follows by the induction hypothesis that $X,D \vdash Y$. Similarly $(\pi(d))^a$ is a sound proof shorter than $\pi$, so by 20.4 and the induction hypothesis $X,Z,A \vdash D$, where Z is the set of formulae which occur as premisses of $\pi(d)$. By the deduction theorem in Heyting's calculus $X,Z \vdash A \supset D$, and since both $A \supset . A \supset B \vdash A \supset B$ and $A \supset \sim A \vdash \sim A$, we have that $A \supset D \vdash D$ and hence that $X,Z \vdash D$; and by cut for D that $X,Z \vdash Y$. By 20.4 d is not initial in $\pi(d)$ so that each initial formula vertex c of $\pi(d)$ is free in $\pi$ and not final; hence by 20.6 $\pi_c$ is a sound proof shorter than $\pi$. By the induction hypothesis $X \vdash C,Y$ for each C in Z, and by cut for Z we have $X \vdash Y$ as required.

*Theorem 20.8* If there is a sound proof from X to Y there is a sound normal one.

*Proof* By 20.7 if there is a sound proof from X to Y then $X \vdash Y$ in the calculus L of that theorem. We again leave it to the reader to show that deducibility by sound normal proofs contains the rules for L (the deduction theorem holding as before), and it is evidently closed under overlap and dilution. By 2.16 it remains only to show that it is closed under cut for formulae, and for this we prove a stronger result. We say that a formula vertex is *major* if it adjoins a major edge other than one from a stroke corresponding to the rule 'from A,~A infer B'; in particular we speak of a *major premiss* or *major conclusion*. We say that $\pi$ is *subject to* $\pi_1,\ldots,\pi_i$ *bar* $A_1,\ldots,A_j$ if no formula other than $A_1,\ldots,A_j$ occurs as a major premiss (or major conclusion) of $\pi$ unless it occurs as a major premiss (major conclusion) of one of $\pi_1,\ldots,\pi_i$. We show that if there are (disjoint) sound normal proofs $\pi$ and $\pi'$ from X to C,Y and from X,C to Y respectively, then there is a sound normal proof from X to Y subject to $\pi$ and $\pi'$ bar subformulae of C.

We may assume that $\pi$ has a unique final occurrence c of C, since if there are more they may be identified, and if there are none $\pi$ itself is the required proof. Similarly we assume that $\pi'$ has a unique initial occurrence c′ of C. We may assume also that c is major, since otherwise all major edges to c are from strokes corresponding to the rule 'from A,~A infer C' and according as (i) there is no major edge to c, or (ii) there is a major edge to c but no minor one, or (iii) there is a major and a minor edge to c, we may obtain the required proof by (i) joining replicas of $\pi$ and $\pi'$, or (ii) replacing c in $\pi$ by an occurrence of a formula in Y, or (iii) deleting the major edges to c, selecting the (standard) component of the resulting argument which contains c, joining it and a replica of $\pi'$, and restoring an edge from each final stroke (which must be free in $\pi$ since as c is not major it cannot be $d_b$ for any initial stroke b) to an arbitrary conclusion. By a similar though simpler argument we may take c′ to be major. Since no occurrence of a propositional variable is major, this covers the basis of an induction proof on the complexity of C. For the induction step, C is not a variable, and our main induction hypothesis is that for all X′ and Y′ and all C′ less complex than C, if there are sound normal proofs from X′ to C′,Y′ and from X′,C′ to Y′, then there is a sound normal proof from X′ to Y′ subject to them bar subformulae of C′. We need a subsidiary induction on the sum of the lengths of $\pi$ and $\pi'$ defined as in the proof of 20.7. The basis is trivial, and the subsidiary induction hypothesis is that for all X′ and Y′, if there are sound normal proofs from X′ to C,Y′ and from X′,C to Y′, together shorter than $\pi$ and $\pi'$ together, then there is a sound normal proof from X′ to Y′ subject to them bar subformulae of C. We distinguish two cases according as $\pi'$ does or does not have an initial stroke.

If $\pi'$ has an initial stroke, then by 20.3 there are edges from such a stroke b′ to an occurrence a′ of a formula A′, and to a free occurrence of D′ (which is either ~A′ or of the form A′⊃B′) which discharges b′. If $\pi(c) = \pi$ and $\pi'(d') = \pi'$ then $D' \in Y$, and applying the subsidiary hypothesis to $\pi$ and $(\pi')^{a'}$ we obtain a sound normal proof from X,A′ to D′ subject to $\pi$ and $\pi'$ bar A′ and subformulae of C; and by joining disjoint simple proofs at the initial occurrences of A′, each proof comprising edges from a single stroke to conclusions A′ and D′ (cf. Figure 20.10), and identifying the conclusions of the resulting argument, we obtain the required proof from X to Y subject to $\pi$ and $\pi'$ bar subformulae of C. If on the other hand $\pi(c)$ and $\pi'(d')$ are together shorter than $\pi$ and $\pi'$ together, let S be the set of vertices initial in $\pi(c)$ but not in $\pi$, or in $\pi'(d')$ but not in $\pi'$, and let Z be the set of formulae occurring at the vertices in S. By 20.4 the vertices in S are all intermediate and free in $\pi$ or in $\pi'$ as the case may be. If C is D′ or a member of Z, we obtain the necessary proof from X to Y by applying the subsidiary hypothesis to $\pi$ and

$(\pi')^{d'}$ or for some d in S to $\pi_d$ and $\pi'$ or to $\pi$ and $(\pi')^d$. Otherwise we obtain in a similar way sound normal proofs from X to D,Y for each D in Z, and from X,D′ to Y, all subject to $\pi$ and $\pi'$ bar subformulae of C, since by 20.4 no d in S is a major conclusion of $\pi_d$ or $(\pi')_d$ as the case may be, while d′ cannot be a major premiss of $(\pi')^{d'}$ because $\pi'$ is normal. Similarly, applying the same hypothesis to $\pi(c)$ and $\pi'(d')$ we obtain a sound normal proof from X,Z to D′ subject to $\pi(c)$ and $\pi'(d')$ bar subformulae of C. Using the main hypothesis to deal with those of D′ and the members of Z which are proper subformulae of C, and junction to deal with the rest, we combine these various proofs to provide a sound normal proof from X to Y subject to $\pi,\pi',\pi(c),\pi'(d')$ bar subformulae of C, and further such junctions provide such a proof subject to $\pi$ and $\pi'$ bar subformulae of C, as required.

If on the other hand $\pi'$ has no initial stroke, all its vertices are free. Suppose first that c is $d_b$ for some initial stroke b of $\pi$. Then C is either A⊃B or ~A for some A and B, and there is an edge from b to an occurrence a of the subformula A of C. Since c′ is a major premiss and an occurrence of A⊃B or ~A, there are edges from it and from an occurrence a′ of A to a stroke b′; and by 20.6 $(\pi')_{a'}$ is a sound normal proof from X,C to A,Y shorter than $\pi'$ and subject to it bar A, while $\pi^a$ is a sound normal proof from X,A to C,Y shorter than $\pi$ and subject to it bar A. Combining these with $\pi$ and $\pi'$ respectively, it follows from the subsidiary hypothesis that there are sound normal proofs from X to A,Y and from X,A to Y subject to $\pi$ and $\pi'$ bar subformulae of C, and the result follows by the main hypothesis. Similarly if c is not $d_b$ for any b, an inspection of the rules shows that there are occurrences a and a′ in $\pi$ and $\pi'$ of a proper subformula A of C, and strokes b and b′ in $\pi$ and $\pi'$ such that there are edges from a to b, from b to c, from c′ to b′, and from b′ to a′. Moreover a is not bound by an initial stroke, since otherwise c also would be bound, and as we have noted every conclusion of a sound proof is free. So $\pi_a$ and $(\pi')^{a'}$ are sound normal proofs from X to A,C,Y and from X,C,A to Y, shorter than and subject to π and π′ respectively bar A; and the result follows as before.

*Theorem 20.9* If X ⊢ B in Heyting's calculus, there is a sound proof from X to {B} in which every formula is a subformula of B or of a member of X.

*Proof* If X ⊢ B in Heyting's calculus, then by 20.7 and 20.8 there is a sound normal proof from X to {B}, and this has the required subformula property by 8.8.

Deducibility by sound proofs permits non-trivial multiple-conclusion inferences - for example that from A∨B to A,B - and in this it differs from the essentially single-conclusion treatment of intuitionist logic by Gentzen, 1934. It does not however permit any such inferences from zero premisses:

*Theorem 20.10* If there is a sound proof from Λ to Y, there is one from Λ to B for some B in Y.

*Proof* We show that a sound normal proof from Λ to Y contains a sound proof from Λ to *each* of its conclusions.

*Lemma 20.11* If b is a conclusion of a sound normal proof π without premisses, then π(b) is a subproof of π.

For the lemma, if π(b) is not a subproof, it has by 20.4 a premiss which is a free occurrence c of a formula C. Let π′ be the standard argument based on the union of the directed paths to c. Although π′ may not be a proof by the same rules as π, it contains every edge of π from its initial strokes and to its intermediate strokes, and hence an edge is major in $\pi'$ only if it is major in $\pi$. It follows that $\pi'$ is a normal argument from $\Lambda$ to C, but as we observed in the proof of 20.4, $\pi'$ contains an occurrence of a formula A∨B, where C is either A or B, and this contradicts 8.8.

For the theorem, if there is a sound proof from $\Lambda$ to Y there is a normal one $\pi$ by 20.8, which has no final strokes and so must have a conclusion b which is an occurrence of some B in Y. By 20.4 and 20.11 $\pi$(b) is a sound proof from $\Lambda$ to Y, and as b is its sole conclusion by 20.4, this is the required proof from $\Lambda$ to B.

Theorem 20.10 does not refer to any particular connectives, but the well-known result of Gödel (1932) regarding the provability of a disjunction is an immediate corollary of it. For by Theorem 20.7 and the rule 'from A∨B infer A,B' we have that if ⊢ A∨B in Heyting's calculus there is a sound proof from $\Lambda$ to A,B, and hence ⊢ A or ⊢ B.

# Bibliography

BELL, J.L. and SLOMSON, A.B.
1969. *Models and ultraproducts: an introduction*, Amsterdam.

BULL, R.A.
1962. The implicational fragment of Dummett's LC, *The Journal of Symbolic Logic*, vol.27, pp.189-194.

CARNAP, R.
1937. *The logical syntax of language*, London.
1943. *Formalization of logic*, Cambridge, Mass.

CHURCH, A.
1944. Review of Carnap's 'Formalization of logic', *The Philosophical Review*, vol.53, pp.493-498.
1953. Non-normal truth-tables for the propositional calculus, *Boletin de la Sociedad Matematica Mexicana*, vol.10, pp.41-52.
1956. *Introduction to mathematical logic*, Princeton.

DUMMETT, M.
1959. A propositional calculus with denumerable matrix, *The Journal of Symbolic Logic*, vol.24, pp.97-106.

FITCH, F.B.
1952. *Symbolic logic: an introduction*, New York.

GENTZEN, G.
1932. Über die Existenz unabhängiger Axiomensysteme zu unendlichen Satzsystemen, *Mathematische Annalen*, vol.107, pp.329-350. English translations of this and the following papers are included in *The collected papers of Gerhard Gentzen*, ed. M.E.Szabo, Amsterdam, 1969.
1934. Untersuchungen über das logische Schliessen, *Mathematische Zeitschrift*, vol.39, pp.176-210 and 405-431.
1936. Die Widerspruchsfreiheit der reinen Zahlentheorie, *Mathematische Annalen*, vol.112, pp.493-565.
1938. Neue Fassung des Widerspruchsfreiheitsbeweises für die reine Zahlentheorie, *Forschungen zur Logik und zur Grundlegung der exakten Wissenschaften*, n.s. no.4, pp.19-44.

GÖDEL, K.
1932. Zum intuitionistischen Aussagenkalkül, *Ergebnisse eines mathematischen Kolloquiums*, vol.4, p.40.

HACKING, I.
1979. What is logic?, *The Journal of Philosophy*, vol.76, pp.285-319.

HARROP, R.
1960. Concerning formulas of the types A→B∨C, A→(Ex)B(x) in intuitionistic formal systems, *The Journal of Symbolic Logic*, vol.25, pp.27-32.

HAY, L.S.
1963. Axiomatization of the infinite-valued predicate calculus, *The Journal of Symbolic Logic*, vol.28, pp.77-86.

HENKIN, L.
1949. The completeness of the first-order functional calculus, *The Journal of Symbolic Logic*, vol.14, pp.159-166.

HERMES, H.
1951. Zum Begriff der Axiomatisierbarkeit, *Mathematische Nachrichten*, vol.4, pp.343-347.

HERTZ, P.
1922. Über Axiomensysteme für beliebige Satzsysteme. I, *Mathematische Annalen*, vol.87, pp.246-269.
1923. Über Axiomensysteme für beliebige Satzsysteme. II, *Mathematische Annalen*, vol.89, pp.76-102.

HILBERT, D. and BERNAYS, P.
1934. *Grundlagen der Mathematik*, vol.I, Berlin.
1939. *Grundlagen der Mathematik*, vol.II, Berlin.

JECH, T.J.
1971. Trees, *The Journal of Symbolic Logic*, vol.36, pp.1-14.

KELLEY, J.L.
1955. *General topology*, Princeton.

KLEENE, S.C.
1952. *Introduction to metamathematics*, Amsterdam.

KNEALE, W.C.
1956. The province of logic, *Contemporary British Philosophy*, 3rd series, ed. H.D.Lewis, London.

KNEALE, W. and M.
1962. *The development of logic*, Oxford.

LEMMON, E.J.
1965. *Beginning logic*, London.

LEWIS, C.I. and LANGFORD, C.H.
1932. *Symbolic logic*, New York.

LINIAL, S. and POST, E.L.
1949. Recursive unsolvability of the deducibility, Tarski's completeness, and independence of axioms problems of propositional calculus, *Bulletin of the American Mathematical Society*, vol.55, p.50.

ŁOŚ, J. and SUSZKO, R.
1958. Remarks on sentential logics, *Indagationes Mathematicae*, vol.20, pp.177-183.

ŁUKASIEWICZ, J.
1920. O logice trójwartościowej, *Ruch filozoficzny*, vol.5, pp.169-171. English translation in S.McCall, *Polish logic 1920-1939*, Oxford, 1967.

ŁUKASIEWICZ, J. and TARSKI, A.
1930. Untersuchungen über den Aussagenkalkül, *Comptes rendus des séances de la Société des Sciences et des Lettres de Varsovie*, Classe III, vol.23, pp.30-50. English translation in A.Tarski, *Logic, semantics, metamathematics*, Oxford, 1956.

LYNDON, R.C.
1964. *Notes on logic*, Princeton.

McCALL, S. and MEYER, R.K.
1966. Pure three-valued Łukasiewiczian implication, *The Journal of Symbolic Logic*, vol.31, pp.399-405.

McNAUGHTON, R.
1951. A theorem about infinite-valued sentential logic, *The Journal of Symbolic Logic*, vol.16, pp.1-13.

ORE, O.
1962. *Theory of graphs*, Providence, R.I.

POPPER, K.R.
1947. Logic without assumptions, *Proceedings of the Aristotelian Society*, vol.47, pp.251-292. Popper's subsequent papers on this theme are indexed in *The Journal of Symbolic Logic*, vol.14, 1949.
1974. Lejewski's axiomatisation of my theory of deducibility, *The philosophy of Karl Popper*, ed. P.A.Schillp, La Salle, Illinois, pp.1095-1096.

PRAWITZ, D.
1965. *Natural deduction*, Uppsala.

PRIOR, A.N.
1960. The runabout inference-ticket, *Analysis*, vol.21, pp.38-39.

ROGERS, H.
1967. *Theory of recursive functions and effective computability*, New York.

ROSSER, J.B. and TURQUETTE, A.R.
1952. *Many-valued logics*, Amsterdam.

SCOTT, D.
1971. On engendering an illusion of understanding, *The Journal of Philosophy*, vol.68, pp.787-807.

1974a. Completeness and axiomatizability in many-valued logic, *Proceedings of symposia in pure mathematics, American Mathematical Society*, vol.25, pp.411-435.
1974b. Rules and derived rules, *Logical theory and semantic analysis*, ed. S.Stenlund, Dordrecht, pp.147-161.

SHOESMITH, D.J.
1962. *Axiomatisation of many-valued logics*, Ph.D. dissertation, University of Cambridge.

SHOESMITH, D.J. and SMILEY, T.J.
1971. Deducibility and many-valuedness, *The Journal of Symbolic Logic*, vol.36, pp.610-622.
1973. Proofs with multiple conclusions, *The Journal of Symbolic Logic*, vol.38, p.547.
1979. A theorem on directed graphs, applicable to logic, *The Journal of Graph Theory*, vol.3, to appear.

STONE, M.H.
1936. The theory of representations for Boolean algebras, *Transactions of the American Mathematical Society*, vol.40, pp.37-111.

TARSKI, A.
1930a. Fundamentale Begriffe der Methodologie der deduktiven Wissenschaften. I, *Monatshefte für Mathematik und Physik*, vol.37, pp.361-404. English translation of this and the following paper in A.Tarski, *Logic, semantics, metamathematics*, Oxford, 1956.
1930b. Über einige fundamentale Begriffe der Metamathematik, *Comptes rendus des séances de la Société des Sciences et des Lettres de Varsovie*, Classe III, vol.23, pp.22-29.

THOMAS, I.
1962. Finite limitations on Dummett's LC, *Notre Dame Journal of Formal Logic*, vol.3, pp.170-174.

TURQUETTE, A.R.
1958. Simplified axioms for many-valued quantification theory, *The Journal of Symbolic Logic*, vol.23, pp.139-148.

WÓJCICKI, R.
1969. Logical matrices strongly adequate for structural sentential calculi, *Bulletin de l'Academie Polonaise des Sciences*, Série des Sciences Matématiques, Astronomiques et Physiques, vol.17, pp.333-335.
1974. Matrix approach in methodology of sentential calculi, *Studia Logica*, vol.32, pp.7-39.

# Index

The page number of a definition is placed first in the entry. Where a word is defined in a multiple- and a single-conclusion sense, the latter is indicated by ‘s-c’.